# Introduction to Optimization Practice

# Introduction to

# OPTIMIZATION PRACTICE

BY

## Dr. Lucas Pun

## JOHN WILEY & SONS, INC.

New York · London · Sydney · Toronto

*To Persons who learn under difficult conditions*

Library of Congress Catalog Card Number: 69-19095
SBN 471  70233  1
Printed in the United States of America

# PREFACE

THE PURPOSE OF this book is to introduce graduate students and practicing engineers to optimization techniques with special emphasis on industrial applications. The required mathematics are some basic principles of linear algebra and differential calculus. The book does not pretend to bridge the gap between theory and practice. Instead, my intention is to supply materials for a first step toward such a task.

The book deals primarily with the exposition of deterministic optimization techniques excluding stochastic optimizations and logical optimizations of automata. The reason for this choice is both practical and pedagogical. Deterministic problems, certainly, are urgent at this time. The study of deterministic problems permits us to displace the gravity center toward static optimization problems, and to explain in an easier manner the fundamental concepts, theories, and methods.

Once the overall objective has been defined, the problem arises of matching optimally the author and the reader. This, at least, can be clarified by three interrelated questions.

1. How long must the book be?
2. How difficult must the reading of the book be?
3. In what manner can the book be useful to the reader?

These questions are not in reversed order. The first one is of primary importance.

Because of the ever-accelerating pace of technical progress, all persons engaged in engineering are in a hammer-and-anvil situation. On one hand, they have to read in order to keep abreast of new developments, while on the other hand, they rarely have the time to do so. Thus, it would seem that if the book is not read within a week ($7 \times 8 = 56$ hours), it will never be read. Therefore (a) perhaps a book of about 200 pages is sufficient, and (b) a reasonable amount of difficulty is acceptable if the book is self-contained. Thus I have answered questions *1* and *2*, and am left with question *3*.

Engineers are generally efficient persons, but are sometimes lax and unsatisfied persons. They need exploitable information for dealing with the

problems at hand. Also they need references to the more advanced literature.
In view of this, I have followed these guidelines:

1. I organized the various topics in a progressive manner (see the Table of
Contents).

2. In the main text, I explained the techniques by giving illustrative
examples.

3. In the Appendices (which can be omitted during the first and second
readings), I supplied basic concepts, notations, theories, and arguments
without unnecessary and lengthy mathematical developments. I emphasized
points such as the relation between the convexity and the optimization
techniques, and weak and strong conditions in extremals search.

Once this had been accomplished, my task became relatively simple. But
it has taken six years to write this book. The inescapable conclusion is:
optimal learning is difficult.

*Lucas Pun*

# ACKNOWLEDGMENT

I am deeply indebted to Professor S. S. L. Chang, Chairman of the Electrical Sciences Department, State University of New York at Stonybrook, for his help in the preparation of this book. The essential elements took form during my visit to the State University (January to May, 1965), made possible by Professor Chang under Research Grant No. AF 31-121. I also thank Prefesseur J. Lagasse, Laboratoire du Génie Electrique, Université de Toulouse, Professor P. Dorato, Electrical Engineering Department, Polytechnic Institute of Brooklyn, and Professors A. Sage and M. Cuénod, Electrical Engineering Department, University of Florida, for the opportunity of conducting seminars on this subject. I express my gratitude to Professor Paquet and Professor J-Ch. Gille, Electrical Engineering Department, Laval University, who have invited me to be a visiting Professor of Control (January to April, 1968) and who have given me the chance to use the material in this book for a semester graduate course.

I am grateful to my wife Arlette, who typed the manuscript twice.

L. P.

# CONTENTS

# 1

---

# IMPLICATIONS OF OPTIMIZATION PROBLEMS

## 1.1 THE PROCESS OF OPTIMIZATION

OPTIMIZATION, CAN BE generally defined as getting the best of something under given conditions. There is no doubt today about the importance of optimization problems in various industries. Their practical solution, however, still remains quite difficult for a number of reasons, among which are:

1. The need to define precisely the elements of the optimization problem.
2. The need to organize these elements in a way so that the problem appears solvable.
3. The need to choose a suitable method for solving the problem.

These difficulties are further increased by the possible effects they may have on each other. In some cases, the knowledge of a method constrains the freedom of defining and organizing the elements of the optimization problem, thus restricting the extent of the optimal solution. In other cases, the elements are defined and organized in such a way that no suitable method can be found. All these points show that it is necessary to clearly understand the process of optimization before discussing any optimization techniques. However, we shall first illustrate these techniques by some examples.

### 1.1.1 An Illustrative Example

We shall consider one example: the maximum range problem of a long-range rocket. The example will be treated in four versions, progressively

1

from simple to complex. This is to show how the same problem can be formulated in different ways according to different desired accuracies, and also how the higher desired accuracies lead to greater difficulties in solving the problem. The method of presentation aims to familiarize the reader with various basic concepts and elements of the optimization techniques. The statement of the problem, i.e., nomenclature of symbols and definitions and various assumptions, will be made in some length. This point is important, because it is necessary, at the beginning of an optimization problem, to understand the basis as clearly as possible. The mathematical development, however, will be restricted to indicating relevant points. Specific expressions such as performance function or index, state equations, etc., will be used without formal definition for the moment.

The common part of the four versions will now be explained. The flight trajectory of a long-range rocket is usually divided into two stages. The first stage, from the launching point to the burnout point, is called the powered stage because the rocket motor is active. The burnout point corresponds to the instant when there is no more combustible. The second stage, from the burnout point to the final point, is called the coasting stage. During this part, the motion of the rocket is governed by the laws of exterior ballistics.

For all the versions, the following general nomenclature will be used $\{[\text{G-43, Ch. 4}]$ (Fig. 1-1)$\}^1$

| | |
|---|---|
| $r$ | Radial distance from the earth center to the rocket. |
| $o$ | Subscript indicating the initial launching point. |
| $b$ | Subscript indicating the burnout point. |
| $f$ | Subscript indicating the final point. |
| $\psi_b$ | Angle between $r_b$ and $r_o$. |
| $\psi_f$ | Angle between $r_f$ and $r_b$. |
| $x, y$ | Rectangular coordinates of the rocket position. |
| $v_x, v_y$ | Speed components of the rocket. |
| $g(r)$ | Gravity acceleration at the distance $r$. |
| $v_b$ | Burnout speed (also the initial speed of the coasting flight). |
| $v_s$ | Satellite speed at the distance $r_b$; $\{v_s = [r_b \cdot g(r_b)]^{\frac{1}{2}}\}$. |
| $\gamma_b$ | Angle between the direction of $v_b$ and the vertical line. |
| $\theta_b$ | Angle between the direction of $v_b$ and the horizontal line. |
| $p$ | Speed-ratio between $v_b$ and $v_s$; $(p = v_b/v_s)$. |
| $a(t)$ | Acceleration of the rocket due to thrust only. |
| $t_b$ | Burning time, prescribed. |
| $t_c$ | Time of coasting flight. |
| $R$ | Total range. |
| $\varphi$ | Angle between the thrust direction and the $x$-coordinate (Fig. 1-4). |

---

[1] Bracketed numbers indicate references given at the end of each chapter (e.g., [1–1]) or at the end of the book (e.g., [G–43]).

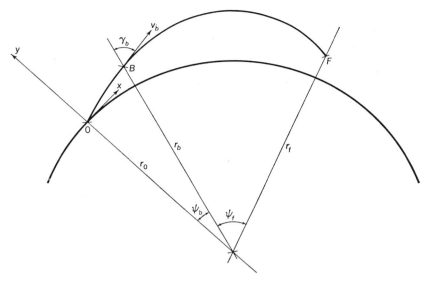

**Fig. 1-1**   Rocket trajectory (spherical earth).

FIRST VERSION.   The following assumptions are made:

1. Flight in the vacuum, atmospheric resistance therefore neglected.
2. The gravitational fields are central, therefore $g$ is independent from $\psi$.
3. Impulsive thrust and the program $v_b$ prescribed.
4. The trajectory is spherical, i.e., max $R$ = max $\psi_f$ (the notation max $X$ means the maximum value of the parameter $X$).
5. Earth rotation neglected.

The control variable, which we can adjust in order to maximize $R$ or $\psi_f$, is the angle $\gamma_b$. We must therefore find the relation connecting $\gamma_b$ to $\psi_f$. To do this we write the usual equations of motion for a particle moving in an inverse square force field [1-1].

$$\ddot{r} - r\dot{\psi}^2 + \left(\frac{GM}{r^2}\right) = 0 \qquad (1\text{-}1)$$

$$\left(\frac{d}{dt}\right)(r^2\dot{\psi}) = 0 \qquad (1\text{-}2)$$

where $G$ is the constant of gravitation, and $M$ is the mass of the earth. The conditions at the burnout point are

$$r = r_b; \qquad \dot{r} = \dot{r}_b = v_b \cos \gamma_b \qquad (1\text{-}3)$$

$$\psi = \psi_b = 0; \qquad \dot{\psi} = \dot{\psi}_b = \frac{v_b}{r_b} \sin \gamma_b \qquad (1\text{-}4)$$

From (1-2), we have the familiar equation for conservation of angular momentum.

$$r^2 \dot\psi = \text{constant} = r_b^2 \dot\psi_b = r_b v_b \sin \gamma_b \tag{1-5}$$

Eliminating $t$ from (1-1) and (1-2) and letting $u = 1/r$, we arrive at the differential equation in $u$ for the trajectory.

$$\frac{d^2 u}{d\psi^2} + u = \frac{GM}{r_b^2 v_b^2 \sin^2 \gamma_b} \tag{1-6}$$

The solution of this equation is

$$u = \frac{1}{r} = A \sin \psi + B \cos \psi + \frac{GM}{r_b^2 v_b^2 \sin^2 \gamma_b} \tag{1-7}$$

The constants of integration $A$ and $B$ are determined by the burnout-point conditions (1-3) and (1-4); the trajectory equation then becomes

$$\frac{r_b}{r} = \frac{1 - \cos \psi}{p^2 \sin \gamma_b} + \frac{\sin (\gamma_b - \psi)}{\sin \gamma_b} \tag{1-8}$$

where the dimensionless parameter $\alpha \equiv r_b v_b^2 / GM = p^2$ is merely twice the ratio of kinetic to potential energy at the burnout point. Now, at the final point, $r = r_f$, $\psi = \psi_f$, and Eq. (1-8) becomes

$$\frac{r_b}{r_f} = \frac{1 - \cos \psi_f}{p^2 \sin \gamma_b} + \frac{\sin (\gamma_b - \psi_f)}{\sin \gamma_b} \tag{1-9}$$

In this equation, all the parameters are known except $\gamma_b$, the value of which is to be determined, and $\psi_f$, the value of which is to be maximized. From (1-9), we form

$$\psi_f = F(\gamma_b) \tag{1-10}$$

Setting then

$$\frac{dF}{d\gamma_b} = 0 \tag{1-11}$$

we obtain the optimum value of $\gamma_b$

$$\tan \gamma_{b\,\text{opt}} = \frac{2}{p} \tan (\tfrac{1}{2}\psi_{f\,\text{max}}) \tag{1-12}$$

and the maximum range

$$\psi_{f\,\text{max}} = 2 \sin^{-1} \left\{ \frac{\left[ 2p \left( \dfrac{r_b}{r_f} + \tfrac{1}{2}p - 1 \right) \right]^{\frac{1}{2}}}{2 - p} \right\} \tag{1-13}$$

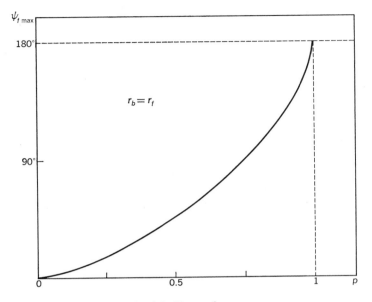

**Fig. 1-2**   Form of $\psi_{f\,\text{max}}$.

In the special case where $r_b = r_f$

$$\psi_{f\,\text{max}} = 2 \sin^{-1}\left(\frac{p}{2-p}\right) \qquad (1\text{-}14)$$

values of $\psi_{f\,\text{max}}$ versus $p$ are plotted in Fig. 1-2.

We note that in this first version, the state equation (1-9), or rather the performance function $\psi_f$ (1-10), is static, i.e., independent from the variable $t$. The maximizing method is relatively simple, and implies only the ordinary theory of maxima and minima. Because of the numerous simplifying assumptions adopted, the value of the solution which we have found is only indicative.

SECOND VERSION.   We make the same assumptions, except that during the powered flight trajectory part from 0 to $B$, we take into account the known and prescribed variations of

$$v_b = v_b(r_b) \qquad (1\text{-}15)$$

$$\gamma_b = \gamma_b(r_b) \qquad (1\text{-}16)$$

with respect to the burnout distance $r_b$. The derivation of the state equation is the same as before. However, instead of considering $v_b(p^2 = v_b^2/v_s^2)$ and $\gamma_b$ as constant parameters, they must be taken as functions of $r_b$. Under this circumstance, Eq. (1-9) becomes

$$\frac{r_b}{r_f} = \frac{1 - \cos\psi_f}{\left[\dfrac{v_b(r_b)}{v_s}\right]^2 \sin^2\gamma_b(r_b)} + \frac{\sin[\gamma_b(r_b) - \psi_f]}{\sin\gamma_b(r_b)} \tag{1-17}$$

As before, we form

$$\psi_f = F[\gamma_b(r_b), v_b(r_b)] \tag{1-18}$$

The maximizing method still implies the ordinary theory of maxima and minima, with the exception that here the compound law $z = f[y(x)]$, $dz = [(df/dy)(dy/dx)\,dx]$ must be used in calculating the derivatives. Omitting the intermediate developments, we shall only mention that in the case where $r_b = r_f$, the maximizing condition is

$$0 = (1 - \cotan^2\gamma_b(r_b) - p^2)\frac{d\gamma_b}{dr_b}$$
$$+ \cotan\gamma_b(r_b)\,\cosec\gamma_b(r_b)\left(\frac{1}{r_b} + \frac{2}{v_b}\frac{dv_b}{dr_b}\right) \tag{1-19}$$

Since $d\gamma_b/dr_b$ and $dv_b/dr_b$ are known, we can extract from (1-19) the values of $\gamma_{b\,opt}$ and $\psi_{f\,max}$.

In this version, the performance index is a functional (function of a function), but its expression is still static. Maximizing work is more laborious, but still simple. Because of the simplifying assumptions, the solution which we have found is still of a qualitative value.

THIRD VERSION.   We shall take as control variable the thrust direction $\varphi$ (Fig. 1-4) instead of $\gamma_b$. We shall use the same assumptions as in the first version with the following exceptions:

1. $t_b$ is prescribed.
2. The acceleration $a(t)$ is prescribed and has the form indicated in Fig. 1-3.
3. Flat earth, i.e., the axis $OX$ is straight instead of being spheric.

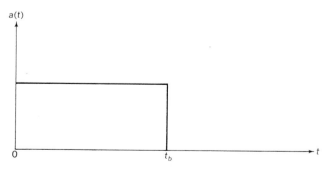

***Fig. 1-3***   Form of $a(t)$.

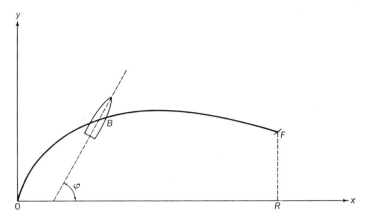

**Fig. 1-4**  Rocket trajectory (flat earth).

The range equation is

$$R = x_b + v_{xb} \cdot t_b = R(x_b, y_b, v_{xb}, v_{yb}) \tag{1-20}$$

where we have

$$t_c = g^{-1}\{v_{xb} + [v_{yb}^2 + 2g(y_b - y_f)]^{\frac{1}{2}}\} \tag{1-21}$$

$$v_{xb} = v_{x0} + \int_0^{t_b} a(t) \cos \varphi \, dt \tag{1-22}$$

$$v_{yb} = v_{y0} + \int_0^{t_b} a(t) \sin \varphi \, dt \tag{1-23}$$

$$x_b = v_{x0} \cdot t_b + \int_0^{t_b} (t_b - t)a(t) \cos \varphi \, dt \tag{1-24}$$

$$y_b = v_{y0} \cdot t_b + \int_0^{t_b} (t_b - t)a(t) \sin \varphi \, dt + \tfrac{1}{2}gt_b^2 \tag{1-25}$$

While Eq. (1-21) is the mathematical formulation of the performance index $R$ which is a functional, Eqs. (1-22) to (1-25) constitute the state equations reflecting the dynamic behavior of the rocket. The extremalizing problem looks variational. However, under the given assumptions, $R$ is a function of $\varphi$ alone: $R = R(\varphi)$. Therefore, setting $dR/d\varphi = 0$ leads to

$$\tan \varphi_{\mathrm{opt}} = \frac{v_{xb}}{[v_{yb}^2 + 2g(y_b - y_f)]^{\frac{1}{2}}} \tag{1-26}$$

and further calculations show that $\tan \varphi_{\mathrm{opt}}$ and $\varphi_{\mathrm{opt}}$ are time-independent.

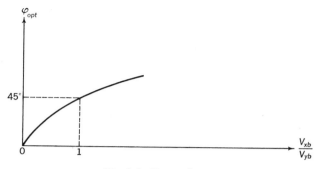

**Fig. 1-5** Form of $\varphi_{\text{opt}}$.

We can then replace $\varphi$ by $\varphi_{\text{opt}}$ in Eqs. (1-22) to (1-25) and integrate by numerical methods. In the special case when

$$y_b = y_f \tag{1-27}$$

we have

$$\tan \varphi_{\text{opt}} = \frac{v_{xb}}{v_{yb}} \tag{1-28}$$

and the diagram of $\varphi_{\text{opt}}$ versus $(v_{xb}/v_{yb})$ is plotted in Fig. 1-5. We note that in this third version, we have obtained mathematical models of a universal type where the performance index is constrained by state equations. These models are dynamic. However, because of its intrinsic properties, the apparent variational problem turns out to be a simple maximizing problem.

FOURTH VERSION.   We always take $\varphi$ as the control variable, but as assumptions, we shall admit that the earth is spheric, that during the powered-flight phase the gravitational force is not negligible, and that $g = g(x, y)$ is approximated by a Taylor expansion, so that

$$g_x = w_x^2 \cdot x \tag{1-29}$$

$$g_y = g_0 - w_y^2 \cdot y \tag{1-30}$$

$$w_x^2 = \tfrac{1}{2}w_y^2 = g_0/r_0 \tag{1-31}$$

From the geometrical representation in Fig. 1-1, we deduce that

$$R = x_b + r_0\psi_f \tag{1-32}$$

$$\frac{r_b}{r_f} = \frac{1 - \cos \psi_f}{p^2 \cdot \sin^2 \gamma_b} + \frac{\sin (\gamma_b - \psi_f)}{\sin \gamma_b} \tag{1-33}$$

The performance index $R$ is therefore a functional

$$R = R_1(r_b, \gamma_b, v_b) \tag{1-34}$$

or

$$R = R_2(x_b, y_b, v_{xb}, v_{yb}) \tag{1-35}$$

The state dynamic equations are the same as in the third version, except that the $g$'s are now variable. After writing the Euler-Lagrange equations.[2]

$$\frac{d}{d\varphi} \frac{\partial R_2}{\partial v_{xb}} - \frac{\partial R_2}{\partial x_b} = 0 \tag{1-36}$$

$$\frac{d}{d\varphi} \frac{\partial R_2}{\partial v_{yb}} - \frac{\partial R_2}{\partial y_b} = 0 \tag{1-37}$$

the following relation is deduced

$$\tan \varphi_{\mathrm{opt}} = \frac{\dfrac{\partial R}{\partial v_{yb}} \cosh w_y(t_b - t) + \dfrac{1}{w_y} \dfrac{\partial R}{\partial y_b} \sinh w_y(t_b - t)}{\dfrac{\partial R}{\partial v_{xb}} \cos w_x(t_b - t) + \dfrac{1}{w_x} \dfrac{\partial R}{\partial x_b} \sin w_x(t_b - t)} \tag{1-38}$$

Replacing $\varphi$ by $\varphi_{\mathrm{opt}}$ in equations in $x_b$, $y_b$, $v_{xb}$, and $v_{yb}$, and then integrating by numerical methods, the value of $R_{\max}$ can be obtained. In this version, we see that the introduction of only an approximation of $g(x, y)$ already leads to a complex extremalizing procedure. The obtained solution has a higher accuracy than the preceding ones, but is insufficient in practice.

In this example, given in four progressive versions:

1. There was no question of limiting values either on the state variables (such as $r_b$) or on the control variables (such as $\varphi$); optimal values $\bar{r}_b$ and $\bar{\psi}$ are supposed to fall into the permissible domain.

2. The variations of the gravitational force $g(x, y)$ are only partially taken into account.

3. The rotation of the earth is not considered. This rotation introduces a velocity vector $V_T(x, y)$ which has to be added to, or subtracted from the velocity of the rocket according to the direction of the movement.

4. Atmospheric resistance has not been considered. This resistance introduces a counterforce, the value of which is function of the altitude and rocket speed.

5. The trajectory is supposed to be contained in a principal plane of the earth. If the problem were three-dimensional, there would be two more state equations in $\dot{z}$ and $z$.

6. The derivatives or variations yield necessary but insufficient conditions of extrema; this question will be treated in subsequent chapters.

---

[2] For the moment, we shall not explain such expressions as: variational method, Euler-Lagrange equations . . .

If all these constraints, limitations, and influences are to be added to the optimization problem, there will be difficulties in adequately formulating them mathematically, as well as difficulties in choosing an extremum-seeking method.

### 1.1.2   Schematic Representation of the Optimization Process

From the preceding example, we may observe that three elements are common to all four versions. These elements are the performance index, the system description and the maximizing method. It is important to emphasize now that the mere definition of these elements does not suffice to completely define the process of optimization; the reason is practical rather than mathematical.

We have seen in the four versions that the difference between them is primarily the degree of simplification (or completeness) in assuming operating conditions. This in turn defines (1) the degree of accuracy with which we mathematically formulate the performance index and the system behavior, and (2) the maximizing method that is to be used. If we now ask why we adopt one degree of simplification over another, we will find the connection of the optimization problem to the practical need. The practical object of optimization, even when the general objective is specified, can bear a variety of forms. We may have, for instance, the following typical cases:

1. A prestudy.
2. A complete study for implementing a prototype.
3. A definitive study for production.

Thus we see the need of defining another element, the *level* of optimization. This element reflects the practical need of the optimization, defines the degree of accuracy in the mathematical formulation, and specifies the desired implementation.

By using the preceding elements, the various steps of the process of optimization, for a given level, can now be formulated. For illustration, we shall use the example of the moon-landing of a spacecraft (Fig. 1-6).

(a) Definition of the overall goal, e.g., moon-landing of a space craft.
(b) Definition of the level of the desired solution. This important and often insufficiently specified point is a two-dimensional problem.
   (b1) Hardware dimensions. Are we concerned with:
      1. Optimum steering under given conditions?
      2. Plus optimum telemetering?
      3. Plus an adequate remote correction control?

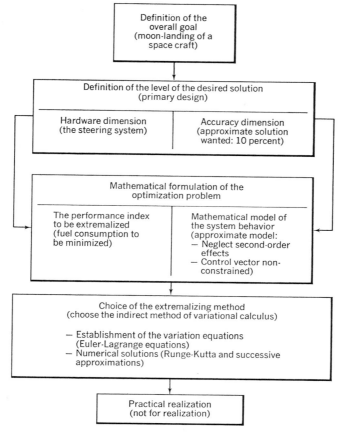

***Fig. 1-6*** Various steps of an optimization process.

(b2) Accuracy dimensions. Are we looking for:

1. An approximate solution?
2. Bounds of the solutions?
3. The exact solution?

(c) Mathematical formulation of the optimization problem which consists of:

(c1) Mathematical formulation of the performance index PI; in optimization problems, the PI is generally extremalized. The form of PI depends on the choice in (b1).

(c2) Mathematical formulation of the system dynamics. The form of the adopted model depends on the choice in (b2).

(d) Choice of the extremalizing method leading to consequent
    (d1) Establishment of variation equations.
    (d2) Numerical solutions.
(e) Practical realizations. In the case of a primary design, no realization
    is asked.

The level consideration does not only help to make the process of optimization more precise; it also helps to emphasize the relationship between the evolution of the optimization levels and the practical approach to optimization problems. When we first consider new problems in industries, we generally do so carefully, by treating them simply. Only when experience is gained do we go further step-by-step. Another point which is quite obvious but worth mentioning is that very often the static version is the simplified form of the dynamic version, just as the linear problem is the simplified form of the nonlinear problem.

## 1.2 SYSTEMS

The object under optimization is called a system, other names often used are plant or process. The basic concept of a system is an object under observation, study, development, or construction. From an industrial point of view, a system may represent a small mechanism or a huge complex operation comprising many subsystems. Since optimization inherently implies control, we shall consider only control systems. Control systems can be partitioned into three main classes [1-5; 1-6; 1-7; G-51, Ch. 1].

1. Control system, formed by the controlled system and the controller.
2. Feedback control system, formed by a forward loop and a feedback loop, in which the control action is a function of the state of the controlled system.
3. Adaptive control system, in which the control action is a function of the parameter variations of the basic control loop.

Optimization of feedback control systems implies the specification of the control as a function of the state, and involves, in general, dynamic optimization techniques. This problem is studied extensively in the literature [G-28, G-34, G-35, G-49, G-54, G-58]. Here, we shall be concerned with both static and dynamic optimizations, but without special emphasis on the feedback feature.

### 1.2.1  State-Space Notation

The mathematical language commonly used to represent the system behavior (other expressions often used are system characteristics and system dynamics) is the state-space notation. This notation is an extension of those used in the theories of vectors, matrices, and sets. In Appendix 1A, at the end of this chapter, the reader can find the basic notations used throughout this book. Complete treatment of the state-space theory can be found in [G-1 and G-57]. Here we give some relevant examples.

***Example 1.***   ***A Production Process.*** Two-products $P_1$ and $P_2$ are manufactured by two machines $M_1$ and $M_2$. The operating times $a_{ij}$ are:

|  | $M_1$ | $M_2$ |
|---|---|---|
| $P_1$ | $a_{11} = 7$ minutes | $a_{12} = 5$ minutes |
| $P_2$ | $a_{21} = 6$ minutes | $a_{22} = 9$ minutes |

The monthly availabilities, in terms of operating time, of the machines are $b_1 = 9200$ minutes for $M_1$, and $b_2 = 8400$ minutes for $M_2$. Let $x_1$ and $x_2$ be the desired monthly number of units of $P_1$ and $P_2$. Clearly, the times needed for producing $x_1$ and $x_2$ can only be *smaller than or equal to* $b_1$ and $b_2$. Thus we have the following constraints or inequalities.

$$a_{11}x_1 + a_{12}x_2 \leq b_1$$
$$a_{21}x_1 + a_{22}x_2 \leq b_2$$

(1-39)

Let $x = (x_1, x_2)$ be the state vector or $x = [x_{kj}] = \begin{bmatrix} x_1 \\ x_2 \end{bmatrix}$; $k = 2, j = 1$ the column state matrix.

$$A = [a_{ik}] = \begin{bmatrix} a_{11} & a_{21} \\ a_{12} & a_{22} \end{bmatrix}; \quad i = 2, k = 2$$

$$B = [b_{ij}] = \begin{bmatrix} b_1 \\ b_2 \end{bmatrix}; \quad i = 2, j = 1$$

Then, in the state-space notation, the system (1-39) is written as a simple inequality:

$$Ax \leq B \tag{1-40}$$

(Each component of the vector $Ax$ is smaller than or equal to the corresponding component of the vector $B$.)

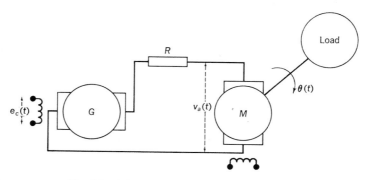

**Fig. 1-7** A D.C. generator and motor system.

## Example 2. DC Generator and Motor System (Fig. 1-7).

Let: $\theta$   Position of motor, radians.
  $v_m$   Voltage applied on the motor, volts.
  $K_m$   Voltage constant of motor, volts/radians/second.
  $T_m$   Motor time constant, seconds.
  $K_g$   Generator voltage constant, volts/field ampere.
  $R$   Series resistance of motor and generator armature circuit, ohms.
  $T_f$   Generator field time constant, seconds.
  $E_c$   Voltage applied to generator control field, volts.
The transfer function from $E_c$ to $\theta$ is

$$\frac{\theta(s)}{E_c(s)} = \frac{K_g/K_mR}{s(T_fs + 1)(T_ms + 1)} \tag{1-41}$$

The corresponding differential equation is

$$T_fT_m\frac{d^3\theta}{dt^3} + (T_f + T_m)\frac{d^2\theta}{dt^2} + \frac{d\theta}{dt} = \frac{K_g}{K_mR}E_c \tag{1-42}$$

If we now define one set of state variables $x_1$, $x_2$, $x_3$, such as

$$x_1 = \theta \qquad x_2 = \dot\theta = \dot x_1 \qquad x_3 = \ddot\theta = \dot x_2 = \ddot x_1$$

(such a set is not unique), then Eq. (1-4) can be expressed as three first-order differential equations

$$
\begin{aligned}
\dot x_1 &= x_2 \\
\dot x_2 &= x_3 \\
\dot x_3 &= \left[\frac{K_g}{K_mR}E_c - x_2 - (T_f + T_m)x_3\right]\frac{1}{T_fT_m}
\end{aligned}
\tag{1-43}
$$

Let the state vector be $x = (x_1, x_2, x_3)$, and the coefficient matrix be

$$A = \begin{bmatrix} 0 & 1 & 0 \\ 0 & 0 & 1 \\ 0 & \dfrac{-1}{T_f T_m} & \dfrac{-(T_f + T_m)}{T_f T_m} \end{bmatrix}$$

and the forcing column matrix

$$f = \begin{bmatrix} 0 \\ 0 \\ \dfrac{K_g E_c}{K_m R T_f T_m} \end{bmatrix}$$

Then, in the state-space notation, the system (1-43) can be simply written as

$$\dot{x} = \frac{dx}{dt} = Ax + f \qquad (1\text{-}44)$$

***Example 3. A Space Vehicle System (Fig. 1-8).***

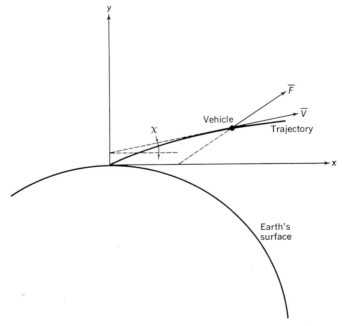

***Fig. 1-8*** Coordinate system for the motion of a space vehicle.

Let:   $F$  Total effective thrust.

       $m$  Vehicle mass.

$X, Y, Z$  Inertial plumb-line coordinate system originating at launching site.

$g_x, g_y, g_z$  Components of gravitational vector in $XYZ$ directions.

      $\chi$  Angle between thrust vector projection into $X$-$Y$ plane and $X$-axis.

According to the conventional theory of general mechanics, the following set of simplified equations of motion (omitting aerodynamic forces) stating the force equilibrium, can be written

$$\ddot{x} = \frac{F}{m} \cos \chi - g_x$$

$$\ddot{y} = \frac{F}{m} \sin \chi - g_y \qquad (1\text{-}45)$$

$$\ddot{z} = -g_z$$

If we now define state variables $x$, $y$, $z$, $p$, $q$, $r$ and the state vector $s = (x, y, z, p, q, r)$ such that $\dot{x} = p$, $\dot{y} = q$, $\dot{z} = r$, then the system (1-45) can be expressed as a system of six first-order differential equations:

$$\dot{x} = p = f_1(x, y, z, p, q, r)$$

$$\dot{y} = q = f_2(x, y, z, p, q, r)$$

$$\dot{z} = r = f_3(x, y, z, p, q, r)$$

$$\dot{p} = \ddot{x} = \frac{F}{m} \cos \chi - g_x = f_4(x, y, z, p, q, r) \qquad (1\text{-}46)$$

$$\dot{q} = \ddot{y} = \frac{F}{m} \sin \chi - g_y = f_5(x, y, z, p, q, r)$$

$$\dot{r} = \ddot{z} = -g_z = f_6(x, y, z, p, q, r)$$

or, in state-space notation

$$\dot{s} = f(s) \qquad (1\text{-}47)$$

### 1.2.2 Mathematical Model of Systems

Optimization techniques generally consist of a mathematical procedure, requiring the mathematical representation of the considered system. Mathematical models are either derived from conventional sciences (mechanics, aerodynamics, electromagnetism, etc.), or formulated according to experimental results. More frequently, they are obtained by a combination of these two methods. The mathematical models, if known, generally have one of the following forms:

1. System of algebraic equations or inequalities (static process).
2. System of integro-differential equations or inequalities (dynamic process).
3. System of partial-derivative algebraic or differential equations or inequalities (static or dynamic processes).

The search for these models is generally progressive. Linear forms are tried first, then transcendental and nonlinear terms are added, and finally higher-order differentials and partial derivatives are used. Discrete forms are used at two occasions: (1) when the considered system is a sampled-data system, or when the system comprises some digital computer, and (2) when numerical integration methods are used for solving equations in continuous form. Stochastic forms are usually introduced when deterministic forms are inadequate to describe the system behavior. Optimization of sampled-data and stochastic control systems will not be specifically considered in this book.

An important note must be made here. The search for the mathematical models (the identification problem) [G-39; G-40; G-41] is not an easy problem. For this reason a number of optimization techniques are developed with the aim of obtaining the extremum without knowing the system behavior. A special chapter has been devoted to this subject.

The main types of mathematical models are summarized below.
Static systems.

$$Ax(\leq, =, \geq) B \tag{1.48}$$

($x = n$-vector, $A = m \times n$ matrix, $B = m$th-order column matrix.)
Dynamic noncontrolled systems (linear).

$$\dot{x}(t) = Ax(t) \tag{1-49}$$

($A = m \times n$ constant matrix.)
Dynamic control systems (linear).

$$\dot{x}(t) = Ax(t) + Bu(t) \tag{1-50}$$

[$u(t) = $ control signal, an $r$-vector, $B = m \times r$ constant matrix.]
Time-varying control systems (linear).

$$x(t) = A(t)x(t) + Bu(t) \tag{1-51}$$

Feedback control systems (linear).

$$\dot{x}(t) = A(t)x(t) + Bu[x(t)] \tag{1-52}$$

Nonlinear control systems (a linearized form).

$$\dot{x}(t) = A(t)x(t) + Bu(t) + g \tag{1-53}$$

[$g = g(x_1, x_2, \ldots, x_n, t)$ is a nonlinear function of its arguments.]

Nonlinear control systems (general form).

$$\dot{x} = g(x, u, t) \tag{1-54}$$

Systems in which the rate of change $\dot{x}$ depends on some hereditary influence (time lags).

$$\dot{x} = g[x(t), x(t - t_1), \ldots, x(t - t_k); t] \tag{1-55}$$

Systems in which the entire past history is involved [G25, p. 5].

$$\dot{x} = g\left[ x(t), \int_{-\infty}^{t} x(s)\, dG_1(s, t), \ldots, \int_{-\infty}^{t} x(s)\, dG_k(s, t); t \right] \tag{1-56}$$

### 1.2.3  Geometrical Representation of Trajectories and Constraints

The state-space language is not only a notation for mathematically representing the behavior of a system; it also permits us to represent this behavior geometrically which gives us an even clearer understanding of the characteristics of optimization. In the form of a mathematical model, such as $Ax \leq B$ we use expressions such as variable ($x$), parameters ($A$, $B$), equation, and inequality. In geometric representation, we use expressions like points, trajectories, and constraints. We shall now look at its use in static and dynamic processes.

STATIC PROCESSES.  Let us take, for example, the mathematical model $Ax \leq B$, where $x$ is an $n$-vector, $A$ is an $m \times n$ matrix, and $B$ an $m$-th-order column matrix. In the $n$-dimensional space $E^n$, any $n$-tuple $x = (x_1, x_2, \ldots, x_n)$ will define a point in $E^n$. This point represents the state of the considered system. In many problems, as we shall see in subsequent chapters, $x$ generally represents physical quantities, and is thus restricted to take on nonnegative values. Such a restriction is written as $x \geq 0$, and is called a constraint. If $n = 2$, then $(x_1, x_2) \geq 0$ means that $x$ must lie on the positive quadrant defined by $x_1 \geq 0$, $x_2 \geq 0$. In the general $n$-dimensional case, we shall say that $x$ is restricted to lie on the nonnegative orthant. The relation $Ax \leq B$ is also called a constraint. The hypersurface $Ax = B$ divides the space $E^n$ into two half-spaces. In Fig. 1-9, an example of $n = 2$ is given. In such a case, $Ax = B$ represents a system of two linear relations $a_{11}x_1 + a_{12}x_2 = b_1$, $a_{21}x_1 + a_{22}x_2 = b_2$. The boundary $Ax = B$ is thus represented by a broken line, dividing $E^2$ into two half-spaces. In the negative half-space (shaded part), we have $Ax < B$. The motivation for using the expression "constraint" is that if $Ax \leq B$ is to be satisfied, then the point-state $x$ is required to lie either on the hyperplane or inside the negative half-space. Similar concepts are also defined for nonlinear relations such as $g(x) \leq B$.

DYNAMIC PROCESSES.  Let us consider the model $\dot{x} = Ax$ where $x$ is an $n$-vector, and $A$ an $m \times n$ matrix. The inclusion of the control $r$-vector $u$

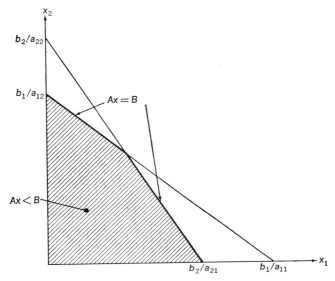

**Fig. 1-9**  State constraints in a two-dimensional space $x \geq 0$   $Ax \leq B$.

does not change the concepts if we assume $a(r + n)$-space instead of $E^n$. Because of the presence of the time variable $t$, the state of the considered system is now represented by a trajectory $\Gamma$ in $E^n$ (Fig. 1-10). The points on $\Gamma$ are the values of $x$ at $t$, $0 \leq t \leq T$, $x_A = x(t = 0)$ at $A$ is the initial state, while $x_B = x(t = T)$ at $B$ is the final state. Both $A$ and $B$ are called terminal or end points. (Note that terminal or end do not necessarily mean final.)

1. The trajectory is called fixed-final-time if $T$ is prescribed, otherwise it is called open-final-time.

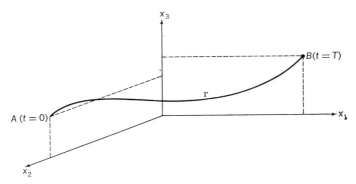

**Fig. 1-10**  Representation of a trajectory in a three-dimensional space.

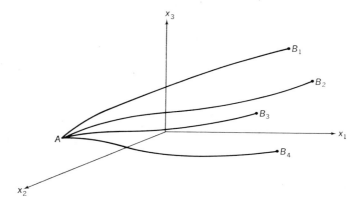

**Fig. 1-11**   Open final-state trajectories.

2. It is called fixed-final-state if the point $B$ is prescribed, otherwise it is called open-final-state (Fig. 1-11). Similar expressions apply for the initial state.

3. Constraints of the form $x \geq 0$ restricts $x$ to lie on the nonnegative orthant for all $t$.

4. Constraints of the form $x_{\min} \leq x \leq x_{\max}$ restrict $x$ to belong to a closed hypercube. The constants $x_{\min}$ and $x_{\max}$ are called boundary values (the

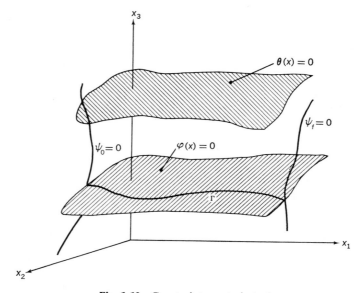

**Fig. 1-12**   Constraints on trajectories.

same applies to $u$; bounded controls are of a special interest in dynamic optimization problems).

Just as in the case of static processes, the equation $\dot{x} = Ax$ or $\varphi(x) = \dot{x} - Ax = 0$ is no longer called "system dynamics," but a "constraint." If this equation is to be satisfied, and a solution of this differential system found, then $x$ is restricted to remain on the hypersurface defined by this solution for all $t$. (Fig. 1-12). Other constraints are:

1. $\psi_f[x(t)] = 0$ at $t = T$, constraint defined by a surface or a curve on which $x$ is restricted to lie at the final time.

2. $\psi_0[x(t)] = 0$ at $t = 0$, constraint represented by a surface or a curve on which $x$ is restricted to lie at the initial time.

3. $\theta(x) \leq 0$ for all $t$, constraints represented by a hypervolume in which (including the boundary) the whole trajectory is restricted to remain.

Note that $\psi_f = 0$, $\psi_0 = 0$ are generally algebraic equations; $\theta(x) = 0$ is an algebraic system, while $\varphi(x) = 0$ is a system of differential equations.


## 1.3  PERFORMANCE INDEX

In optimization problems, performance indices play a bridge role between the general statement of the objective and the mathematical treatment of the problems. We shall now discuss their choice and their formulation.

### 1.3.1  Choice of a Performance Index

When we try to formulate an optimization problem, we must clearly understand the difference between its objective and its performance index. The objective is a literal statement of what to optimize, and of the conditions under which the optimization is to be effected. These conditions comprise the level of the desired solution, the description of the system, and more specifically the limitations and constraints on the system. From an industrial point of view, these conditions must even comprise the practical computing means (how the solutions can be calculated numerically) and the hardware possibilities (how the results will be exploited). The performance index, on the other hand, is a precise mathematical figure which permits quantitative computations and comparisons. Various names are used in the literature: *criterium* (control); *value function* or *cost function* (economic problems); *return function* (dynamic programming language); *functional* or *payoff function* (variational methods).

In the daily engineering practice, to locate or to isolate a performance index out of its technical and economical context is not an easy task. The reasons are (1) there is often some freedom of choice, and (2) the decision influences many other parameters. We shall illustrate this point by a practical example.

Let us consider the problem of optimum thrust programming of electrically powered rocket vehicles [1-8]. It is known that electrically powered rocket engines, such as iron rocket and plasma rocket engines, are characterized by a power source which is separate from the propellant. For a fixed power setting, the rate of propellant expenditure and the exhaust velocity can be varied over wide ranges. As a consequence, a variety of thrust programs can be achieved. Assuming that the total time of powered flight is a specified parameter, and that the thrust is always in the tangential direction, we have as possibilities for a performance index:

$\Delta V$   The change in vehicle velocity.
$\Delta s$   The change in position or range.
$m_p$   Propellant mass.

Noting that each of these parameters may be either maximized or minimized, specified, or left arbitrary, we have at least seven possibilities of choice. The

| Case | $\Delta V$ | $\Delta S$ | $m_p$ |
|------|-----------|-----------|-------|
| 1 | *Maximum* | Arbitrary | Specified |
| 2 | Specified | Arbitrary | *Minimum* |
| 3 | Arbitrary | *Maximum* | Specified |
| 4 | Arbitrary | Specified | *Minimum* |
| 5 | *Maximum* | Specified | Specified |
| 6 | Specified | *Maximum* | Specified |
| 7 | Specified | Specified | *Minimum* |

difficulty of choice is even increased if we state that one can also minimize the time of flight in lieu of one of the parameters $\Delta V$, $\Delta s$, or $m_p$.

### 1.3.2   Formulation of Performance Indices

The difficulty of choice and hence of determination of the mathematical formulation of a performance index is fortunately lightened by the experience gained in control. Generally, we can always assign the performance index to one of two categories.

1. A time (a flight time to be minimized, a chemical-plant operating time to be maximized).

2. An amplitude (a range or a profit to be maximized, a fuel-consumption or an error to be minimized).

For example, mathematical forms for static processes are:

$$J_1 = \sum_i c_i x_i \qquad (x_i: \text{variables, } c_i: \text{constants.})$$

$$J_2 = \sum_i c_i x_i^2 \qquad (\text{A quadratic form.})$$

$$J_3 = \sum_i c_i x_i \qquad (\text{Where the variables } x_i \text{ can only take on}$$
integral values.)

$J_4 = \phi_1(x_1) + \phi_2(x_2) + \cdots \phi_i(x_i) + \cdots$ (Where each of the functions, $\phi_i(x_i)$, linear or nonlinear, depends only upon one variable $x_i$).

$$J_5 = \phi(x) \qquad (\text{Where } \phi \text{ is a nonlinear function in } x.)$$

For dynamic processes, we have:

$$J_6 = \int_0^T dt = T \qquad \text{To be minimized (minimum-time problems).}$$

$$J_7 = \int_0^T (\bar{x} - x)^2 \, dt \qquad \text{To be minimized (minimum-error problems,}$$
$\bar{x}$ is the reference state or the optimal state).

$$J_8 = \int_0^T dt + \mu \int_0^T (\bar{x} - x)^2 \, dt \qquad \text{To be minimized (a combined}$$
performance index, $\mu$ is a weighting constant).

$$J_9 = \int_0^T L(x, u, t) \, dt \qquad (\text{Where } x(t) \text{ is the state vector, } u \text{ is the}$$
control vector, and $L$ is a general nonlinear
functional.)

$$J_{10} = [G(x, u, t)]_{t_0}^{t_f} + \int_{t_0}^{t_f} L(x, u, t) \, dt \qquad (\text{General form commonly used in}$$
dynamic optimization problems.)

The following form has a very great flexibility (rocket problems).

$$J_{11} = \int_\rho^{t+\tau} \{\mu_1(\xi)[\bar{x}(\xi) - x(\xi)]^2 + \mu_2(\xi)[\bar{u}(\xi) - u(\xi)]^2\} \, d\xi$$

If $\rho = 0$, the optimization is effected on the time interval $t + \tau$ after rocket departure. If $t = \rho$, the optimization is effected on the undetermined time interval $\tau$ seconds after the real time. The weights on the state and control errors can be adjusted by time functions $\mu_1(\xi)$ and $\mu_2(\xi)$.

## 1.4   METHODS

Two questions will be discussed here. How to choose an optimizing method and what properties do we demand of it.

### 1.4.1   Choice of an Optimizing Method

We have seen in the illustrative example of Section 1.1.1 that the formulation of the optimization problem depends mainly on the adopted assumptions. However, once the problem is mathematically formulated, the optimizing method is also defined. The problem formulation comprises the definition of the performance index and of the constraints, and this is done if the characteristics of the system under study are known. Of course if these characteristics cannot be formulated mathematically, then there is no possibility of mathematical treatment. Some techniques nevertheless exist. They are grouped under the general title of Extremum Seeking Methods, and are discussed in Chapter 3.

When the performance index and the constraints are formulated, the method to be chosen will depend on whether

1. They are static or dynamic.
2. The performance function is constrained or not.
3. They are linear or nonlinear.
4. They are one-dimensional or multidimensional.

The relations between problem formulations and methods are shown below.[3]

| Problem Formulation | Method |
|---|---|
| *Static processes* | |
| 1. Unconstrained problem<br>$J = J(x)$ performance index<br>$x$: an $n$-vector | Ordinary theory of maxima and minima |
| 2. Constrained problems<br>$J = J(x)$ performance index<br>$g(x) = 0$ constraints | Lagrange-multipliers method |
| 3. Linear programs (programming)<br>$J = \Sigma\, c_i x_i \qquad i = 1, 2, \ldots, n$<br>$\Sigma\, a_{ij} x_i (\leq,\ =,\ \geq) b_i \quad j = 1, 2, \ldots, m$<br>$x_i \geq 0 \qquad i = 1, 2, \ldots, n$ | Simplex method |
| 4. Nonlinear programs (programming)<br>$J = J(x)$<br>$g_j(x)(\leq,\ =,\ \geq) b_i$<br>$x_i \geq 0$<br>$i = 1, 2, \ldots, n \quad j = 1, 2, \ldots, m$ | Convex programming method |

*Dynamic processes*

| | |
|---|---|
| 1. Linear problems | Variational methods |
| $J = J(x, u, t)$ linear | |
| $\varphi = \dot{x} - Ax - Bu = 0$ | |
| $x \in X \quad u \in U$ | Dynamic programming |
| $x$: state $n$-vector | methods |
| $u$: control $r$-vector | |
| $A$: $m \times n$ coefficient matrix | Method of gradients |
| $B$: $m$th-order coefficient column | |
| matrix | |

2. Linear time-varying problems
   $J = J(x, u, t)$ linear
   $\varphi = \dot{x} - A(t)x - B(t)u = 0$
   $x \in X \quad u \in U$
   $A(t)$: $m \times n$ matrix function
   $B(t)$: $m$th-order column matrix
        function

3. Nonlinear problems
   $J = J(x, u, t)$
   $\varphi = \dot{x} - g(x, u, t) = 0$
   $x \in X \quad u \in U$
   $J, g$: nonlinear functions

## 1.4.2  Required Characteristics

As we shall see in subsequent chapters, the optimum conditions in all the techniques are obtained by solving a system of algebraic or differential equations. The techniques are therefore distinguished by two categories of characteristics: (1) mathematical (2) computational.

Mathematical characteristics are those well-known in analysis.

1. Existence of the solution: is there a solution?
2. Uniqueness of the solution: is there only one solution?
3. Necessary conditions: conditions which must be satisfied in all cases.
4. Sufficiency conditions: conditions that, if satisfied, guarantee an extremum.
5. Local or absolute extremum: is the solution valid over a small area of the state space or over the whole state space?
6. Weak and strong extremum: (see Appendix 2A).

Concerning these tracts, the convexity property of constraints and of performance functions (such as defined in Sections 1.2.3 and 1.3.2) plays an important role. This mathematical question is discussed in several Appendices (Appendices 1D, 2B, 4B).

Computational characteristics are of a practical importance.

1. Existence of a numerical computing method.
2. Type and size of the needed computer.
3. Convergence, if the method asks for an iterative procedure.
4. Computing time.

In optimization techniques, computational problems are important *per se*. In static problems, some linear programs may have several hundred variables and constraints. In dynamical problems, as we shall see in Chapter 4, the mathematical basis of the various techniques can be derived from one another, the difference between them being essentially computational.

# Matrices and Vectors: Algebraic Representation

We summarize here a number of notations, operations, and properties. Those concerning vectors, matrices, and simultaneous linear equations are used to represent the systems and those concerning the derivatives and the convex sets are more specially used in the search for the maxima and minima. [G-18, Chs. 2 and 3; G-54, Chs. 2 and 3; G-22; G-59, App. E.]

*1A.1 Basic Matrices.* We define a matrix as a rectangular array of numbers, variables, or functions.

$$A = [a_{ij}] = \begin{bmatrix} a_{11} & a_{12} & \cdots & a_{1n} \\ a_{21} & a_{22} & \cdots & a_{2n} \\ \cdot & & & \cdot \\ \cdot & & & \cdot \\ \cdot & & & \cdot \\ a_{m1} & a_{m2} & \cdots & a_{mn} \end{bmatrix} \tag{1}$$

The numbers, variables, or functions are called the *elements* of the matrix. The matrix (1) is called an $m \times n$ matrix since it has $m$ rows and $n$ columns. The general element $a_{ij}$ has two subscripts: the first one, $i$, indicates the row, and the second one, $j$, indicates the column. If $m = n$, the matrix is said to be *square* and is called an $n$th-order matrix. If $m = 1$, we have an $n$th-order *row* matrix. If $n = 1$, we have an $m$th-order *column* matrix. Row and column matrices are often referred to as row and column vectors.

An $n$-component vector is generally written as

$$a = (a_1, a_2, a_3, \ldots, a_n) \tag{2}$$

which must not be confused with the notation of an $n$th-order row matrix

$$[a_j] = [a_1, a_2, \ldots, a_n] \tag{3}$$

*1A.2 Basic Matrix Operations.* The addition and subtraction of two matrices $A$ and $B$ can be accomplished if and only if they have the same number of rows and columns. To do this, one simply adds or subtracts the corresponding elements:

$$A + B = [a_{ij} + b_{ij}] \tag{4}$$

The matrix addition or subtraction obeys the associative and commutative laws:

$$A + B + C = A + (B + C) = (A + B) + C \tag{5}$$

$$A + B = B + A \tag{6}$$

The multiplication of a matrix $A$ by a scalar $\lambda$ is effected by multiplying each element of $A$ by $\lambda$

$$\lambda A = A\lambda = [\lambda a_{ij}] \tag{7}$$

A matrix $A$ can be multiplied by another matrix $B$ if and only if the number of columns in $A$ is equal to the number of rows in $B$. If $A$ is $m \times r$ and $B$ is $r \times n$, then the product $C = AB$ is $m \times n$. The general element of $C$ is

$$c_{ij} = \sum_{k=1}^{n} a_{ik} b_{kj} \tag{8}$$

For example,

$$A = \begin{bmatrix} 3 & 1 \\ 2 & 2 \\ 1 & 3 \end{bmatrix} \qquad B = \begin{bmatrix} 1 & 2 & 3 & 4 \\ 4 & 3 & 2 & 1 \end{bmatrix}$$

$$C = \begin{bmatrix} 3 \cdot 1 + 1 \cdot 4 & 3 \cdot 2 + 1 \cdot 3 & 3 \cdot 3 + 1 \cdot 2 & 3 \cdot 4 + 1 \cdot 1 \\ 2 \cdot 1 + 2 \cdot 4 & 2 \cdot 2 + 2 \cdot 3 & 2 \cdot 3 + 2 \cdot 2 & 2 \cdot 4 + 2 \cdot 1 \\ 1 \cdot 1 + 3 \cdot 4 & 1 \cdot 2 + 3 \cdot 3 & 1 \cdot 3 + 3 \cdot 2 & 1 \cdot 4 + 3 \cdot 1 \end{bmatrix}$$

$$= \begin{bmatrix} 7 & 9 & 12 & 13 \\ 9 & 10 & 10 & 10 \\ 14 & 11 & 9 & 7 \end{bmatrix}$$

The matrix multiplication is associative

$$ABC = (AB)C = A(BC) \tag{9}$$

but generally not commutative

$$AB \neq BA \tag{10}$$

The *division* of matrices can only be performed by inverting the divisor and multiplying. The *inversion* of a matrix can be performed only for square matrices. Given a square matrix $A$, if there exists a square matrix $B$ which satisfies the relation $AB + BA = I$, then $B$ is called the *inverse* of $A$. ($I$ is the identity matrix, see next section for its definition.) The symbol is the superscript $-1$: $B = A^{-1}$.

The general element of the inverse is

$$b_{ij} = \frac{(-1)^{(i+j)} \begin{vmatrix} \text{cofactors of} \\ A_{ji} \end{vmatrix}}{|A_{ij}|} \tag{11}$$

Note that the elements of $B$ exist only if the determinant $|A_{ij}|$ is not zero. The transpose of an $m \times n$ matrix $A$ is an $n \times m$ matrix $A^T$ (sometimes denoted by $A'$) obtained from $A$ by interchanging rows and columns. If $A^T = [a_{ij}^T]$, then $a_{ij}^T = a_{ji}$. Some interesting properties are: $(A + B)^T = A^T + B^T$; $(A^T)^T = A$; $(AB)^T = B^T A^T$. The transpose of a column matrix is a row matrix, and vice versa.

*1A.3 Special Matrices.*   A square matrix $A$ is said to be symmetric if $A = A^T$, that is $a_{ij} = a_{ji}$. A square matrix is said to be skew-symmetric if $A = -A^T$, that is $a_{ij} = -a_{ji}$, $i \neq j$; $a_{ij} = 0$, $i = j$.

A submatrix of an $m \times n$ matrix $A$ is obtained by deleting all but $t$ rows and $u$ columns; the result is a $t \times u$ matrix. In linear programming problems, it is desirable to partition an $m \times n$ matrix $A$ into a row of column vectors.

$$A = [A_1, A_2, \ldots, A_j, \ldots, A_n] \tag{12}$$

with

$$A_j = \begin{bmatrix} a_{1j} \\ a_{2j} \\ \cdot \\ \cdot \\ \cdot \\ a_{mj} \end{bmatrix} \tag{13}$$

Since it takes too much room to print a vertical column of numbers, we have

C A S E  1.   The $m$-component vector $a_{ij} = (a_{1j}, a_{2j}, \ldots, a_{mj})$ is defined basically as a row matrix $A_j = [a_{1j}, a_{2j}, \ldots, a_{mj}]$. The transpose of the $A_j$'s must be used, and one must write

$$A = [A_1^T, A_2^T, \ldots, A_j^T, \ldots, A_n^T] \tag{14}$$

C A S E  2.   The notation $A_j = (a_{1j}, a_{2j}, \ldots, a_{mj})$, written horizontally, is understood as the column-matrix representation of the $m$-component vector $a_{ij} = (a_{1j}, a_{2j}, \ldots, a_{mj})$, and the form (12) must be used.

The distinction between the row matrix and the column matrix representation of a multicomponent vector must be clearly understood, because the results will be different when multiplication and inversion are performed. A square matrix is said to be *diagonal*, if the nondiagonal elements are all zero, thus $a_{ij} = 0$ for $i \neq j$. Some of the diagonal elements $a_{ii}$ may be zero. A diagonal matrix is called a *scalar matrix* if all the nonzero $a_{ii}$ are identical, and equal to a scalar $\lambda$. The diagonal matrix $I = [\delta_{ij}]$ is called *identity matrix* where $\delta_{ij}$ is the *Kronecker delta*, defined to be zero if $i \neq j$ and unity if $i = j$.

# Matrices and Vectors: Geometrical Representation

*1B.1 Definitions.* The *Euclidian space* of $n$ dimensions is denoted by $E^n$. The space $E^n$ is referred to as the set of all $n$-tuples of real numbers. An element of $E^n$ may be thought of as a *point* whose coordinates are the $n$ real numbers, and $E^n$ as a *point set*. It may also be thought of as a *vector* emanating from the origin, whose components along the coordinate axes are the $n$ real numbers; in this case, $E^n$ is considered as a *vector space*. If the $n$ components of the vector characterize the state of some physical object, then $E^n$ is referred as a *state space*.

Upper-case letters, for example, $X$, $S$, etc., are generally used to denote point sets. If $x$ is an element of the set $X$, we write $x \in X$. The intersection of two sets $X$ and $Y$, written, $X \cap Y$, is the collection of all points common to $X$ and $Y$. The union of $X$ and $Y$, written $X \cup Y$, is the set of all points in either $X$ or $Y$ or both. A point set $X$ is often defined by some property or properties $\rho(x)$ satisfied by its elements $x$ and we write $X = \{x / \rho(x)\}$. A vector $y \in E^n$ is greater than a vector $x \in E^n$ and we write $y > x$ if each component of $y$ is greater than the corresponding components of $x$, that is $y = x$ implies $y_i > x_i$, $i = 1, 2, \ldots, n$. A line in $E^n$ is a set of all points that satisfy a relation

$$\frac{x_1 - a_1}{b_1} = \frac{x_2 - a_2}{b_2} = \cdots = \frac{x_n - a_n}{b_n} \tag{15}$$

where $a$ and $b$ are members of $E^n$. The line includes the point $a$ and is parallel to the direction vector $b$.

A *cone* is a set of points in $E^n$ such that if $x$ is in the set, then $y = \lambda x$ is also in the set for any nonnegative number $\lambda$. A hyperplane in $E^n$ is a set of all points $x$ in $E^n$ that satisfy a relation

$$a^T x = \beta \tag{16}$$

where $a$ and $x$ are column matrices; $a \in E^n$ is the normal vector of the hyperplane and $\beta$ is a scalar. A half-space in $E^n$ is the set of all points $x$ that satisfy a relation:

$$a^T x \geq \beta \tag{17}$$

where $a \in E^n$ and $\beta$ is a scalar. The hyperplane $a^T x = \beta$ is called the bounding hyperplane of the half-space. For a given point $a$ and any $\epsilon > 0$, the set $X = \{x / \|x - a\| = \epsilon\}$ is called a hypersphere in $E^n$ with center at $a$ and radius $\epsilon$. The set $X = \{x / \|x - a\| < \epsilon\}$ is called the interior of the hypersphere. An $\epsilon$-neighborhood of a point $a$ is the interior of a hypersphere with center at $a$ and with radius $\epsilon$. A point $a$ is called an interior point of a set if there exists an $\epsilon$-neighborhood about $a$ that contains only points of the set. A point $a$ is called a boundary point if every $\epsilon$-neighborhood about $a$ contains points that are in the set and points that are not in the set. A set that contains all its boundary points is said to be closed. A set that does not contain any of its boundary points is said to be open. An $\epsilon$-neighborhood of a point $a$ is an open set.

The distance between two points (vectors) $a$ and $b$, written $|a - b|$, is defined as

$$|a - b| = [(a - b)^T(a - b)]^{\frac{1}{2}} = \left[ \sum_{j=1}^{n} (a_j - b_j)^2 \right]^{\frac{1}{2}} \tag{18}$$

This definition is a direct generalization of the definition in two and three dimensions. The length or magnitude of a vector $a$, written $|a|$, is the distance from the origin to $a$: $|a| = |a - 0|$.

*1B.2 Properties.*  *Scalar product.* Given two $n$-component column vectors

$$a = \begin{bmatrix} a_1 \\ a_2 \\ \cdot \\ \cdot \\ \cdot \\ a_n \end{bmatrix} \qquad b = \begin{bmatrix} b_1 \\ b_2 \\ \cdot \\ \cdot \\ \cdot \\ b_n \end{bmatrix}$$

the scalar product of these two vectors $a$ and $b$, and denoted by $\langle a, b \rangle$ is

$$\langle a, b \rangle = a^T b = a_1 b_1 + a_2 b_2 + \cdots + a_n b_n$$

$$= \sum_{i=1}^{n} a_i b_i$$

SCHWARZ INEQUALITY.    For any two $n$-component vectors $a$ and $b$, it is true that $|a^T b| \leq |a| |b|$ where $|a^T b|$ is the absolute value of the scalar product. If $a, b \neq 0$, the angle $\theta$ between $a$ and $b$ is given by

$$\cos \theta = \frac{a^T b}{|a| |b|} \tag{19}$$

and $|\cos \theta| \leq 1$.

ORTHOGONALITY.    Two vectors $a$ and $b$ are said to be orthogonal if $a^T b = 0$. Thus if $a, b \neq 0$, they are orthogonal provided the angle between them is $\pi/2$. In $E^n$, the unit vectors

$$e_1 = (1, 0, 0, \ldots, 0) \quad e_2 = (0, 1, 0, \ldots, 0) \quad \cdots \quad e_n = (0, 0, 0, \ldots, 1)$$

are orthogonal: $e_i^T e_j = 0$, $i \neq j$. The orthogonality property is very often used in the derivation of extremal conditions.

LINEAR DEPENDENCE AND INDEPENDENCE.    A set of vectors $a_1, \ldots, a_k$ is said to be linearly dependent if there exist scalars $\lambda_j$ not all zero such that

$$\sum_{j=1}^{k} \lambda_j a_j = 0 \tag{20}$$

The unit vectors are linearly independent since $\sum_{j=1}^{n} \lambda_j e_j = [\lambda_1, \lambda_2, \ldots, \lambda_n] = 0$ implies $\lambda_j = 0$, $j = 1, 2, \ldots, n$. Given a set of vectors $a_1, \ldots, a_k$, a vector $a$ is said to be a linear combination of these vectors if there exist scalars $\lambda_j$ such that $a = \sum_{j=1}^{k} \lambda_j a_j$. Considering the unit vectors, any $n$-component vector can be written

$$a = [a_1, a_2, \ldots, a_n] = a_1 e_1 + a_2 e_2 + \cdots + a_n e_n \tag{21}$$

where $a_1, a_2, \ldots, a_n$ are scalars and $e_1, e_2, \ldots, e_n$ are vectors. A linearly independent subset of vectors $a_1, a_2, \ldots, a_r$ from $E^n$ is said to be a basis for $E^n$ if every vector in $E^n$ can be represented as a linear combination of $a_1, \ldots, a_r$. The set of unit vectors is a basis for $E^n$. Any set of basis vectors can be thought of as defining a coordinate system (not necessarily orthogonal) for $E^n$. There exists an infinite number of bases for $E^n$.

RANK OF A MATRIX.    The rank of a matrix $A$, denoted by $r(A)$, is defined to be the maximum number of linearly independent columns in $A$, and this number equals that of linearly independent rows in $A$. The rank of $AB$ is not uniquely determined by $r(A)$ and $r(B)$; however, $r(AB)$ is equal to or smaller than the minimum of $r(A)$ and $r(B)$

$$r(AB) \leq \min [r(A), r(B)] \tag{22}$$

SINGULARITY.    A matrix $A$ has an inverse if there exists a matrix $B$ so that $BA = AB = I$. Only square matrices have inverses. An $n$th-order matrix $A$ will have an inverse if and only if $r(A) = n$. If $r(A) = n$, $A$ is called *nonsingular*. A square matrix $A$ is nonsingular if and only if its determinent $|A|$ is different from zero. If $A$ has no inverse, it is said to be singular.

# Simultaneous Equations, Quadratic Forms, and Derivatives

**1C.1 Linear Algebraic Equations.** A system of $m$ linear equations in $n$ unknowns $x_j, j = 1, 2, \ldots, n$ can be written in matrix form

$$Ax = B \quad \text{or} \quad \sum_{j=1}^{n} a_{ij}x_j = b_j \tag{23}$$

where $A$ is an $m \times n$ rectangular matrix, $x$ is an $n$th order column matrix, and $b$ an $m$th-order column matrix (note that the number of equations is *not* necessarily equal to the number of unknowns). Such a system of equations may have (a) no solution, (b) a unique solution, or (c) an infinite number of solutions.

Consider the matrix $A_b = [A \quad b]$. We have either

$$r(A_b) = r(A) \quad \text{or} \quad r(A_b) > r(A)$$

1. If $r(A_b) = r(A) = k$, then every column of $A_b$, and in particular $b$, can be written as a linear combination of $k$ linearly independent columns of $A$. The system (23) has, in this case, at least one solution.

2. If $r(A_b) > r(A)$ and $r(A) = k$, every set of $k$ linearly independent columns from $A_b$ must contain $b$, and therefore $b$ cannot be written as a linear combination of the columns of $A$. In this case, the system (23) has no solution.

3. If $r(A) = r(A_b) = k < n$, then there exists an infinite number of solutions so that $n - k$ of the variables may be given arbitrary values.

4. If $r(A) = r(A_b) = n$, there exists a unique solution.

5. If $r(A) = r(A_b) = k < m$, then $m - k$ equations are redundant.

HOMOGENEOUS EQUATIONS. If in (23) $b = 0$, the system $Ax = 0$ is called homogeneous. The solution $x = 0$ is called a trivial solution. Nontrivial solutions exist if and only if $r(A) < n$.

BASIC SOLUTIONS. Let $n > m$, let $B$ be an $m$th-order submatrix form $A$, and $R$ the $(n - m)$th-order submatrix from $A$ containing the remaining $n - m$ columns. Let $x_B$ and $x_R$ be the vectors containing the variables associated with the columns of $B$ and $R$. Then for any $x_R$, a solution to the system will be obtained if $x_B = B^{-1}b - B^{-1}Rx_R$. A solution $x_R = 0$ and $x_B = B^{-1}b$ is called a *basic solution*. The maximum number of basic solutions to $Ax = B$ is $n!/m!\,(n-m)!$ corresponding to the number of combinations of $m$ columns of $A$ which can be selected to form $B$.

*1C.2 Characteristic Values and Quadratic Forms.* When finding solutions to simultaneous linear differential equations, we meet the characteristic value problem. This problem is to find vectors $x \neq 0$ and scalars $\lambda$ so that for a given square matrix $A$ (the number of equations equals the number of unknowns)

$$Ax = \lambda x \qquad (24)$$

If $x \neq 0$ satisfies (24), then $(A - \lambda I) = 0$. This is possible only if $A - \lambda I$ is singular, so that $|A - \lambda I| = 0$. This, in turn, is possible if and only if $f(\lambda) = |A - \lambda I| = 0$, $f(\lambda)$ is an $n$th-degree polynomial in $\lambda$.

1. $f(\lambda)$ is called the characteristic polynomial of $A$.
2. $f(\lambda) = 0$ is called the characteristic equation for the matrix $A$.
3. The roots of $f(\lambda) = 0$ are called the characteristic values or eigenvalues of $A$.
4. When $\lambda$ is an eigenvalue of $A$, there exists one or more linearly independent vectors $x_i \neq 0$ satisfying (24). Such an $x_i$ is called a characteristic vector or eigenvector of $A$. The number of eigenvectors is $n - r(I - \lambda A)$.

SIMILAR MATRIX.    If $S$ is a nonsingular matrix, then $S^{-1}AS$ has the same characteristic values as $A$. The matrix $S^{-1}AS$ is said to be *similar* to $A$.

ORTHONORMAL BASIS AND VECTORS.    If $A$ is symmetric, the characteristic values of $A$ are real, and the corresponding characteristic vectors are orthogonal. The unit-length orthogonal characteristic vectors denoted $u_j$, $j = 1, 2, \ldots, n$, form an orthonormal basis. Note that $u_i^T u_j = \delta_{ij}$. The matrix $Q = [u_1, u_2, \ldots, u_n]$ where $u_1, u_2, \ldots, u_n$ are column matrices, is called an orthogonal matrix, and is nonsingular. Then we have

$$Q^{-1}AQ = Q^T AQ = Q^T[Au_1, Au_2, \ldots, Au_n]$$

$$= Q^T[\lambda_1 u_1, \lambda_2 u_2, \ldots, \lambda_n u_n] = [\lambda_j \, \delta_{ij}] \qquad (25)$$

Thus any symmetric matrix $A$ is similar to a diagonal matrix whose elements on the main diagonal are the eigenvalues of $A$. This operation is called diagonalization. The matrix $Q^T AQ$ is said to be *congruent* to $A$.

QUADRATIC FORMS.    A quadratic form in $n$ variables $x_1, \ldots, x_n$ is a *numerical function* of these variables which can be written $z = \sum_{j=1}^{n} \sum_{i=1}^{n} \alpha_{ij} x_i x_j$ or $z = x^T Ax$. The matrix $A$ can always be rendered symmetric by defining another matrix $B = [b_{ij}]$ with

$$b_{ij} = b_{ji} = \frac{a_{ij} + a_{ji}}{2}, \qquad \text{all } i, j \qquad (26)$$

if $a_{ij} \neq a_{ji}$.

1. The quadratic form is called positive definite if $x^T A x > 0$ for every $x$ except $x = 0$.

2. $x^T A x$ is called positive semidefinite if $x^T A x \geq 0$ for every $x$ and there exists $x \neq 0$ for which $x^T A x = 0$.

3. $x^T A x$ is called negative definite (semidefinite) if $-x^T A x$ is positive definite (semidefinite).

4. $x^T A x$ is called indefinite if it is positive for some points and negative for others.

Let $x = Qy$, then $z = y^T Q^T A Q y = y^T B y$ or $z = \sum_{j=1}^n \lambda_j y_j^2$. In terms of $y_j$, the quadratic form involves only the square of the variables, and is a diagonal matrix. Every quadratic form can be diagonalized by an orthogonal similarity transformation, and, in addition, the coefficients of the $y_j^2$ are the eigenvalues of $A$.

*1C.3 Derivatives.* (see [G-18], pp. 43–45) FUNCTION OF A SINGLE VARIABLE. The derivative of a function $f(x)$ at the point $x_0$, is defined in terms of a limiting process as

$$\left[\frac{df}{dx}\right]_{x=x_0} = \lim_{x \to x_0}\left[\frac{f(x) - f(x_0)}{x - x_0}\right] = \lim_{h \to 0}\left[\frac{f(x_0 + h) - f(x_0)}{h}\right] \tag{27}$$

provided that the limit exists. Geometrically, $f(x_0)$ is the slope of the curve $z = f(x)$ at $x_0$. The derivative hence provides information about the rate of change of the function at $x_0$.

CLASSES OF CONTINUOUS FUNCTIONS. To indicate that a function $f$ is continuous over some subset of $E_n$ we write $f \in C$. If $f$ and its first derivative are continuous, we write $f \in C^1$. We say that $f(x)$ is differentiable at $x_0$ if $f \in C^1$ at $x_0$. If $f$ and its first and second derivatives are continuous, we write $f \in C^2$. We say that $f(x)$ is twice differentiable at $x_0$ if $f \in C^2$ at $x_0$, etc. These concepts are extendable to multivariable functions and to functionals.

MULTIVARIABLE FUNCTIONS. For these functions it is possible to define $n$ partial derivatives. The partial derivative of $f$ with respect to $x$, at the point $x^0 = (x_1^0, x_2^0, \ldots, x_n^0)$ is

$$\left[\frac{\partial f}{\partial x_j}\right]_{x=x^0} = \lim_{h \to 0}\left[\frac{f(x_1^0, \ldots, x_{j-1}^0, x_j^0 + h, x_{j+1}^0, \ldots, x_n^0) - f(x^0)}{h}\right]$$

$$= \lim_{h \to 0}\left[\frac{f(x^0 + he_j) - f(x^0)}{h}\right] \tag{28}$$

provided the limit exists. This derivative provides information about the rate of change of the function at the point $x^0$ in the direction of the $e_j$ coordinate axis. To obtain the partial derivatives $\partial f / \partial x_j$ we simply determine ordinary derivative of $f$ treating all variables except $x_j$ as constants.

GRADIENT. If $f \in C^1$, and $f$ is an $n$-variable function, when $f$ is differentiable with respect to all variables $x_1, x_2, \ldots, x_n$, we can define an $n$-component row vector $\nabla f$ by

$$\nabla f = \left( \frac{\partial f}{\partial x_1} \frac{\partial f}{\partial x_2} \cdots \frac{\partial f}{\partial x_n} \right) \tag{29}$$

The vector $\nabla f$ is called the gradient vector of $f$, and the symbol $\nabla f$ is read "del $f$." If we wish to indicate that $\nabla f$ is to be evaluated at a point $x_0$, we write $\nabla f(x_0)$.

HESSIAN MATRIX FOR $f$.    If the $n$ partial derivatives of each of the first partial derivative exist, we have $n^2$ second partial derivatives of $f$. The matrix

$$H_f = \left[ \frac{\partial^2 f}{\partial x_i \, \partial x_j} \right] \tag{30}$$

is called the Hessian matrix for $f$. If $f \in C^2$, then

$$\frac{\partial^2 f}{\partial x_i \, \partial x_j} = \frac{\partial^2 f}{\partial x_j \, \partial x_i} \tag{31}$$

it follows that $H_f$ is a symmetric matrix.

TANGENT HYPERPLANE AND NORMAL.    Let $f \in C^1$ at $x_0$. Write $z_0 = f(x_0)$. Then the hyperplane

$$z - z_0 = \nabla f(x_0)(x - x_0)$$

or

$$\nabla f(x_0)x - z = \nabla f(x_0)x_0 - z_0 \tag{32}$$

is called the tangent hyperplane to the surface $z = f(x)$ at $x_0$. The vector $(\nabla f(x_0), -1)$ is normal both to the tangent hyperplane and to the surface $z = f(x)$ and is often referred to as a normal vector.

JACOBIAN MATRIX.    Let us have a set of $m$ linear equations in $n$ variables $x_1,$ $x_2, \ldots, x_n$:

$$g_i(x) = 0 \qquad i = 1, 2, \ldots, m \tag{33}$$

For each function $g_i$, we can determine $n$ first partial derivatives $\partial g_i / \partial x_j$, $j = 1,$ $2, \ldots, n$. For the $m$ functions $g_i$, we obtain a total of $mn$ first-order partial derivatives which can be arranged into an $m \times n$ rectangular matrix $G = [\partial g_i / \partial x_j]$, the functional matrix. From this matrix we can form different square submatrices of order $m$. A typical submatrix of this kind can be written

$$J_x(j_1, \ldots, j_m) = \begin{bmatrix} \dfrac{\partial g_1}{\partial x_{j1}} & \cdots & \dfrac{\partial g_1}{\partial x_{jm}} \\[2ex] \dfrac{\partial g_2}{\partial x_{j1}} & & \cdot \\[1ex] \cdot & & \cdot \\ \cdot & & \cdot \\ \cdot & & \cdot \\[1ex] \dfrac{\partial g_m}{\partial x_{j1}} & \cdots & \dfrac{\partial g_m}{\partial x_{jm}} \end{bmatrix} \tag{34}$$

Such a matrix is called the Jacobian matrix of the functions $g_i$ with respect to the variables $x_{j1}, x_{j2}, \ldots, x_{jm}$. The subscript $x$ on $J$ indicates the point in $E^n$ at which the elements in $J$ are to be evaluated. If $m = n$, the Jacobian matrix becomes unique.

FUNCTIONALS. To give a rigorous mathematical definition of a functional would imply many concepts and definitions (see [G-54, pp. 108–110]). For an easy understanding, we shall say only that the function of a function is a functional.

Let $x$ be a function of $y$, then $f[x(y)]$ is a functional. If $f \in C'$ and $x \in C'$ then $df/dy$ exists and is given by the chain rule

$$\frac{df}{dy} = \frac{df}{dx} \cdot \frac{dx}{dy} \tag{35}$$

If $f$ is a function of one function $x$, but $x$ is a function of $m$ variables $y_1, y_2, \ldots, y_m$, then

$$\frac{\partial f}{\partial y_i} = \frac{df}{dx} \cdot \frac{\partial x}{\partial y_i} \tag{36}$$

if it exists.

If $f$ is a function of $n$ functions $x_1, x_2, \ldots, x_n$, and if each $x_j$ is function of $m$ variables $y_1, y_2, \ldots, y_m$, then each partial derivative has the form

$$\frac{\partial f}{\partial y_i} = \sum_{j=1}^{n} \cdot \frac{\partial f}{\partial x_j} \cdot \frac{\partial x_j}{\partial y_i} \tag{37}$$

if it exists.

All the notations concerning gradient, Hessian matrix, and Jacobian matrix are extendable to functionals.

APPENDIX 1D

# Convex Sets and Functions

*1D.1 Convex Sets.* CONVEX SET. A convex set in $E^n$ is one in which, given any two points $x_1$ and $x_2$ in the set, and $0 \leq \alpha \leq 1$, all points on the line joining $x_1$ to $x_2$ and defined by

$$x = (1 - \alpha)x_1 + \alpha x_2 \tag{38}$$

are also in the set. The, *convex hull* of a given set is the intersection of all convex sets that contain the given set. An *extreme point* of a convex set is a point that does not lie on a line segment joining two distinct points of the set. A *polyhedron* is a convex set; it is the convex hull of its extreme points which are its vertices.

CONVEX POLYHEDRAL SET. A convex polyhedral set in $E^n$ is the intersection of a finite number of half-spaces. Its points satisfy the relation

$$A^T x \geq b \tag{39}$$

where $A$ is an $m \times n$ matrix and $b$ an $m$th-order column matrix. The set of points $x$ which satisfy the constraints of a linear program is a convex polyhedral set. Such a set may not be bounded, and this is why the term polyhedron is not used.

CONVEX POLYHEDRAL CONE. This is the intersection of a finite number of half-spaces whose bounding hyperplanes contain the origin. It is the set of points $x$ satisfying

$$A^T x \geq 0 \tag{40}$$

A convex polyhedral cone is a convex set, a cone, and is the convex hull of a finite number of line segments which emanate from the origin.

*1D.2 Convex and Concave Functions.* CONVEX FUNCTION. The function $f(x)$ is considered to be convex over a convex set $X$ in $E^n$ if for any two points $x_1$ and $x_2$ in $X$ and for all $\alpha$, $0 \leq \alpha \leq 1$

$$f[\alpha x_2 + (1 - \alpha)x_1] \leq \alpha f(x_2) + (1 - \alpha)f(x_1) \tag{41}$$

The same function $f(x)$ is said to be *strictly* convex if the $\leq$ sign is replaced by the $<$ sign. Geometrically, (41) means that the hyper-surface $z = f(x)$ is convex if the line segment joining any two points $(x_1, z_1)$, $(x_2, z_2)$ on the surface lies on or above the surface. (Note the difference between the definitions of a convex set and of a convex function.) The sum of convex functions is a convex function.

CONCAVE FUNCTION. The function $f(x)$ is considered to be concave over $x$ (strictly concave) if $-f(x)$ is convex (strictly convex).

LINEAR FUNCTION. $z = cx$ is a convex (or concave) function over all of $E^n$, since

$$c[\alpha x_2 + (1 - \alpha)x_1] = \alpha c x_2 + (1 - \alpha)c x_1 \tag{42}$$

It is, however, neither strictly convex nor strictly concave. A positive (negative) semidefinite quadratic form $z = x^T A x$ is a convex (concave) function over all of $E^n$. A positive (negative) definite quadratic form is a strictly convex (concave) function over all of $E^n$.

# REFERENCES

[1-1]  Blitzer, L. and A. D. Wheelon,"Maximum Range of a Projectile in Vacuum on a Spherical Earth," *Am. J. Phys.*, **25** (1957), 21.

[1-2]  Lawden, D. F., "Optimum Laucnhing of a Rocket into an Orbit about the Earth," *Astronaut. Acta*, **1** (1955), 185.

[1-3]  Fried, B. D. and J. M. Richardson, "Optimum Rocket Trajectories," *J. Appl. Phys.*, **27** (1956), 955.

[1-4]  Leitmann, G., "Optimum Thrust Direction for Maximum Range," *J. Brit. Interplanet. Soc.*, **16** (1958), 503.

[1-5]  Pun, L., "Automatics, from 1964 Onward," *Control* (February, March, April 1964).

[1-6]  Pun, L.. "Comments on Adaptation Concepts and on Analysis of Control Systems," *IEEE Tr*, **AC-8**, n 3 (July, 1963).

[1-7]  Chang, S. S. L., "On the Relative Time of Adaptive Systems," *IEEE Tr*, **AC-10**, n 1 (1965).

[1-8]  Faulders, C. R., "Optimum Thrust Programming of Electrically Powered Rocket Vehicles in a Gravitational Field," *ARS J.* (October 1960), 954–960.

[1-9]  Womack, B. F. and J. N. Dashiel, "A Weighted Time Performance Index for Optimal Control," *IEEE Tr*, **AC-10**, n 2 (April 1965), 201–202.

# 2

---

# STATIC OPTIMIZATION
# TECHNIQUES

IN THIS CHAPTER we shall discuss optimization techniques for static systems, the mathematical models of which are assumed known. In principle, static systems are those whose parameters do not change with time, however, systems whose parameters vary slightly within a reasonable range of time will also be considered static. In Section 2.1, we shall discuss the techniques of ordinary maxima and minima to be used in unconstrained problems. Section 2.2, deals with the Lagrange-multiplier techniques, which are used in equality-constrained problems. These two methods form the basis of all optimization techniques, both static and dynamic. In Section 2.3, we shall discuss linear-programming techniques, and specifically, the simplex method, used for solving linear programming problems (or, linear programs). In Section 2.4, we shall discuss two kinds of techniques for solving nonlinear constrained problems. The first kind comprises those techniques in which the nonlinear problems are amenable to linear programs. The second kind is the convex programming method used for solving general nonlinear problems. In Appendix 2A, some mathematical definitions about maxima and minima are given which are useful throughout this book. In Appendix 2B, we shall discuss shortly the theory of general programming which shows how to relate the state constraints, the equality- and inequality-constraints, and the Lagrange-multiplier technique to the formulation of programming problems within a general and coherent framework. In Appendix 2C, we shall show the mathematical basis of the duality concept. Until now, programming and optimization have been considered distinct techniques aimed at different problems. Important works, however, are presently being undertaken to unify these two branches G-77, G-78.

## 2.1  ORDINARY LOCAL MAXIMA AND MINIMA

In this section, we shall consider the problem of finding the maxima or the minima of functions of one or several independent variables.

$$\underset{x}{\text{Max}} \quad \text{or} \quad \underset{x}{\text{Min}} \; J_1 = f_1(x)$$

$$\underset{x_1, x_2, \ldots, x_n}{\text{Max}} \quad \text{or} \quad \underset{x_1, x_2, \ldots, x_n}{\text{Min}} \; J_2 = f_2(x_1, x_2, \ldots, x_n)$$

where $f_1$ and $f_2 \in C^2$, are necessarily *nonlinear* functions of their arguments. Referring to Sections 1.2.3 and 1.3, we call $J_1$ and $J_2$ unconstrained performance indices, if $x$ does not have to satisfy any other relations. Current notations concerning maxima and minima are:

1. Maximize or minimize the function $J$ with respect to $x$, using an upper-case M.

$$\text{Max } J(x) \text{ or Min } J(x), \quad [\text{Max}] \, J(x) \text{ or } [\text{Min}] \, J(x), \quad \underset{x}{\text{Max}} \, J(x) \text{ or } \underset{x}{\text{Min}} \, J(x)$$

2. The maximum or minimum value of the function $J(x)$, using a lowercase m.

$$\text{max } J(x) \text{ or min } J(x), \text{ max } [J(x)] \text{ or min } [J(x)]$$

3. The maximum or the minimum of the values $x$, $y$, and $z$, using a lowercase m.

$$\text{max } (x, y, z) \text{ or min } (x, y, z)$$

### 2.1.1  One-Dimensional Problem

Let the problem be

$$\underset{x}{\text{Max}} f(x) = a_0 + a_1 x + a_2 x^2 + \cdots + a_n x^n \qquad (2\text{-}1)$$

The *direct* approach is to calculate and compare the values of $f$ corresponding to a finite and previously chosen number of $x$-values. The maxima will be obtained by inspection (Fig. 2-1). The *indirect* approach is as follows.

1. Calculate the first derivative or the rate of variation of $f$:

$$\frac{df}{dx} = f'(x) = a_1 + 2a_2 x + \cdots + n a_n x^{n-1} \qquad (2\text{-}2)$$

and find the $(n - 1)$ roots $x_a, x_b, \ldots, x_g$, equating $f'(x)$ to zero. These points are called stationary points (maximum, minimum or inflexion points).

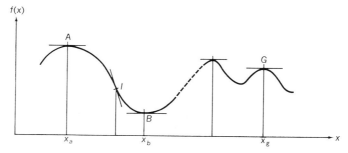

***Fig. 2-1***   One-dimensional extrema.

Let the word extremum signify a maximum or a minimum. The condition

$$f'(x) = 0 \tag{2-3}$$

is a *necessary condition* for the existence of a local extremum. Between two successive maxima, there is always one minimum. Between one maximum and one minimum, there can be one or several inflexion points.

2. Calculate the second derivative or the rate of $f'(x)$ at the points $x_a$, $x_b, \ldots, x_g$ found previously.

$$\frac{df'(x)}{dx} = \frac{d^2 f(x)}{dx^2} = f''(x) = 2a_2 + 6a_3 x + \cdots + n(n-1)a_n x^{n-2} \tag{2-4}$$

When $\alpha$ is any one of the subscripts $a, b, \ldots, g$, then the following three cases are possible:

(a)  For minimum points     $f''(x_\alpha) > 0$       (2-5)

(b)  For inflexion points    $f''(x_\alpha) = 0$       (2-6)

(c)  For maximum points    $f''(x_\alpha) < 0$       (2-7)

The conditions (2-5), (2-6), and (2-7) are *sufficient conditions* for a local extremum. They are insufficient for global extrema, since one considers only the points $x_\alpha$ and their neighborhood. This can be seen also from the curve of Fig. 2-1; the preceding conditions might only supply a maximum point like $G$ which gives a value of $f$ smaller than that at $A$. There is no analytical method which gives the global nature of an extremum.

The sets of two conditions (2-3) and (2-5), or (2-3) and (2-7), respectively, form groups of necessary and sufficient conditions for the existence of a local minimum or a local maximum. Note that the necessary condition type (2-3) defines points like $x_\alpha$ while the sufficient conditions type (2-5) or (2-7), which are inequalities, define areas where $x$ can be located.

It is important to note here that the preceding method does not apply to linear static functions, such as $f(x) = ax + b$, since the derivative $f(x) = a$

is not a function of $x$. This is the reason why the linear programming problems cannot be solved by the ordinary theory of maxima and minima, even when the linear inequalities are amenable to linear equalities.

### 2.1.2   Two-Dimensional Problem

Let the problem be

$$\text{Max } J = f(x_1, x_2) \tag{2-8}$$
$$\scriptstyle x_1, x_2$$

$f$ nonlinear and $\epsilon C^2$. In the vicinity of a point $(a, b)$, $f$ may be represented by a Taylor series expansion.

$$f(x_1, x_2) = f(a, b) + f'_{x_1}(a, b)(x_1 - a) + f'_{x_2}(a, b)(x_2 - b)$$
$$+ \tfrac{1}{2}(f''_{x_1 x_1}(a, b)(x_1 - a)^2 + 2f''_{x_1 x_2}(a, b)(x_1 - a)(x_2 - b)$$
$$+ f''_{x_2 x_2}(a, b)(x_2 - b)^2) + \cdots \tag{2-9}$$

If the point $(a, b)$ is a stationary point of $f(x_1, x_2)$, the two first-order terms will be zero at this point. The existence of an extremum point will depend on the sum of the second-order terms being bigger or smaller than zero for all values of $(x_1 - a)$ and $(x_2 - b)$. A detailed analysis [G-5] will show that this can only happen if the expression $f''_{x_1 x_1} f''_{x_2 x_2} - (f''_{x_1 x_2})^2 > 0$. Then the point $(a, b)$ is a local maximum if $f''_{x_1 x_1} < 0$; a local minimum if $f''_{x_1 x_1} > 0$. Let us summarize as follows.

*Necessary conditions for local extrema:*

$$f'_{x_1}(x_1, x_2) = 0 \tag{2-10}$$

$$f'_{x_2}(x_1, x_2) = 0 \tag{2-11}$$

*Sufficient conditions:*

(a) For a local minimum,

$$f''_{x_1 x_1} > 0 \tag{2-12}$$

$$f''_{x_1 x_1} f''_{x_2 x_2} - (f''_{x_1 x_2})^2 > 0 \tag{2-13}$$

(b) For a local maximum,

$$f''_{x_1 x_1} < 0 \tag{2-14}$$

$$f''_{x_1 x_1} f''_{x_2 x_2} - (f''_{x_1 x_2})^2 > 0 \tag{2-13 repeated}$$

### 2.1.3   Multidimensional Problem

Let the problem be

$$\text{Max } J = f(x_1, x_2, \ldots, x_n)$$
$$\scriptstyle x_1, x_2, \ldots, x_n$$

where $f \in C^2$ for all its arguments is a nonlinear function. The conditions for the existence of a local extremum are given below.

*Necessary conditions*:

$$f_i'(x_1, x_2, \ldots, x_n) = 0, \qquad i = 1, 2, \ldots, n \qquad (2\text{-}15)$$

*Sufficient conditions*:

| | | | |
|---|---|---|---|
| (a) For a local minimum | $D_i > 0$ | $i = 1, 2, \ldots, n$ | (2-16) |
| (b) For a local maximum | $D_i > 0$ | $i = 2, 4, 6, \ldots$ | (2-17) |
| | $D_i < 0$ | $i = 1, 3, 5, \ldots$ | (2-18) |

$D_i$ are determinants of the form:

$$D_i = \begin{vmatrix} f''_{x_1 x_1} & f''_{x_1 x_2} & \cdots & f''_{x_1 x_i} \\ f''_{x_2 x_1} & f''_{x_2 x_2} & \cdots & f''_{x_2 x_i} \\ f''_{x_3 x_1} & & & \cdot \\ \cdot & & & \cdot \\ \cdot & & & \cdot \\ \cdot & & & \cdot \\ f''_{x_i x_1} & \cdots & \cdots & f''_{x_i x_i} \end{vmatrix} \qquad (2\text{-}19)$$

## 2.2   METHOD OF LAGRANGE MULTIPLIERS

In this section, we shall discuss the Lagrange-multiplier technique used for solving equality-constrained problems of the form:

$$\underset{x}{\text{Max }} J = f(x) \quad x = (x_1, x_2, \ldots, x_n) \qquad (2\text{-}20)$$

$$g_i(x) = 0, \qquad i = 1, 2, \ldots, m \qquad (2\text{-}21)$$

where $g_i \in C^2$ are linear or nonlinear functions and where $m \ (\leq, =, \geq) \ n$.

The existence of solutions for the various cases of $m \ (\leq, =, \geq) \ n$ is discussed in Appendix 1C. Here, we shall consider the case where $m < n$. In such a case, we can look at the problem as if there are $(n - m)$ degrees of freedom. Supposing that $n = 5$, $m = 3$, two of the five $x$-values can be chosen to maximize $f$. The other three are automatically determined by the three equations $g_i(x) = 0$.

### 2.2.1   Two-Dimensional Problems

We shall illustrate the Lagrange-multiplier method by a simple two-dimensional problem, and we shall also indicate the important concept of equality of slopes. The problem formulation is

$$\underset{x,y}{\text{Max}} \, J = f(x, y) = 2x + 3y - 1 \tag{2-22}$$

$$g(x, y) = x^2 + 1.5y^2 - 6 = 0 \tag{2-23}$$

Ordinarily, this problem can be solved by the substitution method. From (2-23), extract $x = \pm(6 - 1.5y^2)^{\frac{1}{2}}$, substitute these values of $x$ into (2-22), and maximize $f$ with respect to $y$ by using the ordinary techniques. The Lagrange-multiplier method is preferable, however, for higher-dimensional problems. Let $\lambda$ be some scalar multiplier the value of which is undetermined for the moment. We form a new function

$$F = f + \lambda g = (2x + 3y - 1) + \lambda(x^2 + 1.5y^2 - 6) \tag{2-24}$$

The necessary conditions for $F$ to have stationary points are

$$\frac{\partial F}{\partial x} = \frac{\partial f}{\partial x} + \lambda \frac{\partial g}{\partial x} = 2 + 2\lambda x = 0 \tag{2-25}$$

$$\frac{\partial F}{\partial y} = \frac{\partial f}{\partial y} + \lambda \frac{\partial g}{\partial y} = 3 + 3\lambda y = 0 \tag{2-26}$$

Instead of solving two equations for two variables as in the unconstrained case, we have here three variables $x$, $y$, and $\lambda$. The third necessary condition is supplied by $g$, Eq. (2-23). In other words, we have exactly three equations (2-23) (2-25) (2-26), and three variables. From Eq. (2-25) and Eq. (2-26) we have

$$x = y = -\frac{1}{\lambda} \tag{2-27}$$

Substituting these values in Eq. (2-23):

$$2.5 - 6\lambda^2 = 0 \qquad \lambda = \pm 0.645 \tag{2-28}$$

The second-order partial derivatives are

$$F''_{xx} = 2\lambda \qquad F''_{xy} - F''_{yx} = 0 \qquad F''_{yy} = 3\lambda \tag{2-29}$$

In Table 2.1, we summarize the values of $x$, $y$, $f$, $F''_{xx} F''_{yy}$ for each value of $\lambda$.

The necessary conditions (2-25) and (2-26) can be derived by some simple geometric interpretations. In Fig. 2-2, it is shown that the extremum of the function $J = f(x, y)$ consistent with the constraint $g(x, y) = 0$ occurs at point $P$. At this point, the slope $dy/dx$ of the two curves (see also Appendix 2B, Section 3):

$$f(x, y) = \text{constant} \tag{2-30}$$

$$g(x, y) = 0 \tag{2-31}$$

*Table* **2.1**

| | −0.645 | +0.645 |
|---|---|---|
| Stationary point $\lambda$ | −0.645 | +0.645 |
| Solution $x$ | 1.55 | −1.55 |
| $y$ | 1.55 | −1.55 |
| Performance index $f$ | 6.75 | −8.75 |
| Sufficient conditions $\quad f''_{xx}$ | (−1.29) <0 | (1.29) >0 |
| $f''_{xx}f''_{yy} - (f''_{xy})^2$ | (2.5) >0 | (2.5) >0 |
| Conclusion | Maximum | Minimum |

must be equal, leading to

$$\frac{dy}{dx} = \frac{\partial f/\partial x}{\partial f/\partial y} = \frac{\partial g/\partial x}{\partial g/\partial y} \tag{2-32}$$

Equation (2-32) can be rewritten as

$$\frac{\partial f/\partial x}{\partial g/\partial x} = \frac{\partial f/\partial y}{\partial g/\partial y} = -\lambda \tag{2-33}$$

from which the two conditions (2-25) and (2-26) can be derived immediately. This concept of equality of slopes can be easily generalized to multidimensional problems.

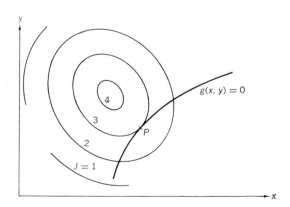

*Fig.* **2-2**   Geometric derivation of the Lagrange-multiplier method.

### 2.2.2  Multidimensional Problems

The general formulation of the multidimensional equality-constrained problem is the following:

$$\text{Max}_{x_i} J = f(x_i) \qquad i = 1, 2, \ldots, n \tag{2-34}$$

$$g_j(x_i) = 0 \qquad j = 1, 2, \ldots, m \quad m < n \tag{2-35}$$

Form the new function

$$F = f + \sum_{j=1}^{m} \lambda_j q_i(x_i) \tag{2-36}$$

The necessary conditions for stationary points are

$$\frac{\partial F(x_i, \lambda_j)}{\partial x_i} = 0 \qquad i = 1, 2, \ldots, n \tag{2-37}$$

The introduction of $m$ multipliers $\lambda$ has increased the number of variables from $n$ variables $x$ to $(n + m)$ variables $x$ and $\lambda$. The simultaneous solution of $m$ equations (2-35) and $n$ equations (2-37) will yield the values of the $(n + m)$ variables.

A typical industrial application of the method of Lagrange multipliers is found in the economic dispatching of interconnected power systems where a set of so-called "coordination equations" is obtained [G-46, G-47, 2-1].[1] As indicated in Fig. 2-3, interconnected power systems consist mainly of three parts: the generators which produce the electrical energy, the transmission lines which transmit it, and the loads which use it. Such a configuration applies for all interconnected networks (regional, national, and

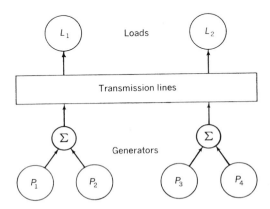

***Fig. 2-3***  A simplified configuration of interconnected power systems.

international), where the numbers of the elements vary. Since the sources of energy are so diverse (coal or gas, riverwater, marine tide, radioactive matter, sun-power), the choice of one or the other is made on economic, technical, or geographic bases. The unit cost of the energy production depends on the category of the plant, and varies even from one plant to another of the same category. The transmission of the energy causes losses in the lines. For a given load, these losses depend on which plant or plants the energy is coming from. In other words, the losses depend on the complexity of the network between the generators and the load. The economic dispatching problem is then: for a prescribed schedule of loads, define the production level of each plant and the paths of transmission to the loads so that the total cost of production and transmission is minimum.

Let:   $P_i$   Power produced by the plant $i$.
     $F_i$   Production cost of $P_i$ (known data).
    $P_{L_i}$   Losses for transmitting $P_i$ to the loads.
  $dF_i/dP_i$   Incremental cost of generation of the plant $i$ (known data).
    $P_L$   Total losses in the transmission lines.
 $\partial P_L/\partial P_i$   Incremental losses in the transmission, for source $i$.
    $L_i$   loss penalty factor of the plant $i$.
    $P_R$   total load (known).

In the language of the optimization problem, we have as:
Performance index

$$F = \sum_{i=1}^{n} F_i \tag{2-38}$$

Constraint

$$H = \sum_{1}^{n} P_i - \sum_{1}^{n} P_{Li} - P_R = 0 \tag{2-39}$$

Using the method of Lagrange multipliers, we have the following necessary conditions for stationary points.

$$\frac{\partial}{\partial P_i}(F + \lambda H) = 0 \tag{2-40}$$

We therefore have $n$ equations (2-40) plus one equation (2-39) for $n$ variables $P_i$ plus one variable $\lambda$. Eq. (2-40) can be written as

$$\frac{\partial}{\partial P_i}\left(\sum_{1}^{n} F_i + \lambda \sum_{1}^{n} P_i - \lambda \sum_{1}^{n} P_{Li} - \lambda P_R\right) = 0 \tag{2-41}$$

Since $P_R$ is independent of $P_i$

$$\frac{\partial P_R}{\partial P_i} = 0 \tag{2-42}$$

On the other hand, for each plant, obviously

$$\frac{\partial P_j}{\partial P_i} = 1 \quad j = i; \qquad \frac{\partial P_j}{\partial P_i} = 0 \quad j \neq i \tag{2-43}$$

$$\frac{\partial F_j}{\partial P_i} = \frac{dF_i}{dP_i} \quad j = i; \qquad \frac{\partial F_j}{\partial P_i} = 0 \quad j \neq i \tag{2-44}$$

Equation (2-41) becomes

$$\frac{dF_i}{dP_i} + \lambda - \lambda \frac{\partial P_L}{\partial P_i} = 0 \tag{2-45}$$

or

$$\frac{dF_i}{dP_i} L_i = \lambda \tag{2-46}$$

where

$$L_i = 1 \bigg/ \left( \frac{\partial P_L}{\partial P_i} - 1 \right).$$

Equations (2-45) or (2-46) are called coordination equations which state that the levels $P_i$ must be chosen so that all the weighted slopes $(dF_i/dP_i) L_i$ of the various plants must be equal to the same quantity $\lambda$. (Fig. 2-4). For each value of $\lambda$, there is a combination of values of $P_i$. Since the $P_i$'s must satisfy the constraint (2-39) (total production = total loads plus total losses), the value of $\lambda$ depends on the total demand. The coordination equations permit the realization of an on-line economic dispatching control, such as shown in Fig. 2-5 for solving Eqs. (2-45) and (2-39). To do this, a computer elaborates the value of $\lambda$ according to the scheduled demand of loads. The signal is sent to each plant where the signal $\lambda/L_i$ is generated. Equations (2-45) are satisfied by implementing function generators of $dF_i/dP_i = f(P_i)$ which deliver two signals, the constant part $P_{i0}$ and the variable $P_{iD}$. Comparison of $P_{iD}$ and

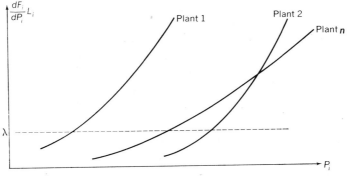

**Fig. 2-4**   Nonlinear functions $dF_i/dP_i = f(P_i)$.

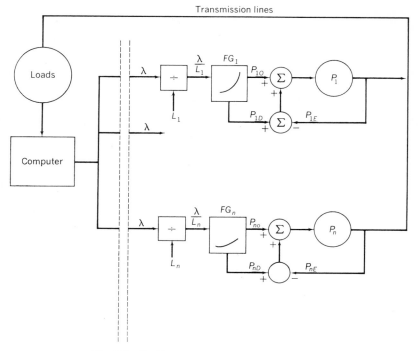

*Fig. 2-5*   On-line economic dispatching control.

the feedback effective production level $P_{iE}$ ensures the desired production level.

Practical difficulties arise in obtaining the values of $dF_i/dP_i$ and $L_i$ considered as known quantities in the preceding treatment. The costs $F_i$ and the cost rates $dF_i/dP_i$ are generally nonlinear functions of $P_i$, (Fig. 2-5), and can only be obtained by laborious experimentations. The values of $L_i$ depend on the configuration of the overall network and on the knowledge of the instantaneous values of $P_i$. The general method of this computation is to assume

$$P_L = \sum_m \sum_n P_m \cdot B_{mn} P_n \qquad (2\text{-}47)$$

where $P_m$ and $P_n$ are production levels, and $B_{mn}$ coefficients to be determined according to the network configurations. The computation of $L_i$ is also very tedious [G-46, Ch. 3]; in practice it can be aided by some network simulators.

### 2.2.3   Computational Difficulties in Multidimensional Nonlinear Problems

The utilization of the method of Lagrange multiplier becomes complex when the performance index $f$ and the constraints $g_i$ are both nonlinear and multidimensional. The application example treated in Section 2.2.2 has been facilitated by the use of experimental curves $dF_i/dP_i = f(P_i)$ and the use of on-line control elements. Generally, analytical difficulties will arise in nonlinear cases.

Let us therefore consider the following problem where $f$ and $g_i$ comprise high-degree terms of the variables.

$$f(x, y, z) = x^4 + 4x^3z^2 - y^2z^3 - 4y^5$$

$$g_1(x, y, z) = 2x^4y + 3z^2y^3 - xy^3 - 4z^4x^2 + 1 = 0 \tag{a}$$

$$g_2(x, y, z) = 7x^6y + 2y^4z^2 - 5y^2xz^3 - y^5 = 0 \tag{b}$$

After the introduction of two Lagrange multipliers $\lambda_1$ and $\lambda_2$, the satisfaction of the conditions

$$\frac{\partial}{\partial x}(f + \lambda_1 g_1 + \lambda_2 g_2) = 0$$

$$\frac{\partial}{\partial y}(f + \lambda_1 g_1 + \lambda_2 g_2) = 0$$

$$\frac{\partial}{\partial z}(f + \lambda_1 g_1 + \lambda_2 g_2) = 0$$

leads to

$$42\lambda_2 yx^5 + (4 + 8\lambda_1 y)x^3 + 12x^2z^2 - 8\lambda_1 z^4 x - 5\lambda_2 y^2z^3 - \lambda_1 y^3 = 0 \tag{c}$$

$$7\lambda_2 x^6 + 2\lambda_1 x^4 - y(3\lambda_1 y + 10\lambda_2 y^3)x - 5(4 + \lambda_2)y^4 + 8\lambda_2 z^2 y^3 + 9\lambda_1 z^2 y^2 = 0 \tag{d}$$

$$8zx^3 - 16\lambda_1 z^3 x^2 - 15\lambda_2^2 yz^2 x + 4\lambda_2 zy^4 + 6\lambda_1' zy^3 - 3z^2 y^2 = 0 \tag{e}$$

There are five equations (a), (b), (c), (d), and (e) for five unknowns $x$, $y$, $z$, $\lambda_1$, $\lambda_2$. However, because of the high degrees of the equations, there will not be one, but $4 \times 6 \times 5 \times 6 \times 4 = 2880$ theoretical solutions without talking about the difficulties for finding these solutions.

## 2.3  LINEAR PROGRAMMING METHOD

In this section, we shall consider the following type of optimization problem.

$$\text{Min } z = \sum_{1}^{n} c_i x_i \tag{2-48}$$

$$\sum_{j=1}^{n} a_{ij} x_j \subseteq b_j \qquad i = 1, 2, \ldots, m \tag{2-49}$$

$$x_i \geq 0 \qquad i = 1, 2, \ldots, n \tag{2-50}$$

where

$$x_i \quad \text{The state variables.}$$

$$c_i, a_{ij}, b_j \quad \text{Arbitrary known constants.}$$

Such a formulation is called a linear program because (2-48) and (2-49) are linear relations [G-14, G-15, G-17, G-20]. Furthermore, inequalities (2-49) are independent of each other. Contrary to the Lagrange-multiplier method, there is no condition imposed on the relative value of $m$ and $n$. The case $m < n$ (more variables than constraints) is normal. But if $m > n$ (more constraints than variables), it is possible to use slack variables to increase $n$ (see Section 2.3.2). This is possible because of the inequality nature of the constraints. The linear program problem is basically treated by the Simplex method, established originally by G. Dantzig. In the linear programming language, the performance index is also called cost function, value function, and economic function, because linear programs usually arise in economic fields. Furthermore, the variables are produced or stocked materials, always measured in nonnegative quantities (zero or positive quantities), and this explains the reason of the constraints (2-50).

There are a number of concepts in the linear programming methods such as basic solution, polyhedron, convex set, and duality, which are interesting for the understanding of other optimization techniques. We shall first expose them by using a simple example and then introduce the Simplex techniques so that the reader will know at least this fundamental tool. Finally, we shall discuss the variety of practical problems which can be formulated as a linear program.

### 2.3.1  Basic Concepts

To introduce the basic concepts, we shall use a blending problem [G-15]. There are a number of blending problems in practice: food, alloy, cement, etc. Here we shall take a food blending problem which can be stated as follows: Develop an economical food for certain animals. The food must contain four kinds of components $A$, $B$, $C$, and $D$. The daily consumption per animal would be at least 0.4 kg of $A$, 0.6 kg of $B$, 2 kg of $C$; 1.7 kg of $D$. From the market, it is possible to buy two products $M$ and $N$ containing these components. One kilogram of $M$, which costs ten dollars, contains 0.1 kg of $A$, 0.1 of $C$, and 0.2 of $D$. One kilogram of $N$, which costs four dollars, contains 0.1 kg of $B$, 0.2 kg of $C$, 0.1 kg of $D$. What are the most economic daily purchases of $M$ and $N$ per animal?

This literal statement is represented in Table 2.2.

This table is called a linear programming model, in which two new concepts are introduced: the expression "activity" denoting elements such as $M$, $N$;

*Table* **2.2**

| Nutritive Components | Market Product M | N | Daily Prescribed Quantity per Animal |
|:---:|:---:|:---:|:---:|
| A | 0.1 | 0 | 0.4 |
| B | 0 | 0.1 | 0.6 |
| C | 0.1 | 0.2 | 2 |
| D | 0.2 | 0.1 | 1.7 |
| Cost | 10 | 4 | Min |

and "item" denoting elements such as $A$, $B$, $C$, $D$. The first step in formulating a linear programming model is to isolate "activities," "items," "cost function," and "constraints" from the considered daily problem. Let $x_1$ and $x_2$ be the quantity of $M$ and $N$ respectively to be purchased per animal per day. The performance index and the constraints are written as[2]

$$[\text{Min}]z = 10x_1 + 4x_2 \tag{2-51}$$

$$0.1x_1 \geq 0.4 \qquad\qquad x_1 \geq 4 \tag{2-52}$$

$$0.1x_2 \geq 0.6 \qquad\qquad x_2 \geq 6 \tag{2-53}$$

$$\text{or}$$

$$0.1x_1 + 0.2x_2 \geq 2 \qquad x_1 + 2x_2 \geq 20 \tag{2-54}$$

$$0.2x_1 + 0.1x_2 \geq 1.7 \qquad 2x_1 + x_2 \geq 17 \tag{2-55}$$

$$x_1 \geq 0 \qquad x_2 \geq 0 \tag{2-56}$$

Relations (2-51) to (2-56) constitute a linear program the form of which corresponds to the general mathematical formulation of relations (2-48) to (2-50). In the case of the illustrative example, inequalities (2-56) are absorbed by inequalities (2-52) and (2-53).

This simple blending problem can be solved graphically. The constraints (2-56) mean that only the first quadrant of the $x_1 - x_2$ plane has to be considered. (Fig. 2-6). The constraints (2-52) to (2-56), if the equal sign is taken, are represented respectively by straight lines $\Delta_A \Delta_B \Delta_C \Delta_D$. The "bigger than" sign in these constraints means that any feasible optimal solution $(x^*, y^*)$ must lie at the right of all the $\Delta$-lines, or, in other words, in the nonshaded part of the $x_1 - x_2$ plane. Clearly, there are an infinite number of feasible solutions.

This infinite number can be reduced by the following consideration. The cost function

$$z = 10x_1 + 4x_2 = \text{constant} \tag{2-57}$$

is represented by a family of straight lines $Z$ parallel to the straight line

$$10x_1 + 4x_2 = 0 \tag{2-58}$$

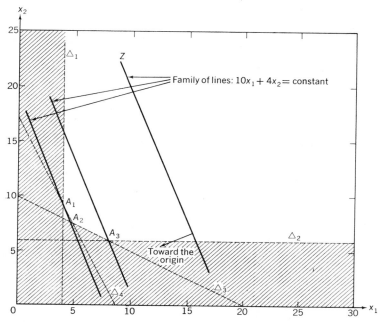

***Fig. 2-6*** Graphical solution of the simple blending problem.

passing through the points $(0, 0)$ and $(-4, 10)$. It is known that the distance $d$ between the origin and a straight line of the form

$$ax + by = c \qquad (2\text{-}59)$$

for given value of $a$, $b$ is proportional to $c$:

$$d = \left| \frac{c}{\sqrt{a^2 + b^2}} \right| \qquad (2\text{-}60)$$

Therefore the value of $z$ is proportional to the distance between the origin 0 and the $Z$-lines. The boundary of the permissible zone, consisting of portions of $Z$-lines, forms an open polygon. The points $(x_1, x_2)$ corresponding to smallest permissible values of $z$ lie necessarily on the polygon. The final solution is therefore the intersecting point of a $Z$-line and the polygon. Since the polygon is convex, such an intersection is always situated at one of the vertices $A_1$, $A_2$, or $A_3$. In our case, it is the point $A_1$, the 2 coordinates of which are $x_1 = 4$, $x_2 = 9$ leading to

$$z = 10.4 + 4.9 = 76 \text{ dollars}$$

The next-to-minimum vertex is $A_2$, where $x_1 = 4.6$, $x_2 = 7.8$, $z = 77.2$ dollars.

From this simple example, we can draw a number of important conclusions which are valid not only for two-dimensional problems, but also for multidimensional problems.

1. The constraints form a *polygon* (in a two-dimensional case), or a *polyhedron* (in a multidimensional case) which is convex from or convex to the origin, according to whether the "bigger than" or the "smaller than" condition is imposed.

2. The $z$-constant equations form a family of straight *lines* or *hyperplanes*.

3. Among the infinite feasible solutions, there is a special category of feasible solutions of a finite number. These solutions, called *basic solutions*, are situated geometrically at the *vertices* of the polygon or the polyhedron. The number of basic solutions is automatically determined by the number of coordinates or variables $N$ and the number of constraints $M$, according to the combinatorial formulas

$$C_N^M = N!/M!\,(N - M)!$$

Table 2.3 shows several examples.

*Table* **2.3**

| Number of Variables | Number of Constraints | Number of Basic Solution |
|:---:|:---:|:---:|
| 2 | 1 | 2 |
| 3 | 1 | 3 |
| 3 | 2 | 3 |
| 4 | 1 | 4 |
| 4 | 2 | 6 |
| 4 | 3 | 4 |
| 5 | 1 | 5 |
| 5 | 2 | 10 |
| 5 | 3 | 10 |
| 5 | 4 | 5 |
| 7 | 5 | 21 |
| 10 | 7 | 120 |

4. The extremum solution is given by the remotest or the nearest intersecting points between the $z$-hyperplane and the polyhedron. Unfortunately, the graphical prodecure is only practical for two-, or perhaps three-dimensional problems. For higher-dimensional problems, nongraphical methods must be used, the basic one being the Simplex method. But, before discussing this method, we shall introduce the concept of duality. (See Appendix 2C.)

Table 2.4

|  | A | B | C | D | Cost |
|---|---|---|---|---|---|
| M | 0,1 | 0 | 0,1 | 0,2 | 10 |
| N | 0 | 0,1 | 0,2 | 0,1 | 4 |
| Prescribed quantities | 0,4 | 0,6 | 2 | 1,7 | → [Max] g |

Instead of restating the problem used in the beginning of this section, we shall state the problem for the competitor of the manufacturer of the products $M$ and $N$. The competitor manufactures the components $A$, $B$, $C$, $D$. His problem is how to define the unit prices of the components $A$, $B$, $C$, $D$, knowing the same constraining conditions, so that his profit is maximum?

Let $y_1$, $y_2$, $y_3$, $y_4$ be these unit prices; the corresponding linear program is

$$[\text{Max}]\, g = 0.4y_1 + 0.6y_2 + 2y_3 + 1.7y_4 \tag{2-61}$$

$$0.1y_1 + 0.1y_3 + 0.2y_4 \le 10 \tag{2-62}$$

$$0.1y_2 + 0.2y_3 + 0.1y_4 \le 4 \tag{2-63}$$

$$y_1, y_2, y_3, y_4 \ge 0 \tag{2-64}$$

The linear programming model is shown in Table 2.4. It is found by the Simplex method that the solution is

$$x_1 = 20 \qquad x_2 = 0 \qquad x_3 = 0 \qquad x_4 = 40 \qquad g = 76$$

The maximum profit of the competitor is equal to the minimum expenses of the buyer.

$$\max g = \min z = 76$$

This property of *duality* does not only stand for the result, but also for the statement (Table 2.5). It can happen, in practice, that it is easier to and the optimal solution of the dual program.

Table 2.5

| [Max]$g = \sum_{j=1}^{n} c_j y_j$ | is | [Min]$z = \sum_{i=1}^{m} b_i x_i$ |
|---|---|---|
| $\sum_{j=1}^{n} a_{ij} y_j \le b_i$ | → dual → | $\sum_{i=1}^{m} a_{ji} x_i \ge c_j$ |
| $i = 1, 2, \ldots, m$ | to | $j = 1, 2, \ldots, n$ |
|  | and max $g$ = min $z$ |  |

## 2.3.2   Simplex Method—Useful Definitions

In Section 2-1.1, we mentioned the difference between *direct* and *indirect* methods in optimization techniques. The direct method consists of prospecting *all* permissible solutions and finding the extremum by comparison; in indirect methods, we first find a set of extremalizing equations, the solution of which, if obtainable, yields the extremum point. The method of Lagrange multipliers is such an indirect method. Both the graphical method, sketched in Section 2.3.1, and the Simplex method which will be described in this section, are neither direct nor indirect. In effect, as can be seen from Fig. 2-6, the permissible solutions are points in the whole nonshaded area including the limiting polygon, while the prospection is only made on three points located at the vertices of the polygon. Such a method can therefore be called a *semidirect method*.

It was seen in the previous section that for solving linear programming problems, it is necessary to inspect only the basic solutions consisting of the vertices of the polyhedron. Unfortunately, the number of these vertices increases quickly with the dimension of the problem. It can be proven that this number has as superior limit the number of combinations of $N$ objects $M$ by $M$. If $N = 17$, $M = 8$, this limit is 24, 310; and $10^{26}$ if $N = 100$, and $M = 38$. In practice, there are linear programming problems having 500 variables, and more than 100 constraints. The complete prospection of even only the basic solutions is impossible.

The Simplex method of Dantzig provides means of optimum walks from one vertex to another. For example:

1. Means to find a vertex of the polyhedron.
2. Criterion for judging quickly if the corresponding value of the cost function is minimum (or maximum) or not.
3. If not, means to find another vertex where the cost function has a smaller (or bigger) value.

The method is not difficult, but is complex. A number of definitions and steps of the operating procedure must be examined and understood carefully before one knows how to use it. In this section, we shall present the definitions, and in the next section, the operating procedure. The slack and surplus variables are used for transforming the general form (2-48) (2-49) (2-50) into the standard form which contains only equality constraints. The standard form is then converted into the canonical form for expliciting the basic variables. The other definitions are useful for understanding the Simplex method.

Definitions

### I. *Slack and Surplus Variables*

Artificial variables used for transforming inequalities into equivalent equalities.

(a) An ($\leq$)-inequality of the form

$$\sum_j a_{ij} x_j \leq b_i$$

is transformed into an equality

$$\sum_j a_{ij} x_j + x_{n+i} = b_i$$

by the use of a slack variable

$$x_{n+i} \geq 0.$$

(b) An ($\geq$)-inequality of the form

$$\sum_j a_{ij} x_j \geq b_i$$

is transformed into an equality

$$\sum_j a_{ij} x_j - x_{n+i} = b_i$$

by the use of a surplus variable

$$x_{n+i} \geq 0.$$

Slack and surplus variables are always nonnegative.

(c) It may happen, in some problems, that some variables $x_j$ may be unrestricted in sign. Since the linear programming method is based merely upon nonnegative variables, the unrestricted variable $x_j$ can be transformed into the difference of two nonnegative variables

$$x_j = x_j' - x_j''$$

$x_j$ sign unrestricted

$$x_j', x_j'' \geq 0.$$

Examples

I.

(a) Use $x_3$ to transform

$$x_1 + 2x_2 \leq 5$$

into

$$x_1 + 2x_2 + x_3 = 5$$

$$x_3 \geq 0.$$

(b) Use $x_4$ to transform

$$2x_1 + 3x_2 \geq 7$$

into

$$2x_1 + 3x_2 - x_4 = 7$$

$$x_4 \geq 0.$$

(c) $x_j > 0$ when $|x_j'| > |x_j''|$

$x_j = 0$ when $|x_j'| = |x_j''|$

$x_j < 0$ when $|x_j'| < |x_j''|$

in all cases

$$x_j', x_j'' \geq 0.$$

| Definitions | Examples |
|---|---|

## II. *Standard System*

The constraint system written in the form

$$\sum_{j=1}^{m+n} a_{ij}x_j = b_i, \quad i = 1, 2, \ldots, m \quad (E_1)$$

$$x_j \geq 0, \quad j = 1, 2, \ldots, m+n \quad (E_2)$$

and obtained from the original constraint system

$$\sum_{j=1}^{n} a_{ij}x_j (\geq, \leq)b_i, \quad i = 1, 2, \ldots, m$$

$$x_j \geq 0, \quad j = 1, 2, \ldots, n$$

by using slack or surplus variables. The maximum number of the slack or surplus variables $x_{n+i}$ is $m$, which is the number of the inequality constraints $(E_1)$.

## II. The original constraint set

$$x_1 + 2x_2 \leq 5$$
$$2x_1 + 3x_2 \geq 7$$
$$x_1, x_2 \geq 0.$$

is transformed into the standard form

$$x_1 + 2x_2 + x_3 = 5 \quad (E_3)$$
$$2x_1 + 3x_2 - x_4 = 7 \quad (E_4)$$
$$x_1, x_2, x_3, x_4 \geq 0.$$

## III. *Feasible Solution*

Any set of values $x_j$ satisfying $(E_1)$ and $(E_2)$ of II is called a feasible solution.

## III. In example of II, the set of values

$$x_1 = -1 \qquad x_2 = 3$$
$$x_3 = 0 \qquad x_4 = 0$$

is a feasible solution.

## IV. *Canonical System*

Set of equations derived from the standard system and written in the following form:

$$x_1 \qquad + \tilde{a}_{1,m+1}x_{m+1} \cdots + \tilde{a}_{1m+n}x_{m+n} = \tilde{b}_1$$
$$x_2 \qquad + \tilde{a}_{2,m+1}x_{m+1} \cdots + \tilde{a}_{2m+n}x_{m+n} = \tilde{b}_2$$
$$\vdots \qquad \vdots \qquad \vdots \qquad \vdots$$
$$x_m + \tilde{a}_{m,n+1}x_{m+1} \cdots + \tilde{a}_{mm+n}x_{m+n} = \tilde{b}_m$$

(a) *Basic variables:* any one of the $x_1, x_2, \ldots, x_m$ is called a basic variable.

(b) *Dependent variables:* In the canonical system form, $x_1, x_2, \ldots, x_m$ are called dependent variables.

(c) *Independent variables:* In the canonical system form, $x_{m+1}, x_{m+2}, \ldots, x_n$ are called independent or nonbasic variables.

(d) *Pivotal variables:* Any of $x_1, x_2, \ldots, x_m$ is a pivotal variable. A pivotal variable $x_i$ has a coefficient of unity in the $i$th row equation, and 0-coefficient elsewhere.

## IV.

$$x_1 \quad - 3x_3 - 2x_4 = -1 \quad (E_5)$$
$$x_2 + 2x_3 + x_4 = 3 \quad (E_6)$$

(a) $x_1, x_2$

(b) $x_1, x_2$

(c) $x_3, x_4$

(d) $x_1, x_2$

Blank

### Definitions

(e) *Pivoting:* A pivot operation consists of $m$ elementary operations which replace a standard system by an equivalent canonical system. These elementary operations are simple multiplications and additions such as those done in the solution of linear algebraic equations.

(f) *Number of canonical systems*

With a standard system, of $(m + n)$ variables and $m$ constraints, the number of canonical systems which can be formed is the number of combinations of $(m + n)$ objects $m$ by $m$.

V. *Basic Feasible Solution*

The special solution of a canonical system obtained by setting all the independent variables equal to zero and solving for the dependent variables is called a basic solution. For the system in IV, the basic solution is

$$x_{m+1}, x_{m+2}, \ldots, x_n = 0$$
$$x_i = b_i, \ i = 1, 2, \ldots, m$$

A basic solution is called basic feasible solution if it is feasible.

VI. *Relative Cost Factors*

In the relation

$$z = \sum_{j=1}^{n} c_j x_j$$

the coefficients $c_j$ are called cost factors. When the standard system is transformed into a canonical system, the cost function is written as

$$(-z) + \tilde{c}_{m+1} x_{m+1} + \cdots + \tilde{c}_n x_n = -\tilde{z}_0$$

The coefficients $\tilde{c}_{m+1}, \ldots, \tilde{c}_n$ are called relative cost function.

*Note.* The value of the variable $z$ in the basic solution is

$$z = \tilde{z}_0.$$

### Examples

(e) The system in IV is obtained from the system in II by pivoting $x_2$

$$2(E_3) \to (E_7)$$
$$2(x_1 + 2x_2 + x_3) = 10 \qquad (E_7)$$
$$(E_7) - (E_4) \to (E_6)$$

from $(E_6)$, one obtains

$$x_2 = 3 - 2x_3 - x_4. \qquad (E_8)$$

substitute $(E_8)$ in $(E_3)$ and solve for $x_1$ gives $(E_5)$.

(f) $m + n = 4$

$$m = 2$$
$$C_4^2 = \frac{4!}{2!\,(4-2)!} = 6$$

V. From $(E_5)$ and $(E_6)$: $x_3, x_4 = 0$
$$x_1 = -1$$
$$x_2 = 3$$

Since $x_1 = -1$, the constraints $x_j \geq 0$ are not satisfied. Neither the canonical system, nor the subsequent basic solution, is feasible.

VI. Let, for example,

$$z = 8x_1 + 9x_2 \qquad (c_1 = 8, c_2 = 9)$$

From $(E_5)$ and $(E_6)$

$$x_1 = -1 + 3x_3 + 2x_4$$
$$8x_1 = -8 + 24x_3 + 16x_4$$
$$x_2 = 3 - 2x_3 - x_4$$
$$9x_2 = 27 - 18x_3 - 9x_4$$

So that

$$z = 19 + 6x_3 + 7x_4$$

Or

$$-z + 6x_3 + 7x_4 = -19$$

Then

$$\tilde{c}_{m+1} = 6 \quad \tilde{c}_n = 7 \quad \tilde{z}_0 = 19$$

In the basic solution, $z = 19$

In the preceding section, we have emphasized the graphical role played by the vertices, or the basic solutions, in the search for the minimal solution. Among the definitions given here, the reader has certainly noticed those concerning the mathematical formulation of the same concept. Another important point is the convexity of canonical system forms. The constraints of the linear program, form a convex polyhedron, from or to the origin according to the sign of the inequalities. The various linear transformations do not change the convexity of the polyhedron. The properties concerning the vertices therefore remain the same; they guarantee the necessity and the sufficiency of the final minimal solution, if it is found.

### 2.3.3  Simplex Method—Operating Procedure

Two procedures are attached to the name of Simplex: the Simplex algorithm and the Simplex method. The Simplex algorithm is the method of optimally going from one basic solution to another, and is always initiated with a program of equations which are already in canonical form. The Simplex method comprises two phases. In phase I, an initial basic solution is found; in phase II, the optimal solution is found. In both phases, the Simplex algorithm is used [G-14] and we shall start with it here.

We suppose that the linear program is already written in canonical form:

$$
\begin{aligned}
x_1 \quad & + \tilde{a}_{1,m+1}x_{m+1} + \cdots + \tilde{a}_{1,m+n}x_{m+n} = \tilde{b}_1 \\
x_2 \quad & + \tilde{a}_{2,m+1}x_{m+1} + \cdots + \tilde{a}_{2,m+n}x_{m+n} = \tilde{b}_2 \\
x_3 \quad & \\
& \hspace{2em}\cdot \\
& \hspace{2em}\cdot \\
& \hspace{2em}\cdot
\end{aligned}
$$

$$x_m + \tilde{a}_{m,m+1}x_{m+1} + \cdots + \tilde{a}_{m,m+n}x_{m+n} = \tilde{b}_m \qquad (2\text{-}65)$$

$$(-z) + \tilde{c}_{m+1}x_{m+1} + \cdots + \tilde{c}_{m+n}x_{m+n} = -\tilde{z}_0 \qquad (2\text{-}66)$$

The values of the coefficients $\tilde{a}$, $\tilde{b}$, $\tilde{c}$, $\tilde{z}_0$ depend on the choice of the basic set of variables. It can be stated that a basic feasible solution is a minimal feasible solution with a total cost $\tilde{z}_0$, if all relative cost factors are nonnegative.

$$\tilde{c}_j \geq 0 \qquad (2\text{-}67)$$

In effect, rewrite (2-66) under the form

$$z - \tilde{z}_0 = \sum_{m+1}^{n} \tilde{c}_j \cdot x_j \qquad (2\text{-}68)$$

If all $\tilde{c}_j$ are positive or zero, the smallest value of $\sum \tilde{c}_j x_j$, hence $z - \tilde{z}_0$, is zero for any choice of nonnegative $x_j$. Therefore

$$z - \tilde{z}_0 \geq 0, \qquad z \geq \tilde{z}_0 \qquad (2\text{-}69)$$

In the case of the feasible basic solution

$$z = \tilde{z}_0; \qquad \text{Min } z = \tilde{z}_0 \tag{2-70}$$

To illustrate, we take the problem of minimizing $z$ where [G-14, pp. 95–97]

$$
\begin{aligned}
5x_1 - 4x_2 + 13x_3 - 2x_4 + x_5 &= 20 \\
x_1 - x_2 + 5x_3 - x_4 + x_5 &= 8 \\
z = \quad x_1 + 6x_2 - 7x_3 + x_4 + 5x_5 & \\
x_j &\geq 0
\end{aligned}
\tag{2-71}
$$

Choose $x_1$, $x_5$ and $(-z)$ as pivotal variables; the following canonical form results:

$$
\begin{aligned}
+x_5 - \quad 0.25x_2 + 3x_3 - 0.75x_4 &= 5 \\
+x_1 - \quad 0.75x_2 + 2x_3 - 0.25x_4 &= 3 \\
(-z) + \quad 8x_2 - 24x_3 + 5x_4 &= -28 \\
x_j &\geq 0
\end{aligned}
\tag{2-72}
$$

The basic feasible solution to system (2-72) is

$$x_2 = x_3 = x_4 = 0 \qquad x_1 = 3 \qquad x_5 = 5$$
$$z = 28 \tag{2-73}$$

This solution is not minimal, because one of the coefficients in $z$, namely the one of $x_3$, is not positive.

Now the problem arises as how to decrease the value of $z$. Since the two terms in $z$, namely $8x_2$ and $5x_4$ are positive, the only way is to increase $x_3$, since the corresponding boundary of $z$ will be

$$z = 28 - 24x_3 \tag{2-74}$$

The larger the value of $x_3$, the smaller will be the value of $z$. However, because of the constraints all $x_j \geq 0$, this increase of $x_3$ is limited by

$$x_5 = 5 - 3x_3 \geq 0 \tag{2-75}$$
$$x_1 = 3 - 2x_3 \geq 0 \tag{2-76}$$

The greatest value that $x_3$ can take, without violating the constraints, is given by Eq. (2-76) where we shall take $x_1 = 0$, resulting in

$$x_3 = 3/2 = 1.5 \tag{2-77}$$

This means that we shall choose $x_3$ and $x_5$ as new pivotal variables, instead of

$x_1$ and $x_5$. The new canonical system is

$$x_5 \qquad - 1.5x_1 + 0.875x_2 - 0.375x_4 = 0.5$$
$$x_3 \quad + 0.5x_1 - 0.375x_2 - 0.125x_4 = 1.5$$
$$(-z) + 12x_1 - \qquad x_2 + \qquad 2x_4 = 8 \qquad (2\text{-}78)$$
$$x_j \geq 0$$

Its basic feasible solution is

$$x_1 = x_2 = x_4 = 0 \qquad x_3 = 1.5 \qquad x_5 = 0.5 \qquad z = -8 \qquad (2\text{-}79)$$

This solution is not yet minimal, because the coefficient of $x_2$ in $z$ is negative. Proceeding as before, we see that the maximum admissible value of $x_2$ is given by setting $x_5 = 0$ in the equation

$$x_5 = 0.5 - 0.875x_2 \geq 0 \qquad (2\text{-}80)$$

Using $x_2$ and $x_3$ as pivotal variables, we have the following canonical system:

$$x_2 \qquad - (\tfrac{12}{7})x_1 - (\tfrac{3}{7})x_4 + (\tfrac{8}{7})x_5 = \tfrac{4}{7}$$
$$x_3 \quad - (\tfrac{1}{7})x_1 - (\tfrac{2}{7})x_4 + (\tfrac{3}{7})x_5 = \tfrac{12}{7}$$
$$(-z) + (\tfrac{12}{7})x_1 + (\tfrac{11}{7})x_4 + (\tfrac{8}{7})x_5 = \tfrac{60}{7} \qquad (2\text{-}81)$$
$$x_j \geq 0$$

and the following basic feasible solution:

$$x_1 = x_4 = x_5 = 0 \qquad x_2 = \tfrac{4}{7} \qquad x_3 = \tfrac{12}{7} \qquad z = -\tfrac{60}{7} \qquad (2\text{-}82)$$

Since all the relative cost factors $\tilde{c}_j$ are nonnegative, the value of

$$z = -\tfrac{60}{7} \qquad (2\text{-}83)$$

is minimal.

The simplex algorithm can be summarized as follows.

**Step 1.** Establish a canonical system such as (2-65) (2-66) from the linear program in order to obtain a basic feasible solution[3]

$$x_j = 0, \qquad j = m + 1, \ldots, m + n$$
$$x_i = \tilde{b}_i, \qquad i = 1, 2, \ldots, m$$
$$z = z_0.$$

**Step 2.** Inspect the sign of the coefficients $\tilde{c}_j$. The solution is minimal if all $\tilde{c}_j \geq 0$.

**Step 3.** In case the optimality test fails, choose a new pivotal variable $x_s$. If there is only one $\tilde{c}_j < 0$, take $x_s$ corresponding to that $\tilde{c}_j = \tilde{c}_s$. If there are several $\tilde{c}_j < 0$, the index $s$ is chosen such that

$$\tilde{c}_s = \text{Min } \tilde{c}_j < 0 \qquad (2\text{-}84)$$

(Min $\tilde{c}_j$ = the smallest of the coefficients $\tilde{c}_j$)

***Step 4.***  Choose the pivotal variable $x_r$ to be replaced by $x_s$ by the condition

$$x_s^* = \frac{\tilde{b}_r}{\tilde{a}_{rs}} = \underset{a_{is} > 0}{\text{Min}} \frac{\tilde{b}_i}{\tilde{a}_{is}} > 0 \tag{2-85}$$

among all the old pivotal variables $x_i$, the coefficients $\tilde{a}_{is}$ of which are positive.

***Step 5.***  Establish the new canonical system and the new basic feasible solution. Inspect the sign of the new coefficient $\tilde{c}_j$. Rework Steps 3 and 4 until the optimality test is proven.

We come now to the problem of finding an initial basic feasible solution. Except for the cases where the linear constraints are of the form

$$\sum_{i=1}^{n} a_{ij} x_j \leq b_j \qquad j = 1, 2, \ldots, m$$

it is not always possible to find easily an initial basic feasible solution. This is because the original system may be redundant, inconsistent, or not solvable in nonnegative numbers. G. Dantzig [G-14, pp. 101–103] has suggested a procedure similar to the Simplex algorithm previously described.

***Step A.***  Transform the original inequality system into the standard form, which will then contain $N = m + n$ nonnegative variables and $M = m$ equality constraints. Arrange the $M$ constraint equations so that all constant terms $b_i$ are positive or zero by changing, where necessary, the signs on both sides of any of the equations. For instance, if we had

$$2x_1 + 3x_2 - x_4 = -7$$

by changing the signs, we would have

$$-2x_1 - 3x_2 + x_4 = 7$$

Note that the variables $x_j$, (including the slack and surplus variables) $j = 1, 2, \ldots, N = m + n$, remain nonnegative after the processing.

***Step B.***  Augment the system to include a basic set of artificial or error variables $x_{N+1} \geq 0$, $x_{N+2} \geq 0$, ..., $x_{N+M} \geq 0$, so that it becomes

$$\begin{aligned}
a_{11}x_1 + a_{12}x_2 + \cdots + a_{1N}x_N + x_{N+1} \quad &= b_1 \\
a_{21}x_1 + a_{22}x_2 + \cdots + a_{2N}x_N \quad\quad + x_{N+2} \quad &= b_2 \\
\vdots \quad\quad\quad\quad\quad\quad\quad\quad \vdots \quad\quad\quad\quad \vdots \quad\quad &\quad \vdots \\
a_{M1}x_1 + a_{M2}x_2 + \cdots + a_{MN}x_N \quad\quad\quad + x_{N+M} &= b_M \\
c_1 x_1 + c_2 x_2 + \cdots + c_N x_N \quad\quad\quad (-z) &= 0 \\
b_i \geq 0 \qquad i = 1, 2, \ldots, M \quad\quad\quad &
\end{aligned} \tag{2-86}$$

and

$$x_j \geq 0 \qquad (j = 1, 2, \ldots, N, N+1, \ldots, N+M) \qquad (2\text{-}87)$$

**Step C.** Use the simplex algorithm (with no sign restriction on $z$) to find a solution to (2-86) and (2-87) which minimizes the sum of the artificial variables denoted by

$$w = x_{N+1} + x_{N+2} + \cdots x_{N+M} \qquad (2\text{-}88)$$

Equation (2-88) is called the infeasibility form. The initial feasible canonical system for this search is obtained by selecting as basic variables $x_{N+1}$, $x_{N+2}, \ldots, x_{N+M}$, $(-z)$, $(-w)$ and eliminating these variables (except $w$) from Eq. (2-88) by subtracting the sum of the first $M$ equations of system (2-86) from Eq. (2-88), yielding the following.

| Admissible Variables | Artificial Variables | |
|---|---|---|
| $a_{11}x_1 + a_{21}x_2 + \cdots + a_{1N}x_N + x_{N+1}$ | | $= b_1$ |
| $a_{21}x_1 + a_{22}x_2 + \cdots + a_{2N}x_N \qquad + x_{N+2}$ | | $= b_2$ |
| $\cdot \qquad\qquad\qquad \cdot \qquad\qquad \cdot$ | | $\cdot$ |
| $a_{M1}x_1 + a_{M2}x_2 + \cdots + a_{MN}x_N$ | $+ x_{N+M}$ | $= b_M$ |
| $c_1x_1 + c_2x_2 + \cdots + c_Nx_N$ | $-z$ | $= 0$ |
| $d_1x_1 + d_2x_2 + \cdots + d_Nx_N$ | $-w$ | $= -w_0$ |

where $b_i \geq 0$, and

$$d_j = -(a_{1j} + a_{2j} + \cdots + a_{Mj}) \qquad (j = 1, 2, \ldots, N) \qquad (2\text{-}89)$$
$$-w_0 = -(b_1 + b_2 + \cdots + b_M)$$

**Step D.** If Min $w > 0$, then no feasible solution exists and the procedure is terminated. If Min $w = 0$, initiate the Simplex algorithm of phase II by (1) dropping from further consideration all nonbasic variables $x_j$ whose corresponding coefficients $d_j$ are positive (not zero) in the final modified $w$-form and (2) replacing the linear form $w$ (as modified by various eliminations) by the linear form $z$, after first eliminating all basic variables from the $z$-form.[4]

### 2.3.4  Application Example

For illustrating the application of the Simplex method, we give an example concerning the energy-source choice for production of electrical power [G-15, pp. 41–46; 2-2, 2-3]. The problem consists of minimizing some cost function comprising the total investment cost and the weighted management cost, with the following demand constraints:

1. Guaranteed power in MW.
2. Peak power in MW.
3. Annual energy consumption in GWh.

The following kinds of plants can be used:

1. Thermal plants.
2. River hydraulic plants.
3. Hydraulic plants with reservoir.
4. Hydraulic plants with dam.
5. Sea-tide plants.
6. Nuclear plants.

For each type of plants $i$, the following coefficients are defined:

$a_i$  Guaranteed power.
$b_i$  Peak power.
$c_i$  Annual energy.
$d_i$  Investment expenses.
$e_i$  Annual management expenses.
$x_i$  Number of plants of the type $i$.
$A$  Overall guaranteed power.
$B$  Overall peak power.
$C$  Overall annual energy demand.
$D$  Imposed investment limit.

Any planning of power production must satisfy the following constraints:

$$a_1 x_1 + a_2 x_2 + \cdots + a_n x_n \geq A$$
$$b_1 x_1 + b_2 x_2 + \cdots + b_n x_n \geq B$$
$$c_1 x_1 + c_2 x_2 + \cdots + c_n x_n \geq C$$
$$d_1 x_1 + d_2 x_2 + \cdots + d_n x_n \leq D$$

Introducing the slack variables $x_a$, $x_b$, $x_c$, $x_d$:

$$a_1 x_1 + a_2 x_2 + \cdots + a_n x_n - x_a = A$$
$$b_1 x_1 + b_2 x_2 + \cdots + b_n x_n - x_b = B$$
$$c_1 x_1 + c_2 x_2 + \cdots + c_n x_n - x_c = C \qquad \text{(2-90)}$$
$$d_1 x_1 + d_2 x_2 + \cdots + d_n x_n + x_d = D$$

The cost function consists of:

1. Initial investments
$$d_1 x_1 + d_2 x_2 + \cdots + d_n x_n$$

2. Annual maintenance expenses

$$k(e_1 x_1 + e_2 x_2 + \cdots + e_n x_n)$$

3. Annual coal expenses for the thermal plants (assuming that $i = 1$ for thermal plants). Let $c_1'$ be the annual thermal energy, $c_1'/c_1$ the annual thermal use rate, $p_0$ the coal cost per kWh; then the coal expenses are $k p_0 c_1' x_1$. Let $x_c$ be the annual saved thermal energy, $c_1$ is related to $c_1'$ and $x_c$ by

$$c_1' x_1 = c_1 x_1 - x_c$$

then

$$c_1' x_1 = C - c_2 x_2 - c_3 x_3 \cdots - c_n x_n$$

The cost function is therefore

$$z = \sum_{i=1}^{n} d_i x_i + k \sum_{i=1}^{n} e_i x_i + k p_0 C - k p_0 \sum_{i=2}^{n} c_i x_i$$

The annual management expenses are therefore

$$f_1 = e_1, f_2 = e_2 - p_0 c_2, \ldots, f_n = c_n - p_0 c_n$$

Let

$$g_1 = d_1 + k f_1, g_2 = d_2 + k f_2, \ldots, g_n = d_n + k f_n$$

The cost function becomes

$$z = g_1 x_1 + g_2 x_2 + \cdots + g_n x_n + k p_0 C$$

By letting $z' = z - k p_0 C$, the function to be minimized is

$$z' = g_1 x_1 + g_2 x_2 + \cdots + g_n x_n \tag{2-91}$$

The linear program (2-90) (2-91) contains $(n + 4)$ variables. A feasible optimal solution must contain at least $n$ zero values. Since there are only 4 constraints, an optimum solution can only contain at most 4 types of production plants (for six available types). This important result shows, if a fifth constraint (for instance, some social constraint) is added, it will be possible to find an optimum solution containing a fifth nonzero value (or production plant).

*Example.* In 1955, the French Parliament fixed as an objective:

$$
\begin{aligned}
A: & \quad 1692 \text{ MW} \\
B: & \quad 2307 \text{ MW} \\
C: & \quad 7200 \text{ GWh}
\end{aligned}
$$

and as constraint

$$D \leq \text{some arbitrary value } D_0.^*$$

---

* The choice of the value $D_0$ is related to the political orientation of the national economy.

***Table* 2.6**

|            | $a_i$ | $b_i$ | $c_i$ | $d_i$ (in millions of francs) | $g_i$ (in millions of francs) |
|------------|-------|-------|-------|------|------|
| 1. Thermal   | 1 | 1.15 | 7    | 97  | 136 |
| 2. River     | 1 | 1.10 | 12.6 | 420 | 56  |
| 3. Reservoir | 1 | 1.20 | 1.20 | 130 | 101 |
| 4. Dam       | 1 | 3    | 7.35 | 310 | 104 |
| 5. Sea-tide  | 1 | 2.13 | 5.47 | 213 | 79  |

Table 2.6 gives the values of the coefficients $a$, $b$, $c$, $d$, and $g$ for five types of plants. Other practical values are rate of interest at 8 percent, or $k = 12.5$, and $p_0 = 3F$; $kp_0C = 2700$ billions of francs. The linear program is therefore

$$[\text{Min}]\ z = 136x_1 + 56x_2 + 101x_3 + 104x_4 + 79x_5$$

$$x_1 + x_2 + x_3 + x_4 + x_5 - x_a = 1692$$

$$1.15x_1 + 1.10x_2 + 1.2x_3 + 3x_4 + 2.13x_5 - x_b = 2307 \qquad (2\text{-}92)$$

$$7x_1 + 12.6x_2 + 1.3x_3 + 7.35x_4 + 5.47x_5 - x_c = 7200$$

$$97x_1 + 420x_2 + 130x_3 + 310x_4 + 213x_5 + x_d = D$$

$$x_1, x_2, x_3, x_4, x_5, x_a, x_b, x_c, x_d \geq 0$$

Such a program ($n = 9$, $m = 4$) has at most 126 basic solutions. Some of them have negative values, and therefore, are not feasible. It is interesting to note the following practical discussion.

1. If only thermal plants are used, then the solution is

$$x_2 = x_3 = x_4 = x_5 = x_6 = 0$$

$$x_1 = 2006 \qquad x_a = 314 \qquad x_c = 6842 \qquad x_d = D - 194{,}582$$

necessitating an investment $D$ higher than 195 billions of francs.

2. If only river plants are used, then the solution is

$$x_1 = x_3 = x_4 = x_5 = x_b = 0$$

$$x_2 = 2097 \qquad x_a = 405 \qquad x_c = 19{,}222 \qquad x_d = D - 880{,}740$$

necessitating an investment $D$ higher than 881 billions of francs.

3. If $D$ is allowed to take several values, then for each of these values, the minimal $z$ and the types of plants to be used are investigated. The results of these comparative investigations constitute a precious element for political decisions.

### 2.3.5 · Problem Formulation

The extensive utilization of linear programming methods in economic problems needs no comments. Application of these methods has begun in petroleum refineries, food-processing industries, iron and steel industries, and some metalworking industries. The rate of expansion is rather slow. One major difficulty resides certainly in the formulation of the problems. There is no systematic method for learning how to formulate a linear programming problem. Only experience can help, or in certain circumstances, understanding may come from association of ideas. That is why we summarize here some typical examples of problem formulation.

A. *Blending problems* such as discussed in Section 2.3.1.

B. *Nationwide planning problems* such as discussed in Section 2.3.4.

C. *Warehouse problems.* Determine the optimal selling, storing, and buying program for a one-year period by quarters, assuming that the warehouse has an initial stock of 50 units. In each period (quarter), $t$, there are four types of activities (Table 2.7).

*Table* **2.7**

| Activity | Quantity |
| --- | --- |
| 1. Selling stock | $x_{t1}$ |
| 2. Storing stock | $x_{t2}$ |
| 3. Buying stock | $x_{t3}$ |
| 4. Unused capacity (slack) | $x_{t4}$ |

There are also three types of items: (a) stock, (b) storage capacity, and (c) costs. The resulting linear programming model is shown in Table 2.8 with 16 variables, 8 constraints. As an exercise, the reader may try to write the linear program corresponding to this model.

D. *Transportation problems.* Determine the optimal transportation program so that the transportation cost of 340 tons of material from 3

**Table 2.8**

| Activities / Items | | First Quarter $x_{11}\ x_{12}\ x_{13}\ x_{14}$ | Second Quarter $x_{21}\ x_{22}\ x_{23}\ x_{24}$ | Third Quarter $x_{31}\ x_{32}\ x_{33}\ x_{34}$ | Fourth Quarter $x_{41}\ x_{42}\ x_{43}\ x_{44}$ | Prescribed |
|---|---|---|---|---|---|---|
| $t=0$ | a | 1  1 −1 | | | | 50 |
| | b | 1     1 | | | | 100 |
| $t=1$ | a | −1 | 1  1 −1 | | | 0 |
| | b | | 1     1 | | | 100 |
| $t=2$ | a | | −1 | 1  1 −1 | | 0 |
| | b | | | 1     1 | | 100 |
| $t=3$ | a | | | −1 | 1 1 −1 | 0 |
| | b | | | | 1     1 | 100 |
| | c | −10  1  10 | −12  1  12 | −8  1  8 | −9  1  9 | $z$ (Min) |

factories $A$, $B$, $C$ to five stores 1, 2, 3, 4, 5 is minimized. The five stores must receive 40, 50, 70, 90, and 90 tons, respectively. The availabilities of the 3 factories are 100, 120, and 120, respectively. The transportation costs from factories to stores are shown in Table 2.9. The resulting linear program is

$$[\text{Min}]\ z = 4x_{11} + x_{12} + 2x_{13} + 6x_{14} + 9x_{15}$$

$$+\ 6x_{21} + 4x_{22} + 3x_{23} + 5x_{24} + 7x_{25}$$

$$+\ 5x_{31} + 2x_{32} + 6x_{33} + 4x_{34} + 8x_{35} \quad (2\text{-}93)$$

**Table 2.9**

Stores

| | | (1) | (2) | (3) | (4) | (5) | |
|---|---|---|---|---|---|---|---|
| | A | 4 | 1 | 2 | 6 | 9 | 100 |
| Factories | B | 6 | 4 | 3 | 5 | 7 | 120 |
| | C | 5 | 2 | 6 | 4 | 8 | 120 |
| | | 40 | 50 | 70 | 90 | 90 | |

Availability constraints

Demand constraints

Constraints are as follows:

$$\sum_{i=1}^{5} x_{1i} = 100$$

$$\sum_{i=1}^{5} x_{2i} = 120$$

$$\sum_{i=1}^{5} x_{3i} = 120$$

$$\sum_{j=1}^{3} x_{j1} = 40 \qquad \text{All } x_{ij} \geq 0$$

$$\sum_{j=1}^{3} x_{j2} = 50 \qquad \begin{aligned} i &= 1, 2, \ldots, 5 \\ j &= 1, 2, 3 \end{aligned} \qquad (2\text{-}94)$$

$$\sum_{j=1}^{3} x_{j3} = 70$$

$$\sum_{j=1}^{3} x_{j4} = 90$$

$$\sum_{j=1}^{3} x_{j5} = 90$$

Since the total availability is equal to the total demand, the 8 constraints are not independent. Therefore, the linear program has 15 variables and 7 constraints. As an exercise, the reader may try to establish the corresponding linear programming model.

E. *Traveling salesman problems.* In what order should a traveling salesman visit $n$ cities in order to minimize the total distance covered in a complete circuit? This is a well-known and difficult problem, having at least three mathematical formulations [G-14, pp. 545–7; G-15, pp. 73–5].

FIRST FORMULATION. Let $x_{ijt} = 1$ or 0 according to whether the $t$th-directed step on the route is from city $i$ to city $j$ or not. Letting $x_{ijn+1} \equiv x_{ijn}$, the conditions

$$\sum_{i} x_{ijt} = \sum_{k} x_{j,k,t+1} \qquad (j, t = 1, \ldots, n)$$

$$\sum_{j,t} x_{ijt} = 1 \quad (i = 1, \ldots, n)$$

$$\sum_{i,j,t} d_{ij} x_{ijt} = z[\text{Min}] \qquad (2\text{-}95)$$

express (a) that if one arrives at city $j$ on step $t$, one leaves city $j$ on step $t + 1$, (b) that there is one directed step leaving city $i$, and (c) the length of the tour is minimum.

SECOND FORMULATION. In case of symmetric distance $d_{ij} = d_{ji}$, only two indices are necessary. Letting $x_{ij} \equiv x_{ji} = 1$ or $0$ according to whether the trip from $i$ to $j$ or from $j$ to $i$ was traversed at some time on the route or not. The conditions

$$\sum_i x_{ij} = 2$$

$$\sum_{i,j} d_{ij}x_{ij} = z\,(\text{Min}) \qquad (2\text{-}96)$$

express that the sum of the number of entries and departures for each city is two.

THIRD FORMULATION. Let $x_{ij} = 1$ or $0$, depending on whether the salesman travels from city $i$ to $j$ or not ($i = 0, 1, \ldots, n$). Then an optimal solution can be found by finding integers $x_{ij}$, arbitrary real numbers $u_i$, and Min $z$ satisfying

$$\sum_{i=0}^{n} x_{ij} = 1 \qquad (j = 1, 2, \ldots, n)$$

$$\sum_{j=0}^{n} x_{ij} = 1 \qquad (i = 1, 2, \ldots, n)$$

$$u_i - u_j + nx_{ij} \leq n - 1 \qquad (1 \leq i \neq j \leq n)$$

$$\sum_{i=0}^{n} \sum_{j=0}^{n} d_{ij}x_{ij} = z\,[\text{Min}] \qquad (2\text{-}97)$$

The obvious method is the direct enumeration of all possibilities. As an exercise, the reader may try to solve a symmetric problem of four cities; he will have only three itineraries to examine. If he wishes to try problems with a higher number of cities, he must know that with 12 cities, there will be 19, 958, 400 itineraries in the symmetric case, and 39, 916, 800 itineraries in the nonsymmetric case.

## 2.4  NONLINEAR PROGRAMMING

In this section, we shall discuss static optimization problems in which the performance index and/or the constraints have nonlinear mathematical forms. Like all nonlinear problems, there are no general techniques, but only special ones, each covering a particular class of practical problems. We shall examine several important techniques.

### 2.4.1  Piecewise Linearization Techniques

It is quite natural that the first idea which comes to mind, when a nonlinear problem is encountered, is to linearize it. There are a number of ways to do this. In the Describing Function technique, the fundamental frequency is used to linearize relay-type functions. In the Perturbation theory, small movement around an equilibrium point is considered so that the dynamics can be expressed in linear form. Even the curve smoothing, either using the Least-Square method, or the Chebyshev method, is a kind of linearization. In this section, we shall discuss piecewise linearization techniques [2-4, 2-5].

The class of problems covered by these techniques are those where the nonlinear curves are obtained experimentally. The analytical representation of the curves is unknown, but the values of the coordinates are known at each point of the curves. The techniques are valid for multidimensional problems. For ease of explanation, we shall consider a two-dimensional problem. Let the nonlinear program be:

$$[\text{Min}]\, y = f(x)$$
$$g(y, x) = 0 \qquad\qquad (2\text{-}98)$$
$$y, x \geq 0$$

$f$ and $g$ being nonlinear curves on the $x$-$y$ plane. This program cannot be treated directly by the Simplex method. Therefore, a set of new variables $n_i$ are used so that the curves $g$ and $f$ can be expressed completely by representing $y$ and $x$ as linear combinations of $n_i$, $y_i$, and $x_i$. The values of $y_i$ and $x_i$ are obtained from experimental curves $f$ and $g$. The final result is a linear program with the variables $n_i$.

Consider, for instance, the nonlinear curve, $y = f(x)$ of Fig. 2-7. We assume that this curve can be replaced by five straight line segments having six vertices. If six new variables $n_1, n_2, \ldots, n_6$ are used, the piecewise curve $f$

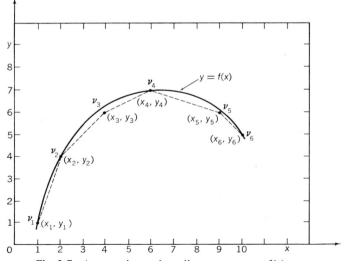

**Fig. 2-7** An experimental nonlinear curve $y = f(x)$.

can then be represented completely by the following set of equations:

$$x = \sum_1^6 n_i x_i$$

$$y = \sum_1^6 n_i y_i \tag{2-99}$$

$$1 = \sum_1^6 n_i \qquad n_i \geq 0$$

It is required that no more than two of the $n_i$'s can be nonzero, and these must be consecutive. Using the numerical values of $x_i$ and $y_i$ of Fig. 2-7, Eqs. (2-99) become

$$x = n_1 + 2n_2 + 4n_3 + 6n_4 + 9n_5 + 10n_6$$

$$y = n_1 + 4n_2 + 6n_3 + 7n_4 + 6n_5 + 5n_6 \tag{2-100}$$

$$1 = n_1 + n_2 + n_3 + n_4 + n_5 + n_6; n_i \geq 0$$

Two solution examples will show that Eqs. (2-100) can represent any vertex, or any point on a straight line segment.

1. $n_1 = n_2 = n_5 = n_6 = 0, n_3 = 0.5, n_4 = 0.5 \, (\Sigma \, n_i = 1, n_3, n_4 \, \text{consecutive})$
then

$$x = 4(0.5) + 6(0.5) = 5$$

$$y = 6(0.5) + 7(0.5) = 6.5$$

this point is on the line segment $v_3 v_4$.

2. $n_1 = n_2 = n_4 = n_5 = n_6 = 0, n_3 = 1.0 \, (\Sigma \, n_i = 1)$ then $x = 4, y = 6$, this is the vertex $v_3$.

In the case of three dimensional problems, the relations are geometrically expressed by surfaces. These surfaces can be approximated by a number of plane triangles. For instance, the surface $f(x, y, z) = 0$ can be approximated by the following set of equations:

$$x = \sum_1^n n_i x_i$$

$$y = \sum_1^n n_i y_i$$

$$z = \sum_1^n n_i z_i \tag{2-101}$$

$$1 = \sum_1^n n_i \qquad n_i \geq 0$$

where $x_i, y_i, z_i$ represent the known coordinates of the vertices of the triangular planes, and $n_i$ represents the new variables. As in the two dimensional case,

Eqs. (2-101) are accompanied by the requirement that no more than three of the $n_i$'s can be nonzero, and these must be associated with a triangle. The reader may easily check that by admitting one, two, or three nonzero $n_i$'s, Eqs. (2-101) will represent a point on a vertex, on a side, or on the surface of a triangle.

The parametric techniques described above can be used for any purposes, not only for programming problems. The conformity between the original nonlinear expressions, and the final linear forms depends on the number of new variables $n_i$ (or vertices, line segments, or triangular planes). The bigger this number is, the higher is the conformity. In some cases, however, the conformity does not only depend on the number of the $n_i$'s, but also on the judicious choice of the location of the vertices. They can be quite remote from each other in flat portions of the nonlinear curves, but they must be close to each other if the curvature is pronounced.

How a linear program is formulated from equations such as (2-99) or (2-101), will now be illustrated by an example taken again from the field of electric power production. In Section 2.3.4, we have discussed how linear programming can be used to establish an optimum planning of energy sources. In Section 2.2.2, we have seen how the Lagrange multipliers method is used to establish an economic dispatching of electric power. Here we shall discuss how to minimize the fuel costs in the process of electric power production by thermal plants.

We consider, as shown in Fig. 2-8, a thermal plant containing three boiler-turbine-generator combinations $v$, $\lambda$, and $\mu$. The efficiency curves of these combinations are obtained experimentally, and shown in Fig. 2-9. The problem is to minimize the overall fuel costs $F_v + F_\lambda + F_\mu$, and find the operating level $G_v$, $G_\lambda$, and $G_\mu$ of each generator so that the total power output of the thermal plant is, say 54 MW (megawatt).

The fuel costs are calculated from the fuel consumptions, which are calculated from the efficiency curves by operating a number of conversions. For example, take the point $v_3$ corresponding to a $G_v$ of 19.5 MW, and to an efficiency of 36.5 percent, the equivalent fuel in MW is:

$$19.5/0.365 = 53.5 \text{ MW}$$

The equivalent Btu's are (1 kW = 3412 Btu/hr): (Btu: British thermal unit)

$$(59.5 \text{ MW}) (3412 \text{ Btu/Mh-kW}) = 1.83 \ 10^8 \text{ Btu/hr}$$

Now, let:

One fuel unit = 40 × $10^6$ Btu.
One pound of fuel oil = 20,000 Btu.
One pound of coal = 13,000 Btu.
One cubic foot of blast furnace gas (BFG) = 92 Btu.
One cubic foot of coke oven gas (COG) = 574 Btu.

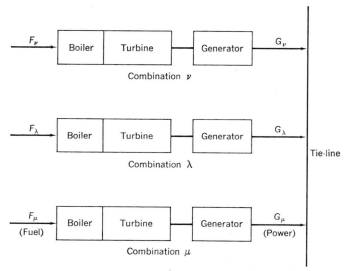

**Fig. 2-8**  A thermal plant containing three boiler-turbine-generator combinations.

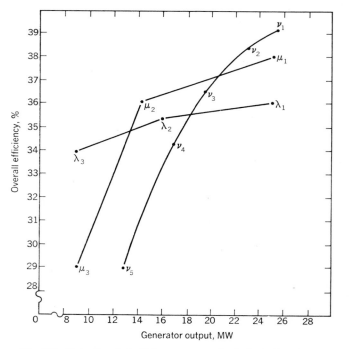

**Fig. 2-9**  Experimental efficiency curves. $\lambda\mu$ is fuel oil and $\nu$ is coal.

One fuel unit will then represent

$$40 \cdot 10^6/20{,}000 = 2000 \text{ pounds} = 1 \text{ ton of fuel oil.}$$
$$40 \cdot 10^6/13{,}000 = 3080 \text{ pounds} = 1.52 \text{ tons of coal.}$$
$$40 \cdot 10^6/92 = 4.35 \cdot 10^5 \text{ ft}^3 \text{ of BFG.}$$
$$40 \cdot 10^6/574 = 6.95 \cdot 10^4 \text{ ft}^3 \text{ of COG.}$$

Assume that the $\nu$ boiler uses coal; then for the point $\nu_3$ we have

$$(1.83 \cdot 10^8/40 \cdot 10^6) \cdot 1.52 = 6.99 \text{ tons of coal/hr}$$

After deciding on the location of the vertices (number of the $n_i$'s) and calculating similar values of all the $\nu_i$'s, we can write equations for $F_\nu$ and $G_\nu$.

$$25n_1 + 23n_2 + 19.5n_3 + 16.7n_4 + 13n_5 = G_\nu$$
$$8.40n_1 + 7.86n_2 + 6.99n_3 + 6.45n_4 + 5.87n_5 = F_\nu \qquad (2\text{-}102)$$
$$n_1 + n_2 + n_3 + n_4 + n_5 = 1.0 \qquad n_i, F_\nu, G_\nu \geq 0$$

Similar calculations can be made for the other two combinations $\lambda$ and $\mu$, after choosing three new variables $l_i$ and three new variables $m_i$. We assume that both of them use fuel oil as the boiler fuel. Note that the $F$'s and $G$'s are also variables. The total number of variables is

$$3(F) + 3(G) + 5(n) + 3(l) + 3(m) = 17$$

In the original problem, we have one constraint

$$G_\nu + G_\lambda + G_\mu = 54.0 \text{ MW}$$

In the linearized formulation, each of the combinations require three constraints; the total number of constraints is, therefore,

$$1(G) + 3(\nu) + 3(\lambda) + 3(\mu) = 10$$

Assuming that the fuel oil costs \$12.6/ton and coal costs \$10/ton, the cost function is

$$12.6F_\lambda + 12.6F_\mu + 10F_\nu = z \text{ [Min]}$$

We then have a 17 variable, 10 constraint linear program. The resulting linear programming model is shown in Table 2.10, including the solution found by using the Simplex method.

### 2.4.2  Parametric Linear Programming

In this section, we shall discuss another class of linearization techniques having great practical interest. These techniques, termed parametric linear programming (PLP) techniques, consist of reducing a multidimensional nonlinear programming problem into a multidimensional linear programming problem with adjustable parameters. These techniques are especially interesting in cases where the number of variables is very great (tens and hundreds) such as in oil and chemical industries.

**Table 2.10**

| Unknowns | Generators | | | Fuel | | | l's | | | m's | | | n's | | | | | |
|---|---|---|---|---|---|---|---|---|---|---|---|---|---|---|---|---|---|---|
| | $G_\lambda$ | $G_\mu$ | $G_\nu$ | $F_\lambda$ | $F_\mu$ | $F_\nu$ | $l_1$ | $l_2$ | $l_3$ | $m_1$ | $m_2$ | $m_3$ | $n_1$ | $n_2$ | $n_3$ | $n_4$ | $n_5$ | |
| Constraints | −1 | | | −1 | | | 25 | 16 | 9 | | | | | | | | | | 0.0 |
| | | | | | | | 5.90 | 3.84 | 2.25 | | | | | | | | | | 0.0 |
| | | | | | | | 1 | 1 | 1 | | | | | | | | | | 1.0 |
| | | −1 | | | −1 | | | | | 25 | 15.5 | 9 | | | | | | | 0.0 |
| | | | | | | | | | | 5.59 | 3.60 | 2.64 | | | | | | | 0.0 |
| | | | | | | | | | | 1 | 1 | 1 | | | | | | | 1.0 |
| | | | −1 | | | −1 | | | | | | | 25 | 23 | 19.5 | 16.7 | 13 | 0.0 |
| | | | | | | | | | | | | | 8.40 | 7.86 | 6.99 | 6.45 | 5.87 | 0.0 |
| | | | | | | | | | | | | | 1 | 1 | 1 | 1 | 1 | 1.0 |
| | 1 | 1 | 1 | | | | | | | | | | | | | | | 54.0 |
| Cost function | | | | 12.6 | 12.6 | 10 | | | | | | | | | | | | |
| Answers | 9.0 | 21.9 | 23.0 | 2.25 | 4.96 | 7.86 | | | 1.0 | | .68 | .32 | 1.0 | | | | | Min z = 1695 |

78

We have assumed, in the preceding section, that the curves relating the dependent variables $y_j$ and the independent variables $x_i$ are known experimentally. Here, we assume that only discrete points of these curves can be known experimentally. Furthermore, we consider the following:

1. $y_i$ and $x_i$ are connected by implicit relations like: $F(x_i, y_i) = 0$ instead of the explicit relations like $y_j = f(x_i)$; $(i = 1, 2, \ldots, n; j = 1, 2, \ldots, m)$ considered in the preceding section.

2. Because of the imperfection of the measuring system, the relations $F$ are only known statistically.

3. The relations $F$ have time-varying (slowly varying, but varying) parameters.

The problems stated above, are clearly very difficult. We can treat these difficulties separately, by using, for example, smoothing techniques, power polynomial expansion techniques, or small perturbations techniques. However, if these difficulties are inter-related we shall have to use the linear programming techniques because these are the only ones available; the treatment then becomes *complex*. The parametric linear programming consists of a procedure (rather than a mathematical algorithm) designed to treat such complex problems [2-7, 2-8, 2-9]. The linearization is obtained by considering only small variations around an operating point; furthermore, the parameters connecting dependent and independent variables are continuously adjusted if the operating point is modified.

Let us consider an industrial process, for instance, a distillation column, (Fig. 2-10), characterized by:

1. The processing plant.
2. The process variables $x_1, x_2, \ldots, x_n$, controlled and regulated (temperatures, feed rates, steam rates).
3. The constrained variables $y_1, y_2, \ldots, y_m$, or dependent variables, because they are not directly controlled. The values of the $y$'s must remain within certain limits for good processing (vapor rates, reflux rates, distillation temperatures, flash and pour points, viscosities).
4. The process information (temperatures, pressures, flows, levels, densities, weights, concentrations) which are either measured directly, or calculated indirectly, and from which the values of the $x$'s and the $y$'s can be obtained.

In a typical distillation column, there are around 160 process informations, approximately 20 $x$'s and 40 $y$'s. The optimization problem is to determine the variations $\Delta x_i$ of the process variables, so that the variation of the cost function (yield, profit) is maximized and the limits $\Delta y_{\max}$ of the variations $\Delta y_j$ of the constrained variables are respected.

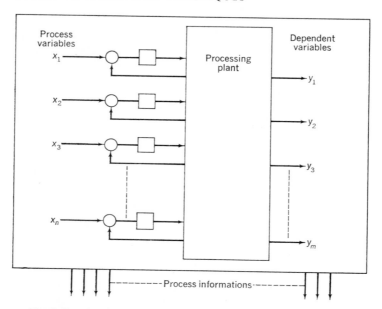

*Fig. 2-10*  Simplified block-diagram of a typical industrial process.

The parametric linear programming procedure consists of the following steps:

1. Determine the state of the process, i.e. what are the $x$'s, the $y$'s, and how can their values be obtained accurately from the numerous process information. At this stage, statistical smoothing and regression techniques can be used.

2. Formulate the cost function $z$ in terms of the process variables. In the general case, we have the nonlinear function (unknown)

$$z = f(x_1, x_2, \ldots, x_n) \qquad (2\text{-}103)$$

Around *an* operating point, we can write the linear equation of the variations

$$\Delta z = c_0 + c_1 \Delta x_1 + c_2 \Delta x_2 + \cdots + c_n \Delta x_n \qquad (2\text{-}104)$$

The parameters $c_1, c_2, \ldots, c_n$ are partial derivatives of $f$

$$c_i = \frac{\partial f}{\partial x_i} \qquad i = 1, 2, \ldots, n \qquad (2\text{-}105)$$

and can be approximated experimentally by $\Delta z / \Delta x_i$.

3. Express the constraint functions from

$$F(x_i, y_j) = 0 \qquad (2\text{-}106)$$

to

$$
\begin{aligned}
y_1 &= g_1(x_i) \\
y_2 &= g_2(x_i) \\
&\phantom{=} \cdot \qquad \cdot \\
&\phantom{=} \cdot \\
&\phantom{=} \cdot \\
y_m &= g_m(x_i)
\end{aligned}
\qquad (2\text{-}107)
$$

Then, as in the case of $z$, linearize around *an* operating point

$$
\begin{aligned}
\Delta y_1 &= a_{10} + a_{11}\,\Delta x_1 + a_{12}\,\Delta x_2 + \cdots + a_{1n}\,\Delta x_n \\
\Delta y_2 &= a_{20} + a_{21}\,\Delta x_1 + a_{22}\,\Delta x_2 + \cdots + a_{2n}\,\Delta x_n \\
&\phantom{=} \cdot \qquad\qquad\qquad \cdot \\
&\phantom{=} \cdot \\
&\phantom{=} \cdot \\
\Delta y_m &= a_{m0} + a_{m1}\,\Delta x_1 + a_{m2}\,\Delta x_2 + \cdots + a_{mn}\,\Delta x_n
\end{aligned}
\qquad (2\text{-}108)
$$

The parameters $a_{ji}$ are partial derivatives

$$\frac{\partial g_1}{\partial x_i}, \frac{\partial g_2}{\partial x_i}, \ldots, \frac{\partial g_m}{\partial x_i} \qquad i = 1, 2, \ldots, n$$

and, like the $c$'s, can be approximated experimentally by $\Delta g_i/\Delta x_i$.

4. Determine the limiting values $y_{j\mathrm{max}}$ of the constrained variables. The constraints in the sense of the linear programming method can then be expressed by inequalities of the form

$$\Delta y = y_{\mathrm{max}} - y = b \qquad (2\text{-}109)$$

Substituting Eq. (2-109) in Eq. (2-108), we have

$$
\begin{aligned}
a_{11}\,\Delta x_1 + a_{12}\,\Delta x_2 + \cdots + a_{1n}\,\Delta x_n &\le b_1 - a_{10} \\
a_{21}\,\Delta x_1 + a_{22}\,\Delta x_2 + \cdots + a_{2n}\,\Delta x_n &\le b_2 - a_{20} \\
\cdot \qquad\qquad\qquad\qquad \cdot \qquad\qquad \cdot& \\
\cdot& \\
\cdot& \\
a_{m1}\,\Delta x_1 + a_{m2}\,\Delta x_2 + \cdots + a_{mn}\,\Delta x_n &\le b_m - a_{m0}
\end{aligned}
\qquad (2\text{-}110)
$$

5. Determine the limiting values $d_i$ and $e_i$ of the variations $\Delta x_i$ of the process variables, so that the solution always remains within the validity of the

linearity. The resulting constraints are

$$d_i \leq \Delta x_i \leq e_i \qquad i = 1, 2, \ldots, n \qquad (2\text{-}111)$$

6. Rewriting Eqs. (2-104), (2-110), and (2-111), we have the linear program

$$[\text{Max}] \, \Delta z = \sum_1^n c_i \, \Delta x_i$$

$$\sum_{i=1}^n a_{ji} x_i \leq b_i - a_{j0} \qquad j = 1, 2, \ldots, n \qquad (2\text{-}112)$$

$$d_i \leq \Delta x_i \leq e_i$$

where $\Delta z$ and $\Delta x$ play the role of $z$ and $x$ in the usual linear programming formulation. The difference is the two-sided constraints on $\Delta x_i$, instead of the usual one-sided constraints. The difficulty can be circumvented by the use of slack variables (see Section 2.3.2). Then solve the program (2-112) by the Simplex method and determine the level of the variations $\Delta x_i$ corresponding to the maximum of $\Delta z$.

7. If, in the optimal solution $\Delta x_i$, some of the constraints $e_i$ and $d_i$ are met, then, instead of the initial operating point $(x_1, x_2, \ldots, x_n)$, use a new operating point $(x_1 + \Delta x_1, x_2 + \Delta x_2, \ldots, x_n + \Delta x_n)$ and recalculate the parameters $a_{ji}, c_i, b_j, d_i$, and $e_i$. The Simplex method is again used for finding the optimal solution in the new conditions.

It is interesting to note that if the linearity constraints are respected, the PLP technique yields an exact solution, although several approximating processes are used. However the practical utility of this method depends on two conditions, the first of which is the existence of a multidimensional optimum point in the bounded (constrained) space. If such a point does not exist, oscillating search will occur. Second, the operating points must be sufficiently steady to allow the evaluations of numerous parameters.

### 2.4.3  Kuhn-Tucker Convex Programming

In Section 2.2, we discussed the Lagrange-multiplier method, which is capable of treating problems where the constraints are nonlinear equations, and the variables are not restricted. In Section 2.3, we discussed the linear programming method, capable of treating problems where the constraints are linear inequalities, and the variables are $(x_j \geq 0)$. In this section, we shall discuss the convex programming method, [2-9, 2-10; G-14, Ch. 24; G-18, Ch. 6], capable of treating more general problems, where the constraints can be linear or nonlinear, equations or inequalities, and the variables can be non-negative or nonpositive.

The mathematical formulation of the convex programming problem is

$$\text{Min } z = f(x)$$
$$
\begin{aligned}
g_i(x) &\leq 0 & i &= 1, 2, \ldots, m \\
h_k(x) &= 0 & k &= 1, 2, \ldots, p \\
x_j &\geq 0 & j &= 1, 2, \ldots, n
\end{aligned}
\tag{2-113}
$$

where $g_i(x)$ and $h_k(x)$ are continuous convex nonlinear functions, and $f(x)$ is a concave nonlinear continuous function.

The convex programming method is stated as a theorem. If there exists $\bar{x} \in R$ such that $g_i(\bar{x}) \leq 0$ and $h_k(\bar{x}) = 0$ for $i = 1, 2, \ldots, m$ and $k = 1, 2, \ldots, p$, then there exist multipliers $\alpha_i \geq 0$ and $\beta_k$ without restrictions of sign, and an $x^*$ which solves Eqs. (2-113) with min $z$ and satisfies

$$F(x^*) = \min F(x) = \underset{x \in R}{\text{Min}} \left[ f(x) + \sum_1^m \alpha_i g_i(x) + \sum_1^P \beta_k h_k(x) \right] \tag{2-114}$$

For applications, this theorem can be translated into the following procedure.

1. Choose arbitrary multipliers $\alpha_i$ and $\beta_k$.
2. Form the function $F$.
3. Write the conditions $\partial F / \partial x_j = 0, j = 1, 2, \ldots, n$.
4. Solve $(m + n + p)$ extremalizing equations: $n$ of which from $\partial F / \partial x_j = 0$, $m$ from $g_i(x) = 0$ and $p$ from $h_k(x) = 0$, for $(n + m + p)$ unknowns: $n \times x$, $m \times \alpha$, and $p \times \beta$.

Note that the solution obtained by this procedure lies on the boundary of the admissible domain, which is defined by the relations: $g_i(x) \leq 0, h_k(x) = 0$, $x_j \geq 0$. This is because we have used $g_i(x) = 0$ instead of $g_i(x) \leq 0$. The true optimal solution $x^*$ may lie inside of this domain ($x^*$ satisfies some or all of the $m$ strict inequality constraints $g_i(x) < 0$). There is no analytical method to find $x^*$. Some iterative method must be used. The interested reader may consult [G-18, G-77]. As an application example, we shall again take the problem of economic dispatching of interconnected systems. This problem has been used for illustrating the Lagrange-multiplier method (see Section 2.2.2). There, we obtained a set of coordination equations which state that all the generators must operate at the same incremental cost.

In practice, this is only possible if the constraints

$$P_i \leq P_{iM} \qquad \text{(superior power bounds of the plant } i) \tag{2-115}$$

$$P_i \geq P_{im} \qquad \text{(inferior power bounds of the plant } i) \tag{2-116}$$

do not exist. Clearly, this is not true in reality. A more exact treatment of the economic dispatching problem must therefore include the boundary constraints Eqs. (2-115) and (2-116). This will exclude the use of the Lagrange-multiplier method, and invalidate the coordination equations.

As can be easily seen, the new formulation of the dispatching problem

$$[\text{Min}] \; F = F(P_1, P_2, \ldots, P_n) \tag{2-117}$$

$$h = \sum_1^n P_i - \sum_1^n P_{Li} - P_R = 0 \tag{2-118}$$

$$g_{iM} = P_i - P_{iM} \leq 0 \tag{2-119}$$

$$g_{im} = P_{im} - P_i \leq 0 \tag{2-120}$$

$$P_i \geq 0 \tag{2-121}$$

falls in the frame of the convex programming. Let $\lambda$, $\alpha_{iM}$ and $\alpha_{im}$ ($i = 1$, $2, \ldots, n$) be multipliers associated respectively with the equation $h$, the $n$ inequalities $g_{iM}$, and the $n$ inequalities $g_{im}$; we then form a new function

$$\mathscr{L} = F + \lambda h + \sum_1^n \alpha_{iM}(P_i - P_{iM}) + \sum_1^n \alpha_{im}(P_{im} - P_i) \tag{2-122}$$

The Kuhn-Tucker theorem will lead to the following optimality conditions

$$\frac{dF_i}{dP_i} L = \lambda + \alpha_{im} - \alpha_{iM} \tag{2-123}$$

$$\alpha_{iM} \geq 0, \qquad \alpha_{im} \geq 0 \tag{2-124}$$

$$\alpha_{iM}(P_i - P_{iM}) = 0, \qquad \alpha_{im}(P_{im} - P_i) = 0 \tag{2-125}$$

Instead of the equal-incremental coordination equations, we shall have nonequal-incremental coordination equations. This can be seen by examining the following three cases derived from Eqs. (2-125).

C A S E   A.   $P_i - P_{iM} \neq 0$, $P_{im} - P_i \neq 0$, the power level of the plant $i$ is neither its maximum, nor its minimum; then

$$\alpha_{im} = \alpha_{iM} = 0$$

$$\frac{dF_i}{dP_i} L = \lambda \tag{2-126}$$

This solution is identical to the one obtained by the Lagrange-multiplier method.

CASE B.    $P_i = P_{iM}$, the power level of the plant $i$ is its maximum; then

$$P_i - P_{im} \neq 0, \qquad \alpha_{im} = 0, \qquad \alpha_{iM} \geq 0$$

$$\frac{dF_i}{dP_i} L = \lambda - \alpha_{iM} \leq \lambda \tag{2-127}$$

$$\frac{dF_i}{dP_i} L \leq \lambda$$

the incremental production cost is inferior to $\lambda$.

CASE C.

$$P_i = P_{im}, \qquad \alpha_{iM} = 0, \qquad \alpha_{im} \geq 0$$

$$\frac{dF_i}{dP_i} L = \alpha_{im} + \lambda \geq \lambda \tag{2-128}$$

$$\frac{dF_i}{dP_i} L \geq \lambda$$

the incremental production cost is greater than $\lambda$.

If the Kuhn-Tucker method belongs to the class of convex programming, the converse is not true. The expression "convex programming" also covers a collection of other programming problems, the principal ones of which are the following.

SEPARABLE CONVEX PROGRAMMING. The cost function is of the form [2-12; G-14, pp. 482–90]:

$$[\text{Min}]z = \sum_{j=1}^{n} \phi_j(x_j) \qquad (0 \leq x_j \leq h_j) \tag{2-129}$$

or

$$[\text{Min}] z = \phi_1(x_1) + \phi_2(x_2) + \cdots + \phi_n(x_n) \tag{2-130}$$

The cost function $z$ is multidimensional in variables $x_1, x_2, \ldots, x_n$, but each convex nonlinear function $\phi_j$ is only a function of the variable $x_j$, and is called convex-separable. One of the methods used for solving these problems is the piecewise linearization method as described in Section 2.4.1.

QUADRATIC PROGRAMMING. Let $x$ be an $n$-vector; the general formulation is [G-14, pp. 490–8; G-18, Ch. 7; 2-13; G-20, Vol. 2, pp. 682–7]

$$\text{Constraints } Ax = b \tag{2-131}$$

$$\text{Cost function [Max] } z = x^T Cx \tag{2-132}$$

$$\text{Boundary constraints } x \geq 0 \tag{2-133}$$

where $C$ is an $n$th-order square matrix, $A$ is an $m \times n$ matrix, $b$ is an $m$th-order column matrix, and $x^T$ the transpose of $x$.

Although this problem can be solved by the Kuhn-Tucker method, the treatment is made easier by the special properties of the quadratic form [see Appendix 1C].

1. Every quadratic form can be expressed by a symmetric matrix associated with its coefficients. For instance, if $n = 3$

$$z(x) = c_{11}x_1^2 + c_{22}x_2^2 + c_{33}x_3^2 + 2c_{12}x_1x_2 + 2c_{23}x_2x_3 + 2c_{13}x_1x_3$$

$$= [x_1 x_2 x_3] \cdot \begin{bmatrix} c_{11} & c_{12} & c_{13} \\ c_{12} & c_{22} & c_{23} \\ c_{13} & c_{23} & c_{33} \end{bmatrix} \cdot \begin{bmatrix} x_1 \\ x_2 \\ x_3 \end{bmatrix} = x^T C x \qquad (2\text{-}134)$$

2. A quadratic form is called *positive definite* if $x^T C x > 0$ for all $x \neq 0$; it is called *positive semidefinite* if $x^T C x \geq 0$ for all $x \neq 0$.

3. $x^T C x$ is convex if and only if it is positive semidefinite, or positive definite.

4. Let $A_j$, $C_j$ denote the $j$th columns of $A$ and $C$, respectively, and let

$$y_j = C_j^T x - \pi A_j \qquad (2\text{-}135)$$

($\pi$ being the multiplier-vector $\pi = \{\pi_1, \pi_2, \ldots, \pi_m\}$). A solution $x = x^0$ is minimal if there exists $\pi = \pi^0$ and $y = y^0$ so that

$$Ax^0 = b, \qquad x^0 \geq 0 \qquad (2\text{-}136)$$

$$y_j^0 = C_j^T x^0 - \pi^0 A_j \geq 0 \qquad (j = 1, 2, \ldots, n) \qquad (2\text{-}137)$$

$$y_j^0 = 0, \quad \text{if} \quad x_j^0 > 0 \qquad (2\text{-}138)$$

As an exercise, the reader may try to translate the mathematical formulation 1 to 4 into a computing procedure as was done in the application example of the Kuhn-Tucker method.

INTEGER PROGRAMMING.   The general formulation [G-14, Ch. 28; G-18, Ch. 8], with the same symbols in (B), is:

$$Ax = b; x \geq 0; \qquad x_j \text{ an integer}, j \in J_1 \qquad (2\text{-}139)$$

$$[\text{Max}]\, z = cx$$

All the variables are restricted to integral values which makes the programming problem nonlinear rather than linear. One might feel that we could simply solve this problem by the linear programming method, ignoring the integrality requirements, and then round out the values in the resulting solution. However, this is not applicable to practical situations because the most interesting integer programming problems are those for which the integer variables must only take the values 0 or 1. These problems are, for instance, sequencing problems, project planning and manpower scheduling problems, traveling salesman problems, etc, generally belonging to economic fields.

# Definitions Relative to Maxima and Minima

In studying optimization techniques, we must recognize clearly the difference between various kinds of maxima and minima.

*Absolute maximum.* The function $f(x)$ defined over a closed set $X$ in $E^n$ is said to take on its absolute maximum over $X$ at the point $\bar{x}$ if $f(x) \leq f(\bar{x})$ for every point $x \in X$.

*Strong relative maximum.* Let $f(x)$ be defined at all points in some $\delta$-neighborhood of $x_0$ in $E^n$. The function $f(x)$ is said to have a strong relative maximum at $x_0$, if there exists an $\epsilon$, $0 < \epsilon < \delta$ such that for all $x$, $0 < |x - x_0| < \epsilon$, $f(x) < f(x_0)$.

*Weak relative maximum.* Let $f(x)$ be defined at all points in some $\delta$-neighborhood of $x_0$ in $E^n$. The function $f(x)$ is said to have a weak relative maximum at $x_0$ if it does not have a strong relative maximum at $x_0$, but there exists an $\epsilon$, $0 < \epsilon < \delta$, so that $f(x) \leq f(x_0)$ for all $x$ in the $\epsilon$-neighborhood of $x_0$.

The definitions of absolute minimum, strong relative minimum, and weak relative minimum are obtained from the corresponding definitions of maxima, by reversing the directions of the inequality signs in $f(x) \leq f(\bar{x})$, or $f(x) < f(x_0)$, or $f(x) \leq f(x_0)$. The classical approach to the theory of maxima and minima does not provide a direct method of obtaining the absolute maximum or minimum of a function. Instead, it only provides a mean for determining the relative maxima and minima by making use of the partial derivatives of $f$. In such a case, it is necessary that $f \in C^1$ over all of $E^n$ or over any part of $E^n$ which is of interest to us. From the first partial derivatives, we derive only necessary conditions for maxima and minima. Sufficient conditions can only be derived from the second partial derivatives of $f$; and in this case, it is necessary that $f \in C^2$ over all of $E^n$ or over any part of $E^n$ which is of interest to us. In ordinary problems, we do not need to distinguish between strong and weak relative maxima or minima. In optimization problems, this distinction becomes necessary because the maxima or minima may occur at the boundaries of the set $X$ over which $f(x)$ is defined. Such boundaries may appear in two ways:

(a) By limiting the variations of $x$, e.g., $x \leq 0$ or $|x| \leq x_M$ ($x_M$ being some prescribed value), then $x = 0$ or $|x| = x_M$ form the boundaries of the set $X$.

(b) By constraining $x$ to be on some hypersurfaces defined by

$$g_i(x) = b_i, \qquad i = 1, 2, \ldots, m$$

In the latter, let $f \in C^1$, $g_i \in C^1$, $Y$ be the set of points $x$ satisfying $g_i(x) = b_i$; the preceding definitions then become the following.

*Constrained absolute maximum.* The function $f(x)$ is said to take on its absolute maximum over the closed set $X$ in $E^n$ for those $x$ satisfying $g_i(x) = b_i$, $i = 1, 2, \ldots$, $m$, at the point $\bar{x} \in X$ for all $x \in X \cap Y$, $f(x) \leq f(\bar{x})$.

*Constrained strong relative maximum.* The function $f(x)$ is said to take on a strong relative maximum at the point $x_0$ for those $x$ satisfying $g_i(x) = b_i$, $i = 1, 2$, $\ldots, m$, if $x_0 \in Y$ and there exists an $\epsilon > 0$ such that for every $x \neq x_0$ in an $\epsilon$-neighborhood of $x_0$ for which $x \in Y$, $f(x) < f(x_0)$.

*Constrained weak relative maximum.* The function $f(x)$ is said to take on a weak relative maximum at the point $x_0$ for those $x$ satisfying $g_i(x) = b_i$, $i = 1, 2, \ldots, m$, if it does not have a strong relative maximum at $x_0$, but $x_0 \in Y$ and there exists an $\epsilon > 0$ such that for every point $x \in Y$ in an $\epsilon$-neighborhood of $x_0$, $f(x) \leq f(x_0)$.

The Lagrange-multiplier method provides a set of necessary conditions for the constrained relative maxima or minima by using the partial derivatives of the Lagrangian function $f(x) - \lambda_i(g_i - b_i) = 0$, $i = 1, 2, \ldots, m$. Sufficient conditions can only be obtained by using second partial derivatives of the Lagrangian functions. There does not exist any general method providing a direct way of obtaining the absolute maximum or minimum of a function or a functional. However, the knowledge of the convexity or the concavity of the functions $f(x)$, or $g_i(x)$ or of the functionals $f[x(t)]$ or $g_i[x(t)]$ can help. Note that a linear function is convex. Note also that there is no special method of recognizing a convex function except the one of checking if it possesses the property stated in Appendix 1A. Let us note now that in the definitions of strong and weak maxima, the emphasis is on the behavior of the function $f(x)$ at the neighborhood of the optimal point $x_0$. In the case of functionals such as $v = v[y(x)]$, the preceding definitions of strong and weak minima can, in principle, be extended to $v$ with respect to $y$. However, in this case, the argument is no more a point $x$, but a curve $y(x)$. Hence it becomes necessary to pinpoint the behavior of the extremal $\bar{y}(x)$ with respect to its neighboring $y(x)$ before stating any property of $v[\bar{y}(x)]$ with respect to $v[y(x)]$.

CLOSENESS. Two curves $\bar{y} = \bar{y}(x)$ and $y = y(x)$ are close or neighboring in the sense of *closeness of order zero*, if the absolute value of the difference $\bar{y}(x) - y(x)$ is small. [Fig. 2A-1, $y(x)$ of both cases $a$ and $b$ has the closeness of order zero.]

The closeness is of *order one*, if the absolute values of the differences $\bar{y}(x) - y(x)$ and $(d\bar{y}/dx) - (dy/dx)$ (the slopes) are small. [Fig. 2A-1, $y(x)$ of case $a$ has the closeness of order one, while $y(x)$ of case $b$ has only the closeness of order zero.)

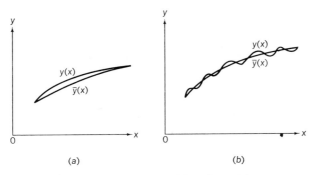

*Fig. 2A-1*  Closeness of functions $y(x)$.

The closeness is of *order $k$*, if the absolute values of the differences

$$\bar{y}(x) - y(x), \quad \frac{d\bar{y}}{dx} - \frac{dy}{dx}, \ldots, \frac{d^k\bar{y}}{dx^k} - \frac{d^ky}{dx^k}$$

are small.

*Strong maximum for functionals.* If the curve $\bar{y}(x)$ which makes a functional $v[y(x)]$ a maximum has a closeness of order zero, then such maximum is called strong.

*Weak maximum for functionals.* If the curve which makes a functional $v(y(x))$ a maximum has a closeness of order one, then such maximum is called weak. Obviously, if a curve $\bar{y}(x)$ makes a functional a weak maximum, then it makes the same functional a strong maximum too. Similar definitions are easily obtained by changing strong or weak maximum into strong or weak minimum, strong or weak relative maximum, or strong or weak absolute maximum.

APPENDIX 2B

# The Theory of General Programming

The general programming problem is stated as follows:
Maximize

$$z = f(x) \tag{1a}$$

with

$$\begin{aligned} g_i(x) &\le b_i & i &= 1, 2, \ldots, u \\ g_i(x) &\ge b_i & i &= u + 1, \ldots, v \\ g_i(x) &= b_i & i &= v + 1, \ldots, m \end{aligned} \tag{1b}$$

$$x_j \ge 0 \qquad j = 1, 2, \ldots, n \tag{1c}$$

This problem (or theory) has several other names: nonlinear, convex or concave programming and Kuhn-Tucker problem (or theory). It is called general or nonlinear, because $f(x)$ and $g_i(x)$ are not restricted to be linear as in the linear programming problem. It is called convex or concave because of the convexity and concavity of $g_i(x)$ and $f(x)$ involved in the proof of the Kuhn-Tucker theorem. It is called Kuhn-Tucker because the basic method has been developed by them.

In order for showing a number of points which are of interest to the optimization theory in general, we shall start the exposition of the general programming theory with the simplest unconstrained case, and then continue by introducing progressively the various kinds of constraints.

1. *Minima of a function.*

Let $f(x)$ be a function of the variable $x$, without any constraint on $x$, i.e., $x$ can be $>$, $=$, $<0$. The necessary conditions for a relative minimum of $f(x)$ are obtained, if $f \in C^1$, by considering

(a) $df(x)/dx = 0$, if $x$ is one-dimensional.
(b) $\nabla f(x) = 0$, if $x$ is $n$-dimensional.

2. *Minima of a function with constraint on the state variable.*

Let us consider the same minimization problem, but this time, let $x$ be constrained to take nonnegative values, i.e., $x \ge 0$. Let us consider first the one-dimensional problem. Two cases are possible:

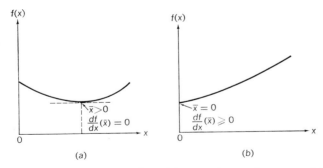

**Fig. 2B-1**   Minima of a function with constraint on the state variable.

(a) The minimum occurs at some value $\bar{x}$, with $\bar{x} > 0$, and at which

$$df(\bar{x})/dx = 0 \qquad \text{(Fig. 2B-1}a\text{)}$$

(b) The minimum occurs at the point $\bar{x} = 0$, in order for this to be possible; it is necessary that the slope of $f(x)$ be nonnegative, i.e.,

$$df(\bar{x})/dx \geq 0 \qquad \text{(Fig. 2B-1}b\text{)}$$

The usual necessary conditions that $\bar{x}$ yields a minimum of $f(x)$ turns out in this case to be

$$df(\bar{x})/dx - \bar{y} = 0 \tag{2a}$$

$$\bar{x} \geq 0 \qquad \bar{y} \geq 0 \tag{2b}$$

$$\bar{x} \cdot \bar{y} = 0 \tag{2c}$$

for some value of $\bar{y}$. The relation (2c) requires that either $\bar{x}$ or $\bar{y}$ be zero. Hence only one of the inequalities (2b) can be satisfied in the strict inequality sense. The relation (2c) is known as a complementary slackness condition, and is introduced into the necessary condition because of the initial constraint $x \geq 0$.

If the problem is $n$-dimensional, the necessary conditions become

$$\nabla f(x) - \bar{y} = 0 \tag{3a}$$

$$\bar{x} \geq 0 \qquad \bar{y} \geq 0 \tag{3b}$$

$$\bar{x}^T \bar{y} = 0 \tag{3c}$$

where $\bar{x}$ and $\bar{y}$ are $n$-vectors. Note that the relation (3c) states that the two vectors $\bar{x}$ and $\bar{y}$ are mutually orthogonal.

3. *Minima of a function without constraint on the state variable but with equality constraints.*

This is the problem treated by the ordinary Lagrange-multiplier method. Let us consider a two-dimensional problem:

Minimize

$$f(x) = f(x_1, x_2) \tag{4a}$$

with

$$g(x) = g(x_1, x_2) = 0 \tag{4b}$$

Forming the Lagrangian equation

$$F(\lambda, x) = f(x) - \lambda g(x) \tag{5}$$

and differentiating, we obtain

$$\partial f(x) - \lambda \, \partial g(x) = 0 \tag{6a}$$

$$g(x) = 0 \tag{6b}$$

(Three equations for three unknowns). The relations (6a) and (6b) are illustrated geometrically in Fig. 2B-2. Note first that the gradients $\partial f(x)$ and $\partial g(x)$ are respectively normal to the curves defined by $f(x) = $ constant and $g(x) = 0$. The solution point $\bar{x}$

Is on the curve $g(x) = 0$, satisfying relation (6b), and

Is on that curve $f(x) = $ constant so that the value of $\partial f(\bar{x})$ is $\lambda$ times that of $\partial g(\bar{x})$.

The problem treated here can be generalized in two ways:

(a) We still have one equality constraint, but $x$ is an $n$-component vector.

(b) $x$ is an $n$-component vector, and furthermore, we have $m$ equality constraints $g_i(x) = 0, i = 1, 2, \ldots , m$.

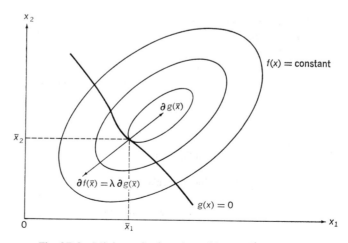

*Fig. 2B-2*  Minima of a function with equality constraint.

We ask the reader to derive the necessary conditions and the geometrical interpretations.

4. *Minima of a function with equality constraints and constraints on the state variable.*

Let us consider a two-dimensional problem:

Minimize

$$f(x) = f(x_1, x_2) \tag{7a}$$

with

$$g(x) = g(x_1, x_2) = 0 \tag{7b}$$

$$x = (x_1, x_2) \geq 0 \tag{7c}$$

By combining the results obtained in (2a), (2b), (2c), (6a), and (6b), we arrive to the following set of necessary conditions:

$$\partial f(\bar{x}) - \lambda\, \partial g(\bar{x}) - \bar{y} = 0 \tag{8a}$$

$$g(\bar{x}) = 0 \tag{8b}$$

$$\bar{x} \geq 0 \qquad \bar{y} \geq 0 \tag{8c}$$

$$\bar{x}^T \bar{y} = 0 \tag{8d}$$

where $\bar{x}$ and $\bar{y}$ are two-component vectors, and $\lambda$ a scalar. If the minimizing $\bar{x} = (\bar{x}_1, \bar{x}_2) \neq 0$, then the complementary slackness condition (8a) requires that $\bar{y}_1 = \bar{y}_2 = 0$ and the problem becomes exactly the one stated by (4a) and (4b). Consider however the situation indicated in Fig. 2B-3 where the minimizing $\bar{x}$ is on the curve $g(x) = 0$, but has the components $\bar{x}_1 > 0$, $\bar{x}_2 = 0$. According to (8d), we have $\bar{y}_1 = 0$, $\bar{y}_2 > 0$. The two component-relations of (8a) become

$$\partial f(\bar{x}_1) - \lambda\, \partial g(\bar{x}_1) = 0 \tag{9a}$$

$$\partial f(\bar{x}_2) - \lambda\, \partial g(\bar{x}_2) - \bar{y}_2 = 0 \tag{9b}$$

The geometrical interpretation of these relations is given in Fig. 2B-3. The component $\partial f(\bar{x}_1)$ is equal to $\lambda\, \partial g(\bar{x}_1)$ in a similar fashion to (6a). The component $\partial f(\bar{x}_2)$, however, is equal to $\lambda\, \partial g(\bar{x}_2)$ plus some value $\bar{y}_2$.

The generalization of the problem formulated by the relations (7a), (7b), and (7c) is the extension to an $n$-component $x$-vector, and to $m$ equality constraints $g_i(x) = 0$, $i = 1, 2, \ldots, m$. The necessary conditions are

$$dg(x)^T \lambda - \partial f(x) + y = 0 \tag{10a}$$

$$g(x) = 0 \tag{10b}$$

$$x \geq 0 \qquad y \geq 0 \tag{10c}$$

$$x^T y = 0 \tag{10d}$$

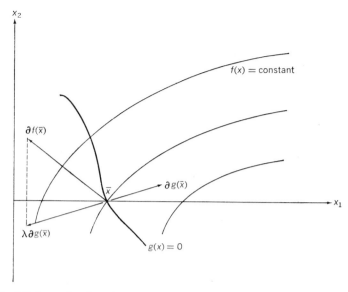

**Fig. 2B-3**  Minima of a function with equality constraint and constraint on the state variables.

where the state variable $x$ and the slack variable $y$ are $n$-component vectors, and the Lagrangian-multiplier $\lambda$ an $m$-component vector. We note now that while the constraint $x \geq 0$ introduces the slack vector $y$, the equality constraints $g_i(x) = 0$ introduce the Lagrangian-multiplier vector $\lambda$.

5. *Minima of a function with inequality constraints and constraints on the state variable.*

Let us now consider the following problem.
Minimize

$$f(x) \tag{11a}$$

with

$$g(x) \geq 0 \tag{11b}$$

$$x \geq 0 \tag{11c}$$

First, we introduce a slack vector $u$ (maximum number of components equal to $m$) so as to transform the inequality (11b) into an equality.

$$g(x) - u = 0 \tag{11d}$$

The problem is now similar to the preceding one, except for the presence of the new vector-variable $u$. A reasoning similar to the one given in (b) concerning $\bar{x}$ and $\bar{y}$ applies here on $\bar{\lambda}$ and $\bar{u}$, and leads to a relation similar to (3c) or (10d).

THEOREM E1.   *Consider the general programming problem: minimize*
$f(x)$, *with* $g(x) \geq 0$ *and* $x \geq 0$, *where* $x$ *is a n-component vector,* $g(x)$ *an m-component vector,* $f \in C^1$ *and* $g \in C^1$. *A necessary condition that* $\bar{x}$ *be a local minimum of the general program is the existence of a solution* $(\bar{x}, \bar{\lambda}, \bar{y}, \bar{u})$ *satisfying the following set of relations:*

$$dg(\bar{x})^T \bar{\lambda} - \partial f(\bar{x}) + \bar{y} = 0 \tag{12a}$$

$$g(\bar{x}) - \bar{u} = 0 \tag{12b}$$

$$\bar{x} \geq 0 \qquad \bar{y} \geq 0 \qquad \bar{u} \geq 0 \qquad \bar{\lambda} \geq 0 \tag{12c}$$

$$\bar{y}^T \bar{x} = 0 \tag{12d}$$

$$\bar{\lambda}^T \bar{u} = 0 \tag{12e}$$

*Now, forming the Lagrangian function*

$$F = f(x) - \lambda g(x)$$

*and noting that from* (12b)

$$\bar{y} = \nabla_x F(\bar{x}, \bar{\lambda}) = \partial f(\bar{x}) - dg(\bar{x})^T \bar{\lambda}$$

*and*

$$\bar{u} = \nabla_\lambda F(\bar{x}, \bar{\lambda}) = g(\bar{x})$$

*we have an alternate way of stating the conditions* (12a) *through* (12e)

$$dg(\bar{x})^T \bar{\lambda} - \partial f(\bar{x}) + \bar{y} = 0 \tag{13a}$$

$$g(\bar{x}) - \bar{u} = 0 \tag{13b}$$

$$\bar{x} \geq 0 \qquad \bar{y} \geq 0 \qquad \bar{u} \geq 0 \qquad \bar{\lambda} \geq 0 \tag{13c}$$

$$\bar{\lambda}^T g(\bar{x}) = \bar{\lambda}^T \nabla_\lambda F(\bar{x}, \bar{\lambda}) = 0 \tag{13d}$$

$$\bar{x}^T [\partial f(\bar{x}) - dg(\bar{x})^T \bar{\lambda}] = \bar{x}^T \nabla_x F(\bar{x}, \bar{\lambda}) = 0 \tag{13e}$$

THEOREM E2.   *Let* $C$ *be the set of points which satisfy* $g(x) \geq 0$, $x \geq 0$. *If in the preceding general programming problem,* $f$ *is a concave function,* $g_1, g_2, \ldots, g_m$ *are convex functions, and* $\bar{x}$ *be a relative minimum, satisfying the set of necessary conditions* (12a) *through* (12e), *then* $\bar{x}$ *minimizes* $f(x)$ *in* $C$. *That is,* $\bar{x}$ *is also an absolute minimum.*

*Proof*

Suppose $\bar{x}$ does not minimize $f(x)$ in $C$. Then let $\overset{\circ}{x}$ be some point in $C$ for which $f(\overset{\circ}{x}) < f(\bar{x})$. The line segment $x = (1 - \alpha)\bar{x} + \alpha\overset{\circ}{x}, 0 \leq \alpha \leq 1$, is entirely contained in $C$ because $C$ is convex. The objective function evaluated on this line segment is

$$f(\alpha) = f[(1 - \alpha)\bar{x} + \alpha\overset{\circ}{x}] \leq (1 - \alpha)f(\bar{x}) - \alpha f(\overset{\circ}{x})$$

Now there is a point on this line segment which is contained in the neighborhood of $\bar{x}$ but is not identical to $\bar{x}$. Let this point be $x = (1 - \delta)\bar{x} + \delta\overset{\circ}{x}$, $0 \leq \delta \leq 1$. Then

$$f(x) = f[(1 - \delta)\bar{x} + \delta\overset{\circ}{x}] \leq (1 - \delta)f(\bar{x}) + \delta f(\overset{\circ}{x})$$
$$= f(\bar{x}) - \delta[f(\bar{x}) - f(\overset{\circ}{x})]$$

by the concavity of $f(x)$. Since $f(\overset{\circ}{x}) < f(\bar{x})$, we have

$$f(x) < f(\bar{x})$$

which contradicts the assumption that $\bar{x}$ is a local minimum.

NOTE.   The truly general programming problem implies that $g(x)$ be $\geq 0$, or $\leq 0$, as it is formulated in the main text. We have only taken the case of $g(x) \geq 0$; otherwise, we would see so many complicated notations that the essence of the demonstration would be hindered.

# APPENDIX 2C

# Duality

It was seen, in Appendix 2B, that in the solution of programming problems, we use the so-called Lagrangian multipliers. As they are used, we might think that they are just mathematical artifices. However, in the study of programming problems, it has been found that they have some physical meanings; that is, if the original optimization problem is formulated in terms of the variables $x$, the *same* problem can be formulated in terms of some other variables $\lambda$ which also have very precise physical interpretation. To distinguish between these two problems, it is usual to call the original formulation the *primal* problem, and the new one the *dual* problem. An interesting point in this primal-dual correspondence is the existence of the *minimax* property. This is shown in the following.

We take the programming problem

$$[\text{Max}] f(x_j) \qquad j = 1, 2, \ldots, n \qquad \text{(1a)}$$

with

$$g_i(x_j) = b_i \qquad i = 1, 2, \ldots, m \qquad \text{(1b)}$$

$$x_j \geq 0 \qquad j = 1, 2, \ldots, n \qquad \text{(1c)}$$

$f \in C^1, g_i \in C^1$, and we denote by $Y$ the set of points $x$ satisfying (1b). Now, we form the Lagrangian function

$$F(x_j, \lambda_i) = f(x_j) - \sum_{i=1}^{m} \lambda_i [g_i(x_j) - b_i] \qquad \text{(2)}$$

We have previously seen that under certain conditions, $f(x_j)$ has a relative maximum at $\bar{x}_j$ for $x_j$ satisfying $g_i(x_j) = b_i$, $i = 1, 2, \ldots, m$. Let $\bar{\lambda}_i = (\bar{\lambda}_1, \bar{\lambda}_2, \ldots, \bar{\lambda}_m)$ a set of Lagrangian multipliers corresponding to $x_j$. Let us now suppose that there exists a $\delta$-neighborhood of $\bar{\lambda}_i$ in $E^m$ such that for any $\lambda_i$ in this $\delta$-neighborhood $F(x_j, \lambda_i)$ has an unconstrained relative maximum with respect to $x_j$ at a point which satisfies

$$\frac{\partial F}{\partial x_j} = 0 = \frac{\partial f}{\partial x_j} - \sum_{1}^{m} \lambda_i \frac{\partial g_i}{\partial x_j} \qquad j = 1, 2, \ldots, n \qquad \text{(3)}$$

Imagine now that Eqs. (3) have a unique solution for every $\lambda$ in the $\delta$-neighborhood of $\bar{\lambda}_i$ and that $\bar{x}_j$ can be written $\bar{x}_j(\lambda)$. Then in the $\delta$-neighborhood of $\bar{\lambda}$ we can

**97**

write

$$\max_{x_j} F(x_j, \lambda_i) = F(\bar{x}_j, \lambda_i) = h(\lambda_i) \tag{4}$$

i.e., the maximum with respect to $x_j$ will be a function of $\lambda_i$. Note that at $\bar{\lambda}_i$, we have

$$h(\bar{\lambda}_i) = F(\bar{x}_j, \bar{\lambda}_i) = f(\bar{x}_j) \tag{5}$$

Consider $F(x_j, \lambda_i)$ for a fixed $\lambda_i$, for $x_j$ satisfying (1b), $x_j \in Y$, Eq. (2) becomes

$$F(x_j, \lambda_i) = f(x_j) \tag{6}$$

Hence

$$\max_{x_j} F(x_j, \lambda_i) = \max_{x_j} f(x_j) = f(\bar{x}_j) = h(\bar{\lambda}_i) \tag{7}$$

Since in (7), we require that $x_j \in Y$, we have restricted the range of variation of $x_j$, and it follows that

$$h(\lambda_i) \geq h(\bar{\lambda}_i) \tag{8}$$

and hence $h(\lambda_i)$ has a minimum at $\bar{\lambda}_i$. Since $h(\lambda_i) = F(x_j, \lambda_i)$ when $x_j$ and $\lambda_i$ are related by (3), it follows that $F(x_i, \lambda_j)$ has a relative minimum at $\bar{\lambda}_i$ with respect to $\lambda_i$. The problem of minimizing $F(x_j, \lambda_i)$ with respect to $\lambda_i$ subject to the constraints (3) is called a dual of the original problem. The primal and dual problems have the following property:

$$f(\bar{x}_j) = h(\bar{\lambda}_i) = \min_{\lambda_i} \max_{x_j} F(x_j, \lambda_i) \tag{9}$$

The two points shown above, namely (1) the existence of a dual problem and (2) the minimax property, apply in principle to all optimization problems. However, the difficulty, in practice, is to derive analytically the dual problem from the primal problem. The reasons are (1) $f(x_j)$ and $g_i(x_j)$ may be nonlinear and (2) $g_i(x_j)$ may be $\geq$ or $\leq b_i$. Even in the simplest case of linear programming problems, where $g_i(x_j)$ are reduced to be $\Sigma\, a_{ij}x_j$, the presence of the inequalities imposes the use of complex mathematical notations and arguments to derive the dual program. Therefore, we shall not do it here.

# REFERENCES

[2-1] Ruge, H. "On the Optimal Control of Hydroelectric Power Systems," Second IFAC Congress, Basles (June 1963).

[2-2] Gibrat, R., "Le problème général des plans de production d'énergie électrique et les usines marémotrices," Document EDF (French Electricity Board) (November 1955).

[2-3] Ventura, E., "Un exemple de recherche opérationnelle, la détermination d'un plan optimum de production d'énergie électrique par la théorie des programmes linéaires," Document SOFRO (Société Française de Recherche Opérationnelle) (1956).

[2-4] Miller, C. E., "Simplex Method for Local Separable Programming," *Proc. Symp. Math. Program.*, Univ. of Chicago (June 1962).

[2-5] Hovanessian, S. A., and T. M. Stout, "Optimum Fuel Allocation in Power Plants," Tr. IEEE on Power Apparatus and Systems (June 1963), 329–335.

[2-6] Kuehn, D. R., and H. Davidson, "Mathematics of Computer Control," *Chem. Eng. Progr.*, **57**, n 6 (June 1961).

[2-7] Lermoyez, "Optimalisation des unités de raffinage," *Automatisme* (French) (May 1962).

[2-8] Conway, F. M., "The Use of Parametric Techniques to enhance the Interpretive Potential of Linear Programming," IBM Seminar on Refinery Engineering and Operation, Poughkeepsie, New York (June 1958).

[2-9] Kuhn, H. W., and A. W. Tucker, "Nonlinear Programming," *Proc. 2nd Berkeley Symp. Math. Statist. Probability*, J. Neyman (Ed.), Univ. of California Press (1951), 481–492.

[2-10] Tucker A. W., "Linear and Nonlinear Programming," *Operations Res.* **5** (1957), 244–257.

[2-11] Carpentier J., "Contribution à l'étude du dispatching économique," *Bull. Soc. Franç. Electriciens* (August 1962), 431–447.

[2-12] Charmes, A., and C. E. Lemke, "Minimization of Nonlinear Separable Convex Functions," *Naval Res. Logist. Quart.*, **1**, n 4 (December 1954).

[2-13] Wolfe, P., "The Simplex Method for Quadratic Programming," *Econometrica*, **27** (1959), 382–398.

[2-14] Hamza, M. H., "Optimum Design of Multi-input Multi-output Control Systems using Linear Programming," *New Techniques* (Swiss), **9**, n A6 (November 1967), 357–360.

[2-15] Jewell, W. S., "Complex: A Complementary Slackness, out-of-kilter Algorithm for Linear Programming," AD-662668 (March 1967), 96.

[2-16] Moore, J. C., "Some Existensions of the Kuhn-Tucker Results in Concace Programming," *NASA* CR-91076, (August 1967), 44.

[2-17] Fath, A. F., and Thomas J. Higgins, "Fixed-time Fuel-optimal Control of Linear State Constrained Systems by Use of Linear Programming Techniques," AD 660064 (1967).

[2-18] Waespy, C. M., "An Application of Linear Programming to Minimum Fuel Optimal Control," *NASA* CR-86654, Rept. 67-16 (May 1967), 159.

[2-19] Bankoff, S. G., "Some Generalization of the Discrete Optimal Control Problem (Nonlinear Programming)," TR-66-104, AD 643280 (October 1966), 33.

[2-20] Vardiya, P. P., "Nonlinear Programming and Optimal Control," *NASA* CR-479 (May 1966), 62.

# 3

---

# EXTREMUM SEEKING
# METHODS

## 3.1 INTRODUCTION

IN CHAPTER 2, we have discussed optimization of static systems, some knowledge of which is available in the form of system-equations (or constraints). Such a situation can be compared to that of a climber seeking the summit of a mountain, given the summit orientation, the category of roads, and the prohibited area. His problem is to make a correct deduction, and, if this is done, he will be led directly to the summit. In this chapter, we shall discuss optimization of static and dynamic systems, the behavior of which is not previously known. Using the same analogy, this situation can be compared to that of a climber seeking the summit of a mountain in foggy weather. Nothing is known to him except the feeling of gravity, indicating whether he has ascended or descended. This category of optimization problems can be adequately referred to as *seeking* problems, as opposed to all the others treated in this book, because of the high degree of ignorance in which the seeker is placed.

With respect to our usual terminology, the seeking problem can be one- or multidimensional, static or dynamic, and the seeking process implemented in an open- or closed-loop form. In the control field, the development of extremum-seeking techniques (the system behavior not known), began in the early 1950's. The study of these techniques parallel that of the optimization techniques (the system behavior known). There is no common theory covering the two branches, although there are many similar concepts useful to both.

In Section 3.2 of this chapter, we shall discuss Fibonacci search,[1] thus introducing, among others, two important concepts: (1) the interval of uncertainty, and (2) the optimum of an extremum search. In Section 3.3, we discuss the Steepest Ascent (or Descent) Search applicable to multidimensional problems. Emphases will be on the concepts of gradient and ridge. The Fibonacci search and the steepest ascent search are applicable to static problems. Methods applicable to dynamic problems are discussed in Sections 3.4 and 3.5. In Section 3.4, the category of peak-holding methods is explained; these methods are useful for quasi-static dynamic problems. In Section 3.5, we discuss the Analogue Extremum Forcing method, applicable to general dynamic systems. Another relevant point is that the static searching techniques discussed here have a convenient theoretical basis, while the two dynamic searching techniques are rather heuristic, and are studied in the on-line form at the beginning of their development.

## 3.2  FIBONACCI SEARCH IN ONE-DIMENSIONAL STATIC PROBLEMS

### 3.2.1  Problem Formulation

In one-dimensional problems, the mathematical model of the system is reduced to a certain performance function $y$ of the control variable $x$. Since we assume that the system behavior is unknown, all we can formulate is some general form

$$y = f(x) \tag{3-1}$$

In order for the optimization problem to be meaningful, it is required that:

1. The value of $y$ can be known by some direct or indirect measurement.
2. The function $f(x)$ exhibits some maxima (or minima) with respect to $x$ such as shown in Fig. 3-1.

For simplicity, we shall assume that, by proper scalings, we only operate in the first quadrant of the $x$-$y$ plane, and that $x$ can only take values between 0 and 1, as shown in Fig. 3-2. The optimization problem can then be formulated as follows. Find the value of the control variable $x$, $0 \leq x \leq 1$, so that the resulting value of the performance $y$ is an extremum. Since the analytical form of $f(x)$ is unknown, the solution of such a problem will necessarily be the result of a trial-and-error process.

However, we can imagine that there may exist a great number of possibilities.

---

[1] We understand *seeking* as an action, and *search* as a policy.

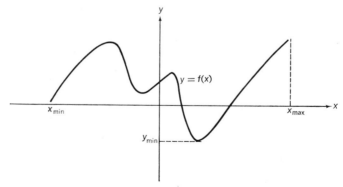

*Fig. 3-1*    Some one-dimensional performance function.

1. We can start from one of the extreme values $x = 0$ or $\alpha = 1$ and then go to the other extreme value.

2. We can divide the whole 0-1 line into two equal parts. We compute $f(x)$ at the three points $x = 0$, $x = .5$, and $x = 1$ yielding $f(0)$, $f(.5)$, and $f(1)$. If $f(x)$ is unimodal, i.e., having only one maximum or minimum, the promising part where the extremum is located can be deduced by comparing the magnitudes of $f(0)$, $f(.5)$, and $f(1)$. For instance, in the case of a maximum, if the magnitude of $f(0)$ or $f(1)$ is not the highest, i.e., $f(.5) > f(0)$ and $f(.5) > f(1)$, then the maximum is between $x = 0$ and $x = .5$ if $f(1) < f(0)$; it is between $x = .5$ and $x = 1$ if $f(1) > f(0)$.* After this step, the promising part is again divided into two equal parts, and so on.

3. We can use simultaneous search policies by choosing several values $x_{k1}, x_{k2}, \ldots$ at the same time $t_k$.

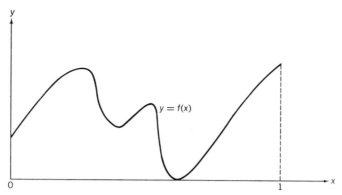

*Fig. 3-2*    Function of Fig. 3-1 after standardization.

* Statements only valid for functions unimodal and with continuous first derivatives. For other unimodal functions, see [G-78, p. 220].

4. We can use sequential search policies by choosing only one value of $x$: $x_1$, $x_2$, ... , at each time $t_1$, $t_2$, ... .

The question then is not only to find a good search policy, but *the* good policy. Obviously, the criterium of the good policy is the search speed. Therefore, the preceding optimization problem must be formulated differently. Find, in a minimum number of steps, the value of $x$, $0 \leq x \leq 1$, so that the resulting $y$ is an extremum.[1] The Fibonacci search is the method which solves this problem, because it reduces to the minimum the search interval between each two subsequent steps, as compared to any other search policy.

### 3.2.2  Demonstration

First of all, let us define the expression "interval of uncertainty." Suppose that we search the maximum of a function $y(x)$ within the interval $x = a$ and $x = b$ (Fig. 3-3c). The interval $a - b$ is called an interval of uncertainty because we do not know the exact location of the maximizing value $x_m$

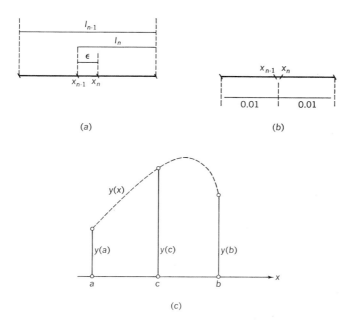

(a)

(b)

(c)

*Fig. 3-3*  Optimum location after the $n$th step.

---

[1] This requirement of "minimum number of steps" forms a second optimization problem distinct from the original one. The minimal case is one step. This happens when $f(x)$ is known. The zeroing of the derivative supplies the extremalizing values of $x$ in one step.

within this interval. Let us now make the first search step. We divide $a - b$ into two intervals $a - c$ and $c - b$. We compute the values of $y$ at the points $a$, $b$, and $c$. Suppose that we find $y(c) > y(b) > y(a)$; we then know that the maximizing $x_m$ is between $x = c$ and $x = b$. However, we still do not know the exact location of $x_m$. The interval $c - b$ is the new interval of uncertainty. Any search method must, in principle, reduce the magnitude of the interval of uncertainty after each search step. The higher the rate of this reduction, the better the corresponding search method. The Fibonacci search gives the highest rate of reduction. We give here a short demonstration of its optimality. Suppose that we have a search of $n$ steps. *At the $(n - 1)th$ step*, the $(n - 1)$th interval of uncertainty $l_{n-1}$ and the position $x_{n-1}$ corresponding to the $(n - 1)$th best outcome $y_{n-1}$ are known. The problem is to place the $n$th choice $x_n$ so that the next interval of uncertainty $l_n$ is the smallest possible.

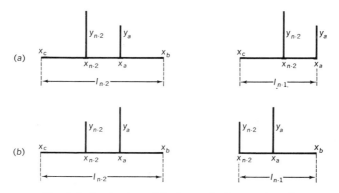

**Fig. 3-4**   Derivation of the interval of uncertainty $l_{n-1}$.

It can be seen that this is obtained if (1) $x_n$ is placed as close to $x_{n-1}$ as possible, so that $x_{n-1} - x_n = \epsilon$, $\epsilon$ being the smallest measurable unity, and (2) the group of the two points $(x_{n-1}, x_n)$ is at the center of $l_{n-1}$ (see Fig. 3-3a) so that

$$l_{n-1} = 2l_n - \epsilon \tag{3-2}$$

To solidify out understanding, we can imagine, with a definition of one out of 100, that $\epsilon = 0$, $l_n = \cdot 01$ (see Fig. 3-3b).

*At the $(n - 2)th$ step*, the $(n - 2)$th interval of uncertainty and the position $x_{n-2}$ corresponding to the $(n - 2)$th best outcome $y_{n-2}$ are known. Now we have to locate a point $x_a$ which will result in an outcome $y_a$. If $y_{n-2} > y_a$, (Fig. 3-4a), we will take $l_{n-1} = x_a - x_c$. However, if it happens that $y_{n-2} < y_a$, then we will take $l_{n-1} = x_b - x_{n-2}$. Note here that at the $(n - 1)$th step, $x_{n-2}$ or $x_a$ becomes the point $x_{n-1}$ according to the case. The problem at this $(n - 2)$th step is therefore to locate $x_a$ so that the resulting interval of uncertainty is as small as possible. Since in either case (a) or (b) its value must be at least the one required by the (n − 1)th step, the best way is to make it equal in either eventuality. As a conclusion, $x_a$ must be located so that $x_a - x_b = x_c - x_{n-2}$.

At the (n − 1)th step, we have shown that the point $x_{n-1}$ whether it replaces $x_{n-2}$ or $x_a$, must be, in both cases, distant of $l_n$ from the opposing end of $l_{n-1}$; the relation between the three intervals of uncertainty is therefore (Fig. 3-5)

$$l_{n-2} = l_{n-1} + l_n \tag{3-3}$$

Substituting Eq. (3-2) into Eq. (3-3), we have

$$l_{n-2} = 2l_n - \epsilon + l_n = 3l_n - \epsilon \tag{3-4}$$

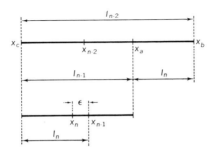

**Fig. 3-5**  Relation between the intervals of uncertainty.

Similar arguments are valid for all intervals with decreasing values of the index. Then, setting $\epsilon$ to zero, and generalizing, we have

$$l_{n-3} = l_{n-2} + l_{n-1} = 5l_n \tag{3-5}$$
$$l_{n-4} = l_{n-3} + l_{n-2} = 8l_n \tag{3-6}$$

$$l_{j-1} = l_j + l_{j+1} = kl_n \qquad \text{for all } 1 < j < n \tag{3-7}$$

Let us now introduce the sequence of numbers $F_k$ defined as

$$F_0 = F_1 = 1$$
$$F_k = F_{k-1} + F_{k-2} \qquad k = 2, 3, \dots \tag{3-8}$$

For historical reasons, $F_k$ is called the $k$th Fibonacci number, and the sequence (3-8) the Fibonacci sequence. It is easy to see the similarity between the expressions (3-7) and (3-8), and that the coefficients of the right member of the Eqs. (3-4) to (3-7) form the Fibonacci sequence. The best extremum-seeking policy is therefore the one which, at each search step, chooses the location $x_j$ so that the previous interval of uncertainty $l_{j-1}$ is divided into two intervals $l_j$ and $l_{j+1}$ and that $l_{j-1}$, $l_j$, and $l_{j+1}$ contain respectively $F_k$, $k_{(k-1)}$, and $F_{(k-2)}$ unities of interval, $F_k$, $F_{(k-1)}$, and $F_{(k-2)}$ being any of three successive Fibonacci numbers.

### 3.2.3   Search Procedure

We shall illustrate the Fibonacci search by an example: find the maximum of a function $y = f(x)$, $0 \leq x \leq 1$, the smallest interval of $x$ distinguishable being .05.

*Search*                                       *Illustration*

**Step 1.**   Rescaling. Divide 1 by .05 to obtain 200. The Fibonacci number nearest to 200 is 233 ($k = 12$). Testing:

for $x = 0$     $y = 10$
$x = 200$     $y = 15$

the extremum seems near the end.

**Step 2.**   Choose $x_a = 144$ ($k = 11$). Outcome $y_a = 17$. The new interval of uncertainty is therefore $0 - 144$.

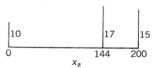

**Step 3.**   Choose $x_b = 89$ ($k = 10$). Outcome $y_b = 19$. The new interval of uncertainty is $0 - 89$.

**Step 4.**   Choose $x_c = 55$ ($k = 9$). Outcome $y_c = 18$. The new interval of uncertainty is $55 - 89$. ($89 - 55 = 34$, $k = 8$.)

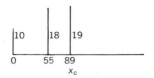

**Step 5.**   Choose

$$x_d = 55 + 21 = 76$$

(21 corresponds to $k = 7$). Outcome $y_d = 22$. The new interval of uncertainty is $55 - 76$.

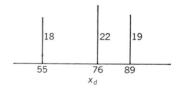

*Search*                                   *Illustration*

**Step 6.** Choose

$$x_e = 55 + 13 = 68$$

(13 corresponds to $k = 6$). Outcome $y_e = 25$. The new interval of uncertainty is therefore $55 - 68$.

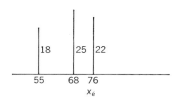

**Step 7.** Choose

$$x_f = 55 + 8 = 63$$

(8 corresponds to $k = 5$). Outcome $y_f = 21$. The new interval of uncertainty is therefore $63 - 68$. ($68 - 63 = 5$, $k = 4$.)

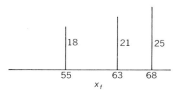

**Step 8.** Choose

$$x_g = 63 + 3 = 66$$

(3 corresponds to $k = 3$). Outcome $y_g = 24$. The new interval is $68 - 66$.

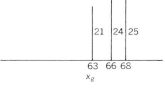

**Step 9.** Choose $x_h = 67$. Outcome $y_h = 25$. The maximum of $y$: $y_{\max} = 25$ occurs at $x = 67$ or $68$ (real values .335 or .34).

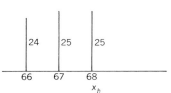

The search is accomplished in 9 steps with 11 measurements. If the dichotomous search was used, 64 steps would be needed. Note that in the normal Fibonacci treatment of this problem, we have $k = 12$, leading to 11 search steps and 13 measurements. Economy of two steps has been obtained in Step 4 and Step 7.

The Fibonacci search suffers from two insufficiencies. First, this search does not extend in any *simple* way to higher dimensions than one. Second, the optimality of the search, in the sense of minimum number of steps in reaching the extremum of $f(x)$, is true only if $f(x)$ is unimodal.

In effect, if $f(x)$ is bimodal or multimodal, i.e., if $f(x)$ has two or more maxima (minima), the result guarantees only a local, but not necessarily absolute maximum (minimum). The interest of the method is, however, great. In many practical situations, we meet one-dimensional unimodal functions. Also, in the case of multimodal or multidimensional functions, this method can be useful for an approximating approach.

## 3.3  STEEPEST ASCENT SEARCH IN MULTIDIMENSIONAL STATIC PROBLEMS

### 3.3.1  Intuitive Concept and Application Areas

In this section, we shall discuss multidimensional optimization problems in which the system behavior is unknown. The problem can be formulated similarly to that of the preceding section. Given a performance index $y = f(x_1, x_2, \ldots, x_n)$, function of several variables $x_1, x_2, \ldots, x_n$, the analytic form of $f$ being unknown, find, in a minimum number of steps, the values of $x_1, x_2, \ldots, x_n$, so that the resulting value of $y$ is an extremum. Geometrically, the function $y$ can be imagined as scenery (Fig. 3-6) furnished with mountains and valleys. The optimization problem is then to find as quickly as possible summits like $A$ or $B$, or a bottom like $C$.

Suppose that initially we are at some point, $D$. The quickest way of getting to the summit $A$ is certainly not the zigzagging path $DEA$, but the path $DFA$ along which we have ascended the most at each step. Likewise, the steepest descending path $GHC$ is also the quickest way to reach the bottom from the initial point $G$. It is therefore intuitive to understand that the steepest ascent (or descent) is the quickest way to reach a maximum (or a minimum). Two remarks must be made here to show the limitations:

1. The applicability depends on the contour shape of the functions. For instance, from $I$ the steepest ascending path $IA$ will contain some descending portion in the vicinity of some point $J$.

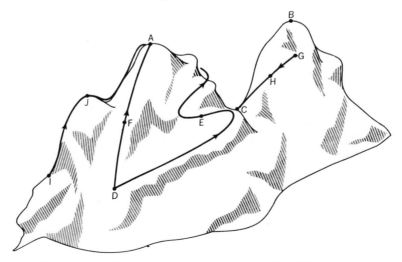

***Fig. 3-6***  Descriptive representation of some multivariable function.

2. The search only leads to local extrema. For instance, from $D$, the steepest ascent search will lead to the summit $A$, but not $B$, which is higher.

The intuitive concept of steepest search has long been mechanized into a mathematical method in connection with the properties of the gradients. [3-1; G-29, Ch. 6; G-23, Chs. 3–5]. The technique is also called the method of gradients. For clarification, we shall use two distinct names for two distinct categories of applications.

1. The name "steepest search" will be used in this section. The problems discussed as those where the analytical form of the performance function is *not known*. The main purpose of the method is to seek for the extremum.

2. The name "method of gradients" is used in the next chapter. The problems discussed are those where the performance function is analytically *known*. The properties of the gradients are used in solving differential equations during the optimization process.

### 3.3.2   Search Techniques

Although based on the same concepts, the method of gradients will be exposed more mathematically in the next chapter; here, the steepest search will be exposed more geometrically.

Let us consider a function of two variables

$$y = f(x_1, x_2) \tag{3-9}$$

the $x_1 - x_2$ plane forming the ground, while $y$ represents the altitude. For ease of discussion, we shall cut the function $f$ with horizontal planes and project the equi-$y$ or equi-altitude contours on the ground (Fig. 3-7). Starting from any point $A$, the problem of reaching the summit $S$ as quickly as possible is twofold; (a) find the good direction $AS$, (b) go to $S$ from $A$ as quickly as possible.

The reader is reminded here that in reality, either the equi-$y$ contours or the location of the point $S$ are completely unknown. The problems (a) and (b) turn out to be (c) how to avoid the tedious work of testing all the 360° directions around $A$ for getting the good one, and (d) once the good direction is obtained, since the location of $S$ is unknown, how to minimize the number of trial-and-error tests for finding the extremum.

To solve the problem (c), it is necessary to assume first that $y$ is linear in $x_1$ and $x_2$ around the point $A$, so that one can take only the first-degree terms of the polynomial expansion, and write

$$y = m_1 x_1 + m_2 x_2 \tag{3-10}$$

and

$$\Delta y = m_1 \, \Delta x_1 + m_2 \, \Delta x_2$$

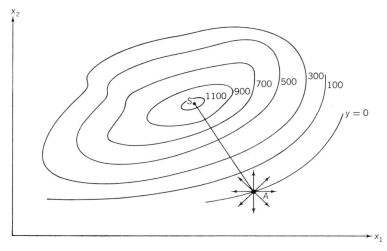

**Fig. 3-7**  Projection of equi-$y$ contours on the $x_1 - x_2$ plane.

if we consider the initial point $A$ as the origin of $y$, $x_1$, and $x_2$. Note that $m_1$ and $m_2$ are the values of the 2 first-order partial derivatives of $f$

$$m_1 = \frac{\partial f}{\partial x_1} \qquad m_2 = \frac{\partial f}{\partial x_2} \qquad (3\text{-}11)$$

These coefficients $m_1$ and $m_2$ can be obtained experimentally. In effect, $\Delta x_1$ and $\Delta x_2$ can be considered as testing values measured from the initial point $A$ along the axis of the $x_1$- and $x_2$-variable, while $(\Delta y)_1$ and $(\Delta y)_2$ will be respectively the resulting values of the outcome that we are supposed to obtain by some measuring device. Suppose, for example,

$$\text{for} \qquad \Delta x_1 = -1 \qquad (\Delta y)_1 = 3$$
$$\Delta x_2 = 1 \qquad (\Delta y)_2 = 3$$

Then

$$m_1 = -3 \qquad m_2 = 3$$

and

$$y = -3\,\Delta x_1 + 3\,\Delta x_2$$

In order to qualify the direction, we introduce the parameter $\theta$ (see Fig. 3-8) and a circle centered at $A$ and, for simplification, of a unit radius. We can write, assuming that the testing values are also unity:

$$\left.\begin{array}{l} x_1 = 1 \times \cos\theta = \cos\theta \\ x_2 = 1 \times \sin\theta = \sin\theta \end{array}\right\} \qquad (3\text{-}12)$$

$$y = m_1 x_1 + m_2 x_2 = m_1 \cos\theta + m_2 \sin\theta \qquad (3\text{-}13)$$

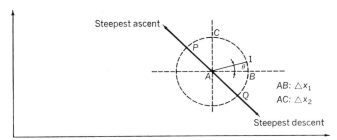

**Fig. 3-8**  Parametric representation of the search direction.

The direction $\theta^*$ leading to the greatest (or the smallest) increment of the function is given by the following condition:

$$\frac{\partial y}{\partial \theta} = -m_1 \sin \theta + m_2 \cos \theta = 0$$

and

$$\tan \theta = m_2/m_1 \qquad (3\text{-}14)$$

Equation (3-14) has two solutions:

$$\left.\begin{array}{c} \theta^* \\ \theta^* + \pi \end{array}\right\} = \arctan \frac{m_2}{m_1} \qquad (3\text{-}15)$$

The one which leads to the positive value of $\Delta y$ corresponds to the steepest ascent, while the other one corresponds to the steepest descent. These two directions, by Eq. (3-15), are necessarily back-to-back. In the case of our example:

$$\frac{m_2}{m_1} = -1 \qquad \theta_a^* = 135°$$

$$\theta_d^* = 315$$

leading to the points $P[\Delta x_1 = -(\sqrt{2}/2), \Delta x_2 = (\sqrt{2}/2)]$ and $Q[\Delta x_1 = (\sqrt{2}/2),$ $\Delta x_2 = -(\sqrt{2}/2)]$. The values of $\Delta y$ at these points are:

$$(\Delta y)_P = \frac{3\sqrt{2}}{2} + \frac{3\sqrt{2}}{2} = 3\sqrt{2}$$

$$(\Delta y)_Q = -\frac{3\sqrt{2}}{2} - \frac{3\sqrt{2}}{2} = -3\sqrt{2}$$

Therefore, $AP$ indicates the steepest ascent direction, while $AQ$ shows the steepest descent.

Once problem (c) is solved, we attack problem (d). For this, we shall consider the line of steepest ascent as a new axis of a parameter $r$ representing the distance from the point $A$ to any point of this axis. The value of $r$, by this definition, is

$$r = (x_1^2 + x_2^2)^{\frac{1}{2}} \qquad (3\text{-}16)$$

From Eqs. (3-12) and (3-14), along the steepest ascent direction,

$$\frac{x_2}{x_1} = \frac{m_2}{m_1} \tag{3-17}$$

The variables $x_1$ and $x_2$ can then be parametrically represented in terms of $r$.

$$r = \frac{x_1}{m_1}(m_1^2 + m_2^2)^{\frac{1}{2}}; \qquad x_1 = \frac{m_1 r}{(m_2^2 + m_1^2)^{\frac{1}{2}}}$$

$$r = \frac{x_2}{m_2}(m_1^2 + m_2^2)^{\frac{1}{2}}; \qquad x_2 = \frac{m_2 r}{(m_1^2 + m_2^2)^{\frac{1}{2}}} \tag{3-18}$$

Likewise, the extrapolated estimate of the performance function is

$$y = m_1 x_1 + m_2 x_2 = r(m_1^2 + m_2^2)^{\frac{1}{2}} \tag{3-19}$$

We note that the original two-dimensional problem has been reduced here to a one-dimensional problem, the new direction being the $r$-dimension. For any given value of $r$, not only the values of $x_1$ and $x_2$ are defined, even the value of $y$ can be predicted. In principle, starting from $A$, positive values of $r$ increase the outcome, and negative values of $r$ decreases it. However, we can not increase the value of $r$ indefinitely in one direction. For instance, in the steepest ascent search, before reaching the summit $S$, increase of $r$ will increase $y$; but after the summit, increase of $r$ will contrarily decrease $y$ (see Fig. 3-9). Consequently there is the need for trial-and-error tests for locating the summit $S$. Fortunately, this is a one-dimensional search, and we know that the best one is the Fibonacci search, if and only if, $y(r)$ is unimodular.

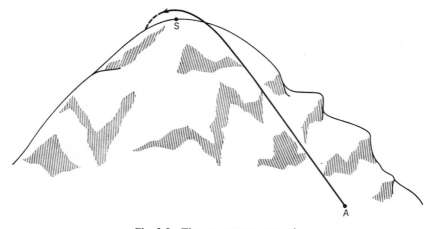

**Fig. 3-9**   The steepest ascent path.

The preceding techniques explained in a two-dimensional case can easily be generalized to multidimensional problems, if the linear approximation

$$\Delta y = \sum_{i=1}^{k} m_i \Delta x_i \qquad (3\text{-}20)$$

is adopted. As the first step, one makes $k$ tests to supply the values of the coefficients $m_i$. As the second step, one introduces an arbitrary parameter $\lambda$, the gradient line is then parametrically represented by:[1]

$$\Delta x_i = m_i \lambda \qquad i = 1, 2, \ldots, k \qquad (3\text{-}21)$$

The outcome, by combining Eq. (3-20) and (3-21), can be written as

$$\Delta y = \lambda \sum_{1}^{k} m_i^2 \qquad (3\text{-}22)$$

and its value can be predicted if the value of $\lambda$ is given. In practice, the location of the multidimensional summit can be found by trial-and-error or iterating procedure. Since Eq. (3-22) is again a function of a single variable $\lambda$, the Fibonacci search can be used. Examples of the steepest ascent search can be found in [3-2, 3-3, and 3-4].

### 3.3.3   Limitations

The linear or first-degree approximation, represented by Eq. (3-10), of the general function $f(x_1, x_2)$ [Eq. (3-9)], means that the curved face of the mountain is replaced by a slant plane. The steepest ascent direction found is, therefore, only valid if the hypothesis is correct. This is seldom the case in reality, and the usefulness of the preceding techniques becomes limited. In order to examine the extent of these limitations, and eventual remedies, we shall go into more explicit details.

Let us consider the three-dimensional Euclidian-coordinate system $y - x_1 - x_2$ centered at $A$ (Fig. 3-10). The plane $x_1 - x_2$ represents the ground plane, and $y$ the altitude; neither the $x_1$-axis nor the $x_2$-axis a priori coincides with the ground-contour. The choice of the testing values $\Delta x_1$ and $\Delta x_2$ fixes the points $B$ and $C$. With the help of $\Delta y_1$ and $\Delta y_2$, we obtain the partial slopes or the coefficients.

$$m_1 = \Delta y_1/\Delta x_1 \qquad \text{(segment } AD\text{)}$$
$$m_2 = \Delta y_2/\Delta x_2 \qquad \text{(segment } AE\text{)} \qquad (3\text{-}23)$$

The line segments $AD$ and $AE$, having a common point $A$, form univocally the slant plane $P$. By virtue of the assumed linearity, the plane $P$ is tangent to the function $f(x_1, x_2)$. Graphically, it means that the plane $P$ touches the mountain only at one point, the point $A$. The ground-trace $MN$ of the plane

---

[1] In Eq. (3-18), the second members are divided by the quantity $m_1^2 + m_2^2$. Since this is a constant coefficient common to all $x$ and $y$, it can be omitted.

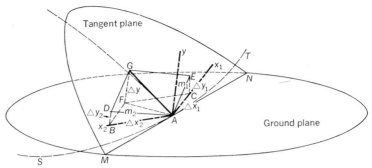

***Fig. 3-10***    The first step of steepest ascent search.

$P$ is also necessarily tangent to the ground-contour $ST$ of the function $f(x_1, x_2)$ at the point $A$. The equation of the ground-tangent $MN$, which is the linearization limit of the ground-contour $ST$ can be found by setting $y$ to 0.

$$y = f(x, y) \simeq m_1 x_1 + m_2 x_2 = 0 \qquad (3\text{-}24)$$

Leading to

$$\frac{x_2}{x_1} = -\frac{m_1}{m_2} \qquad (3\text{-}25)$$

Let us now consider the steepest ascent direction $AF$ defined on the ground. This direction is determined by Eq. (3-15) or Eq. (3-17). The comparison of Eq. (3-25) and Eq. (3-17) shows that the two lines $MN$ and $AF$ are negative reciprocals of each other. They are, therefore, perpendicular, a known fact of analytical geometry. Equation (3-19) gives the value of the slope $AG$, or the unit-increment of the outcome along the steepest ascent direction.

$$\frac{\Delta y}{\Delta r} = (m_1^2 + m_2^2)^{\frac{1}{2}} \qquad (3\text{-}26)$$

Since $G$ is necessarily on the slant plane $P$, $m_1 = AD$, and $m_2 = AE$, Eq. (3-26) means that $AD$ is perpendicular to $AE$. It can be shown also that since $G$ is the projection of $F$ on the ground-plane, and that $AF$ is perpendicular to $MN$ at $A$, $AG$ is also perpendicular to $MN$, at $A$ and on the plane $P$.

The fact that the linear hypothesis does seldom hold in practice leads to two categories of consequences. (a) the optimum search direction found at the first step will not be valid for subsequent steps, (b) the slope of the out-come-increment will not be given by Eq. (3-26). We shall discuss these points first, in the case where the function $f(x_1, x_2)$ is represented by some conic section, and secondly, in the case where $f(x_1, x_2)$ contains terms in $x_1$ and $x_2$ with powers higher than 2.

Supposing that the axis of revolution is perpendicular to the $x_1 - x_2$ plane, the projection of the equi-$y$ contours is of four main types (a) circles, (b)

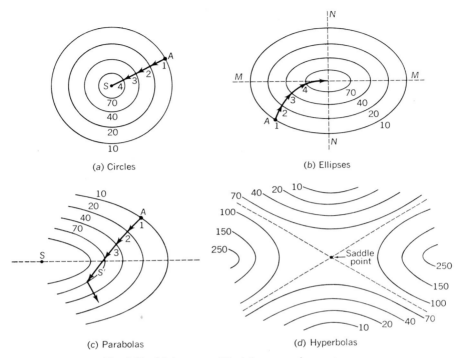

(a) Circles

(b) Ellipses

(c) Parabolas

(d) Hyperbolas

**Fig. 3-11**  Main types of 2nd-degree equi-$y$ contours.

ellipses, (c) parabolas, and (d) hyperbolas (Fig. 3-11). One fact which is common to all these cases is the diminution of the slope $AG$, or the $y$-increment ratio, while we ascend on the steepest path. This happens because at each new search step, the tangent plane $P_1, P_2, \ldots$, to the function $f(x_1, x_2)$ at the points $A_1, A_2, \ldots$, make an angle with the horizontal plane which is always smaller (see Fig. 3-12$a$). The diminution means that for equal

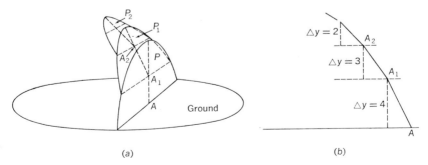

(a)

(b)

**Fig. 3-12**  The diminution of the $y$-increment.

$\Delta r$ at each step, $\Delta y$ becomes smaller (see Fig. 3-12b); or, which is the same thing, for equal $\Delta y$, the distance between the contour projections becomes greater towards the maximum (see Fig. 3-11).

Two questions arise regarding the best search direction: (1) is this direction modified while ascending; (2) what is the influence of the location of the starting point. The case of circles is perfect in both of these respects. Because of the concentricity, the direction found at the initial step 0 is valid all along the subsequent steps 1, 2, ... until the summit; this is true independently from the location of the initial point $A$ (Fig. 3-11a). Unfortunately, this case rarely happens in practice. The case of ellipses, on the contrary, happens quite often. The validity of the initially found direction depends on the location of the starting point. If, with luck, the initial point happens to be on the major axis $MM$ or on the minor axis $NN$, then the same direction leads to the summit. Otherwise, the best direction is modified at each step (see the path $A_0 - A_1 - A_2, \ldots$, of Fig. 3-11b). What is said of the ellipses is also valid for the parabolas, except that there is danger of reaching some false summit such as $S'$ if the path $A_0A_1A_2A_3A_4$ has been followed. The steps $A_2$ and $A_3$ yield the same outcome, and the step $A_4$ yields a smaller outcome than the one obtained at $A_3$, confirming that some summit $S'$ must lie between $A_2$ and $A_3$. The case of hyperbolas is even more troublesome. The axis does not play any facilating role, while the danger of false summits is very great. Furthermore, a very peculiar situation occurs in this case. If the search is vertical, from the up-side or the down-side, one will reach the middle as the highest point. If the search is horizontal from right or left, one will also find an extremum point at the middle, but this time it is the lowest point. This point is called a saddle point.

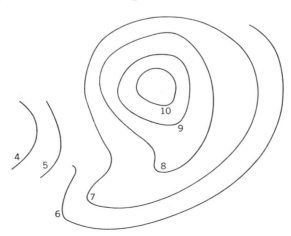

***Fig. 3-13***   Distorted unimodal contours.

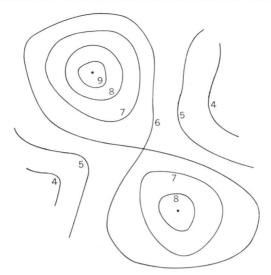

**Fig. 3-14**   Nonlinear bimodal contours.

The four types of contour shown in Fig. 3-11 are in fact very simple. In reality, they will be either distorted because of the effects of higher-degree terms in $x_1$ and $x_2$ (Fig. 3-13), or combined to form a multimodal configuration because of the effects of nonlinearities (Fig. 3-14). The situation of bimodal or multimodal contours is similar to that of one-dimensional functions exhibiting several extrema, there is no quick method. In the case where the two-dimensional functions $f(x_1, x_2)$ are represented by one of the three types of unimodal function (Fig. 3-15): (a) Strongly unimodal, (b)

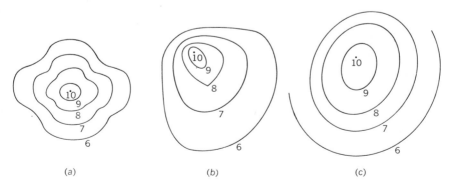

**Fig. 3-15**   Two-dimensional unimodal functions. (*a*) Strongly unimodal, (*b*) linearly unimodal, (*c*) concave down.

Linearly unimodal, and (c) Concave down, the steepest ascent search will be always successful.

For the distorted contours, the basic type of steepest ascent search will not apply; we shall see the remedy together with an application example.

### 3.3.4  Pattern Search

Let us consider the situation represented in Fig. 3-16: the equi-$y$ contours are distorted and exhibit a bimodal behavior. Properly speaking, the pattern search is not a steepest descent search, but an extension of it. The method

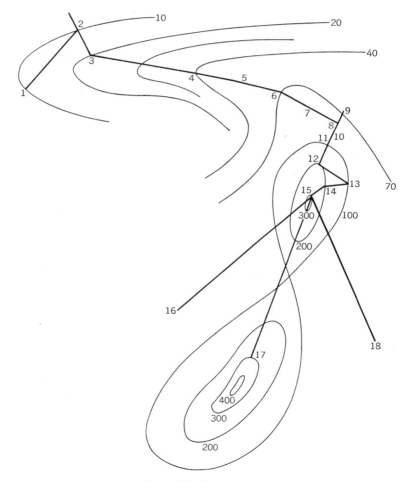

*Fig. 3-16*  Pattern search.

uses the steepest descent search for finding the main direction, *and* experimental patterns for overcoming obstacles which arise from the distortions. Let 1, 2, 3, . . . , represent the successive search steps. For the entire search to be effective, we note that not only the basic steepest ascent strategy (i.e., following the direction normal to contours) must be used, but, at several points this basic strategy must be modified to suit the circumstances more adequately. For instance:

1. Since steps 1 and 2 give the same yield $y = 10$, small steps are necessary to locate a contour with a better yield.

2. After step 3, a change in the direction (with respect to the normal direction) and the use of larger step size are more efficient.

3. After step 5, a smaller step size is more prudent until step 8 to adapt to the modification of contour shapes.

4. After step 8, trials 9 and 10 show the necessity of a 90° change in the search direction.

5. Trial step 13 shows that a smaller angle in the direction change is better.

6. After checking that the maximum yield is 300, large-size trial steps 16, 17, 18 are launched to locate eventual bimodal situation.

The difference between such a strategy and the basic steepest strategy lies in the use of some type of pattern at crucial points. Here a pattern can be defined as a search configuration with various numbers of search steps, sizes, and directions. Since the number of possible pattern configurations is infinite, their use is sensible only if (a) the adequate pattern configuration can be deduced from a reasonably limited number of previous trials, and (b) there is great hope that the same pattern can be used in later steps. These conditions are often met in practice, because of the inherent homogeneity and uniformity of industrial processes. Furthermore, one is never in complete ignorance of the relation between the performance function $y$ and the variables $x_1$, $x_2$, . . . ; theoretical knowledge, and past operating data can help to give at least some idea of the shapes of the contours. As a consequence, crucial points and search patterns can be predetermined and tested.*

### 3.3.5   Application Example with OPCON

The pattern search has been implemented in some industrial equipment. We mention here one application of this kind of equipment, not only for illustrating the principle, but also for emphasizing several practical aspects. The application concerns the optimization of the catalytic dehydrogenation process of an ethylbenzene (Fig. 3-17); the schematic representation is rearranged in the form of a bloc-diagram (Fig. 3-18) so that the roles of the

* See another presentation of pattern search in reference [3-21].

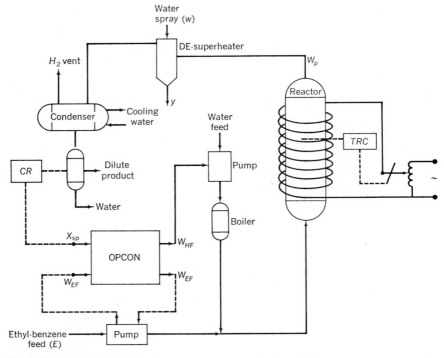

**Fig. 3-17** OPCON optimizing control of the catalytic dehydrogenation of ethyl benzene.

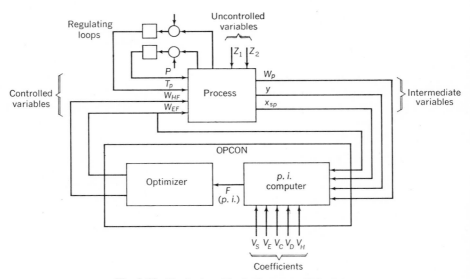

**Fig. 3-18** Equivalent block-diagram of Fig. 3-17.

signals appear more clearly. The streams, water $W$ and ethylbenzene $E$, to be dehydrogenated are fed by means of variable-speed, positive displacement pumps, to an electrically heated fixed-bed catalytic reactor. The water is vaporized in a flash boiler before being mixed with the hydrocarbon; water vapor serves as a heating and vaporizing medium for the feed, as a diluent in the reactor, and as a deterrent to the formation of carbon on the catalyst. The gaseous mixture emerging from the reactor is cooled and condensed. The liquid passes to the separator where the styrene $S$ mixed with unreacted ethylbenzene and small quantities of by-products, is decanted from a water phase. The nomenclature of the symbols used follows.

1. Controlled variables
   $P$    Pressure in the reactor.
   $T_p$   Reaction temperature (determined by the heating coils).
   $W_{EF}$   Flow rate of ethylbenzene feed (mol/hr).
   $W_{HF}$   Flow rate of steam to the reactor (mol/hr).
2. Intermediate variables
   $W_p$   Flow rate of total hydrocarbons leaving the reactor (mol/hr).
   $x_{sp}$   Mol fraction of styrene in the hydrocarbon mixture.
   $y$    Fraction of the total ethylbenzene reacted which forms styrene.
3. Cost coefficients
   $V_s, V_E$   Prices of the ethylbenzene and the purified styrene respectively ($/mol).
   $V_H$   Cost of producing water vapor fed to the reactor ($/mol).
   $V_D$   Cost per mol of distilling the hydrocarbon mixture from the reactor to produce purified styrene ($/mol).
   $V_C$   Price received for by-products recovered in the distillation process to purify styrene, per mol of ethylbenzene to by-products ($/mol).
4. Uncontrolled variables
   $z_1$   Catalyst activity.
   $z_2$   Random fluctuations of temperature distribution, reactant purity, etc.

The relations between the outputs $W_p$, $y$, and $x_{sp}$ of the process and its inputs $T_p$, $P$, $W_{HF}$, and $W_{EF}$ are not known. However, the performance function, or the operating profit $F$, in $/hr, is assumed to be dependent of the variables $W_p$, $y$, $x_{sp}$, and $W_{HF}$ and of the given cost coefficients according to

$$F = V_s \cdot x_{sp} \cdot W_p - V_E \cdot x_{sp} \cdot W_p/y$$

$$+ V_C \cdot (1 - y) \cdot x_{sp} \cdot \frac{W_p}{y} - V_H \cdot W_{HF} - V_D \cdot W_p \quad (3\text{-}27)$$

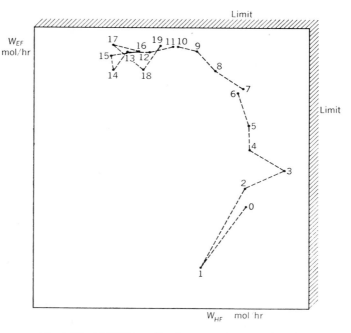

**Fig. 3-19**   OPCON application example—search path.

The OPCON equipment, conceptually, has two parts: the performance-index computer which computes the values of $F$, and the optimizer which decides the moves of the two independent variables $W_{EF}$ and $W_{HF}$. The values of the other two independent variables, $P$ and $T_p$, are kept constant by local regulating loops. A typical operating path is shown in Fig. 3-19, while the corresponding yield is shown in Fig. 3-20. The yield, or rather the production rate of styrene $W_{sp}$ is significant of the operating profit $F$. The extremum region is attained after 10 search steps. After 19 search steps, the summit seems located quite definitely. Other significant points of this application example follow.

1. The performance index, such as $F$, is generally computed indirectly from the given parameters like the $V$'s, some dependent variables like $W_p$, $x_{sp}$, and $y$, and some independent variable like $W_{HF}$.

2. Local regulating loops (for $P$, $T_p$ and also implicitly $W_{EF}$) keep their role. In the application example, other tests have been made by choosing $T_p$ and $W_{EF}$ as controlled variables; the OPCON then adjusts the reference signal of the temperature regulating loop.

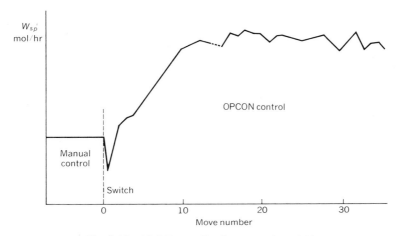

***Fig. 3-20*** OPCON application example—yield.

3. The optimum operating point drifts due to fluctuations of the un-controlled variables $z_1$ and $z_2$. Here, the deterioration of $z_1$, the catalyst activity, which is by far the most influent of all the fluctuations, depends again on the particular catalyst used. Some catalyst becomes insufficient within a period of 2 to 3 hours, while others can be used without regeneration for 4 to 5 weeks. Therefore, care must be taken in this respect, because approx-imately 45 minutes are required for the equilibrium state to be reached again after a change of operating variables.

## 3.4 EXTREMUM FORCING SEARCH IN QUASI-STATIC PROBLEMS

### 3.4.1 Extremum Tracking

In the previous sections, we have discussed search techniques which generally aim to reach the extremum point, starting from some arbitrary initial point. The criterion is usually the speed of search. If the environmental conditions are modified after the extremum has been reached, the same techniques are used for the new searches.

There is another class of search techniques, that we shall call "forcing search techniques," which, although also pertaining to problems where the analytical form of the performance function is not known, operate differently from the preceding techniques. The forcing search techniques also bring the operating state of a system from any arbitrary state to the optimum state.

However, their characteristic feature is their capacity to maintain this optimum state once it is reached. Another adequate way of qualifying these techniques is to call them "extremum tracking techniques" implying:

1. That the object to be controlled consists of some dynamic process with the following special feature: the location of the extremum point varies with time.

2. The "pursuit" nature of these techniques.

In this section, we shall first examine several forcing searches, applicable to one-dimensional problems. In the next section, we shall discuss the influence of the noises and the system lags (time constants and time delays) on the efficiency of extremum tracking. Finally, we shall see how these forcing searches can be extended to multidimensional problems.

A distinct feature of the extremum tracking techniques, such as will be described here, is their heuristic nature. Instead of being based on some analytical arguments, they evolve from intuition, aided by various available technological devices [3-9 to 3-15, 3-20].

### 3.4.2   Peak-holding Method

The earliest extremum tracking technique is the "peak-holding method." This method was applied to find the best "brake mean effective pressure" of an internal combustion engine by adjusting one of the two independent variables: ignition timing and fuel-air ratio. The simplified block-diagram of the process plus the on-line controller is shown in Fig. 3-21a. The nonlinear relation between the performance signals $p$ and the control signal $m$ is supposed to exhibit some extremum. The signal $m$ is varied at a constant rate until the performance signal $p$ drops a preset amount from the peak value measured during the current cycle. The direction of change of the signal $m$ is then reversed and the operation repeated (Fig. 3-21b). Actually, by using this method, the process does not operate at its true extremum state, but at some mean performance state. (Fig. 3-21c). The loss $\delta p$ (Fig. 3-21d) can, in principle, be reduced by restraining the difference between the measured peak value and the reversal level. However, this will meet two major obstacles (a) the various noises prohibiting significant measurements of $\delta p$, and (b) dynamical lags of $p$ from $m$ delaying the knowledge of the needed information.

Various improvements can be introduced in this peak-holding method: (a) use other forms of $m$ (b) make $m$ a function of the slope $dp/dm$ and, (c) make $m$ a function of the phase of $p$, and $dp/dt$ [3-16].

Although basically simple, the peak-holding method can be extended, with difficulty, to multidimensional problems. In principle, different testing frequencies can be used for each of the input signals $m_1, m_2, \ldots, m_n$. However, there is no theoretical justification of how to correlate these

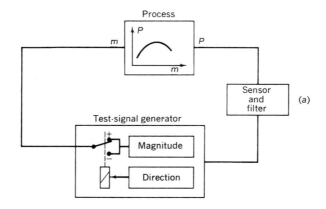

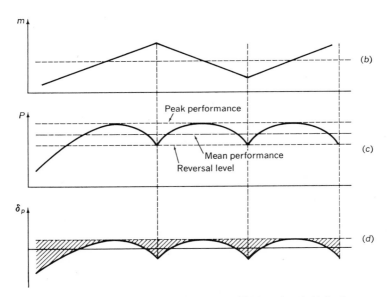

***Fig. 3-21*** Peak-holding method. (*a*) Block diagram. (*b*) Test-signal. (*c*) Performance signal. (*d*) Loss signal.

frequencies so that the resulting loss $\delta p$ is a minimum. Another disadvantage of this method is the intrinsically inevitable oscillations of the input signals, which are not always tolerable to all industrial processes.

### 3.4.3   Divider Method

Another significant extremum tracking technique is the "divider method," which, also very simple, has the advantage of not perturbing the input

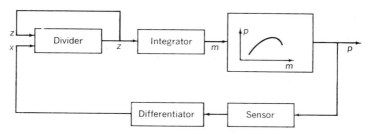

**Fig. 3-22**  Divider method.

signals. This method can best be explained by using the block-diagram of Fig. 3-22. The divider, in effect, generates the derivative of the performance signal $p$ with respect to the input variable $m$. This derivative is integrated with respect to the time needed to yield the input signal $m$. The relations between $p$, $m$, $x$, and $z$ can be established easily.

$$z = \frac{x}{z} \tag{3-28}$$

$$x = T_1 \frac{dp}{dt} \tag{3-29}$$

$$z = T_2 \frac{dm}{dt} \tag{3-30}$$

where $T_1$ and $T_2$ are the differentiator and integrator time constants, respectively. By dividing Eq. (3-29) by Eq. (3-30), and by using Eq. (3-28) and Eq. (3-30) we have

$$\frac{dm}{dt} = \frac{T_1}{T_2^2} \frac{dp}{dm} \tag{3-31}$$

Thus, the rate of change of $m$ is proportional to the slope of the performance curve, and $m$ is forced in the direction to reduce the slope to zero. This method has been effectively applied to a pilot plant fractionating column. Analysis of the dynamics shows that the system is stable for a number of practical applications.

It can be noted that in both methods, although heuristic, the derivative $dp/dm$ is used in one way or another to generate the search signals. The basic concept is, therefore, similar to the steepest descent method. However, here the important zone considered is the region around the extremum point, where the search is operating in a continuous mode. The interest is in the behavior of the tracking process. We shall discuss this question in the next section.

### 3.4.4  Effects of System Dynamics

The process, as it is considered in Fig. 3-21 or Fig. 3-22, is somewhat idealized. In practice, there are a number of factors, belonging to the process or to the instruments, which can affect the location of the extremum point, and hereafter, the behavior of the extremum tracking process. These factors are:

1. The variations of the load of the process (load torque of an engine, steam consumption of a boiler).
2. The variations of the uncontrollable environmental conditions (in the case of an internal combustion engine: temperatures, pressures, humidities of air and fuel).
3. Time constants of the type $1/(1 + \tau_1 p)$ due to the process itself or to instrumentation.
4. Time delays of the type $e^{-\tau_2 p}$ representing various dead-zones either in the process or in the instrumentation.

The variations of the load are random in nature and normally unknown. Their effect is the shift of the optimum point. However, this shift is generally slow so that there is enough time for the tracking process to be efficient. The environmental uncontrolable variations are also random and normally unknown. Their effect is consisting of fast fluctuations of the optimum point, generally of small magnitudes. If the desired accuracy of the problem does not imply the need of stochastic considerations, then the remedy is the introduction of some simple filters. The utilization of a filter makes possible the measurement of smoothed mean values of the concerned signals, but it also increases the number of time constants of the overall loop. The effects of the time constants and time delays reside mainly in the loss of the time-coincidence of the performance state and the state of the consequent controlling signal. The extremum tracking policy must therefore be modified accordingly.

## 3.5  EXTREMUM FORCING SEARCH IN DYNAMICAL PROBLEMS

### 3.5.1  Dynamical Switching Search

Let us consider a typical optimization loop as shown in Fig. 3-23 [3-17, 3-18]. The bloc $C$ represents the steady-state nonlinear unknown relation between the performance index $y$ and the controlled variable $m$. The forcing

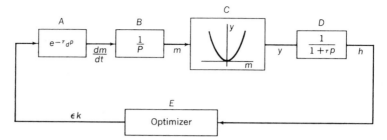

**Fig. 3-23**   A typical optimization loop.

search policy implemented in the optimizer is similar to that of Fig. 3-21, i.e., generate the rate of $m$ in accordance with what is known on the state of $y$, except that, instead of the first derivative $dy/dm$, the second derivative $d^2y/dm^2$ is used. The block $D$ represents the time constant $\tau$ of the process itself, or of some measuring instruments, so that what is known on the state of the performance is not $y(t)$, but $h(t)$. The bloc $A$ represents the time delay $\tau_d$ assumed to be concentrated at the input of the process. The bloc $B$ represents the fictitious integration since the optimizer generates the rate of $m$. The output of the optimizer is $\epsilon K$; $K$ determines the magnitude of $dm/dt$, while $\epsilon = \pm 1$, indicates its direction. Assuming that a predetermined constant value of $K$ is adopted, the minimization problem therefore consists of determining the sign of $dm/dt$ according to the knowledge of $h[m(t)]$, so that the operating point $Q$ in the phase-plane $h - m$, is brought from any initial point $Q_i$ (at initial time $t_i$) to the minimal point $Q_m$ (at final time $t_f$). (See Fig. 3-24a.) However, as we have seen in the preceding section, the actual search trajectory will exhibit some oscillatory behavior (with limit cycles), such as

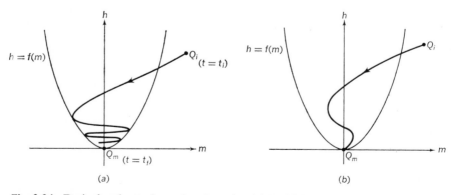

**Fig. 3-24**   Desired and actual search trajectories. (*a*) Desired search trajectory. (*b*) Actual search trajectory.

shown in Fig. 3-24b. The final form of the oscillations will, of course, depend on the adapted search strategy and on the searching conditions. The exposition which follows is not difficult, but implies many symbols characterizing these search strategy and conditions. The reader is asked to carefully read their definition.

$m(t)$   Control variable.

$y(t)$   Performance index.

$\alpha$   Characteristic parameter of the nonlinear steady-state relation $y = f(m) \cong \alpha m^2$.

$h(t)$   Measured value of the performance index $y(t)$, $h(p)/y(p) = 1/(1 + \tau p)$.

$\tau$   Characteristic time constant of the controlled process (sec).

$\tau_d$   Characteristic dead time of the process (sec).

$\lambda$   Integration constant characterizing the initial conditions of the point $Q_i$.

[$m(t)$ through $\lambda$ are known parameters, which characterize the search conditions.]

$\epsilon$   $\pm 1$.

$K$   Constant magnitude of $dm/dt$ (sec$^{-1}$).

$\delta$   Switching threshold (sec$^{-2}$).

$g$   Switching function $[g = (d^2h/dt^2) - \delta]$ (sec$^{-2}$).

$\tau_w$   Switching period (sec).

($\epsilon$ through $\tau_w$ are unknown parameters characterizing the search policy, they are to be chosen at each search condition.)

$H(t)$   Dimensionless value of $h$ ($H = h/2\alpha K^2\tau^2$).

$M(t)$   Dimensionless value of $m$ ($M = m/K\tau$).

$\Delta$   Dimensionless value of $\delta$ ($\Delta = \delta/2\alpha K^2$).

$T$   Dimensionless time ($t/\tau$).

$G$   Dimensionless value of $g$ $[G = (d^2H/dT^2) - \Delta]$.

$D$   Dimensionless value of dead time ($D = \tau_d/\tau$).

$W$   Dimensionless switching period ($W = \tau_w/\tau$).

$\mu$   Coefficient $(1 - \Delta)e^{-R} = \left(1 - \dfrac{\delta}{2\alpha K^2}\right)e^{-\tau_d/\tau}$.

### 3.5.2   Search Strategy

The control signal $m$, during the course of the search, varies with a constant rate positive or negative.

$$\frac{dm}{dt} = \epsilon K \qquad (3\text{-}32)$$

The sign of $\epsilon$ is modified each time the switching function

$$g = \frac{d^2h}{dt^2} - \delta \tag{3-33}$$

or

$$G = \frac{d^2H}{dT^2} - \Delta \tag{3-34}$$

passes from a negative to a positive value. The new state is to continue during a period $\tau_w$ (or $W = \tau_w/\tau$) as long as the function $G$ remains positive.

In this strategy, $d^2h/dt^2$ or $d^2H/dT^2$ is a quantity to be measured or fabricated from $h(t)$ or $H(t)$, the values of $K$ and $\delta$ are to be chosen according to the search conditions. To see how this strategy works, it is necessary to know how the dynamical and steady-state response curves or $H = f(M)$ are represented on the plane $H - M$.

### 3.5.3   Response Curves with Time Constant

Since we assume $y = \alpha m^2$, we have

$$h = \alpha m^2, \quad \text{or} \quad H = \frac{M^2}{2} \tag{3-35}$$

the steady-state response curve, when $dH/dt$, $d^2H/dt^2$, ... all the time derivatives of $H$ are zero, which is represented by a parabola in the $H - M$ plane, symmetrical with respect to the $H$-axis and having the minimum point $Q_m$ at the origin. Dynamically, because of the time constant $\tau$ and the time delay $\tau_d$, the function $H(t) = f[M(t)]$ is no longer represented by the $s - s$ response curve. To find the dynamical response curves, we shall first introduce $\tau$, and then $\tau_d$. If only $\tau$ is considered, we have the following equations:

$$\tau \frac{dh}{dt} + h = \alpha m^2 \tag{3-36}$$

$$\frac{dm}{dt} = \epsilon K \tag{3-37}$$

Dividing Eq. (3-36) by Eq. (3-37), we have

$$\frac{dh}{dm} = \frac{1}{\epsilon K}\left(-\frac{h}{\tau} + \alpha \frac{m^2}{\tau}\right) \tag{3-38}$$

Using the dimensionless variables, the solution of Eq. (3-38) is

$$H = \frac{M^2}{2} - \epsilon M + 1 + \lambda e^{-\epsilon M} \tag{3-39}$$

where $\lambda$, an integration constant, characterizes the initial conditions. From Eq. (3-39), we can deduce

$$\frac{dH}{dM} = M - \epsilon - \epsilon \lambda e^{-\epsilon M} \tag{3-40}$$

$$\frac{d^2H}{dM^2} = 1 + \lambda e^{-\epsilon M} \tag{3-41}$$

Combining Eqs. (3-40), (3-41), and (3-42), we have

$$H = \frac{M^2}{2} - \epsilon \frac{dH}{dM} \tag{3-42}$$

$$H = \frac{M^2}{2} - \epsilon M + \frac{d^2H}{dM^2} \tag{3-43}$$

From these two equations, we can write

$$\text{if } \frac{dH}{dM} = 0; \quad \text{then } H = \frac{M^2}{2} \tag{3-44}$$

$$\text{if } \frac{d^2H}{dM^2} = 0; \quad \text{then } H = \frac{M^2}{2} - \epsilon M \tag{3-45}$$

Eqs. (3-39), (3-44), and (3-45) represent three groups of response curves in various circumstances. The group represented by Eq. (3-39) consists of two networks $L^-$ and $L^+$ of response curves for various values of initial-condition parameter $\lambda$, $L^-$ for $\epsilon = -1$, and $L^+$ for $\epsilon = +1$. These curves are asymptotic to the symmetrical parabolas of equation

$$H = \frac{M^2}{2} - \epsilon M + 1 \tag{3-46}$$

The network $L^-$ is shown in Fig. 3-25. The group represented by Eq. (3-44) is reduced to one parabola $F$, which is, as one can see also from Eq. (3-35), the steady-state response curve. The group represented by Eq. (3-45) consists of two response curves $H^-$ and $H^+$, two parabolas symmetrical with respect to the vertical axis, $H^-$ for $\epsilon = -1$, $H^+$ for $\epsilon = +1$. The response curve $H^-$ is drawn in Fig. 3-25. According to circumstances, the operating point $Q$ will follow one of these response curves.

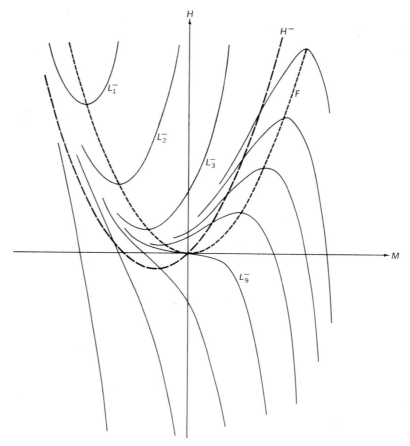

**Fig. 3-25** Response curves $H = f(M)$ with $\tau$.

### 3.5.4 Search Trajectory with Time Constant

It is important to note here that because of the imposed constant rate $K$, [Eq. (3-32)] $dm$ is linearly proportional to $dt$ for both signs of $\epsilon$, ($dm = K\,dt$ and also $dM = K\,dt$). As a consequence, the space derivatives $dH/dM$ and $d^2H/dM^2$ are also linearly proportional to $dH/dT$ and $d^2H/dT^2$. Therefore, conditions implied in Eqs. (3-44) and (3-45) are also satisfied if, respectively, $(dH/dT) = 0$ and $(d^2H/dT^2) = 0$. Assume now: (a) $\epsilon = -1$, $M$ decreases with time, and (b) the initial operating point $Q_i$ is on one of the $L^-$ curves, say $L_4^-$. Since $M$ is decreasing, $Q$ moves on $L_4^-$ in the direction indicated in Fig. 3-26. Assuming for the moment the $\delta$ (or $\Delta$) is equal to zero, the switching condition $G = 0$ [Eq. (3-34)] becomes $d^2H/dT^2 = 0$, and happen when $Q$

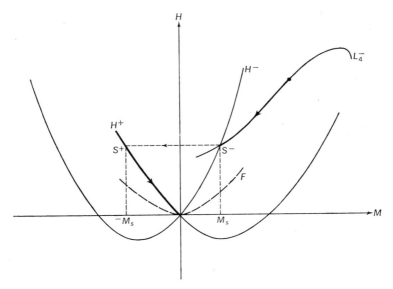

**Fig. 3-26**  Search trajectory with $\tau$.

arrives at $S^-$, intersection of $L^-$ and $H^-$. The sign of $\epsilon$ changes ($\epsilon = +1$ now), and $Q$ jumps to $S^+$ on $H^+$ with the abscissa $-M_s$. Now, since $\epsilon = +1$, $M$ is increasing, therefore $Q$ moves along $H^+$ towards $Q_m$ if the condition $G = 0$ is always satisfied. The point $Q$ will reach $Q_m$ if the condition $dH/dM = 0$ is also satisfied. Or, in other words, according to Eq. (3-40).

$$M = +\epsilon(1 + \lambda e^{-\epsilon M}) = 1 + \lambda e^{-M} \qquad (\epsilon = +1) \qquad (3-47)$$

Now, if the switching threshold $\delta$ (or $\Delta$) is not equal to zero, the response curve $H^-$ is higher or lower according to the sign of $\delta$. As a matter of fact, the choice of $\delta$ belongs to the search strategy, and its value must be such that the initial response curve $L_4^-$ can meet it in the course of the search.

### 3.5.5  Influence of the Dead Time

Assume now that we introduce the dead time $\tau_d$, or its dimensionless value $D$; this will change the location of the switching points $S$, the switching response curves $H^-$ and $H^+$, and will impose several conditions on the choice of strategy parameters $\Delta$ and $W$. We recall that (Fig. 3-26) at the switching point $S^-$, with the abscissa $M_s$, the condition is $G = 0$; therefore from Eq. (3-34)

$$\frac{d^2H}{dM^2} = \Delta \qquad (3-48)$$

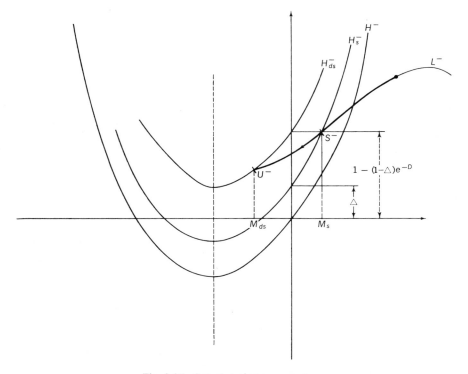

**Fig. 3-27**   Search trajectory $\tau$ and $\tau_d$.

Substituting Eq. (3-48) in Eq. (3-41), we have

$$1 + \lambda e^{-\epsilon M_s} = \Delta \tag{3-49}$$

Because of the dead time, the switching occurs neither at the time when $G = 0$ nor at the abscissa $M_s$, but $\tau_d$ seconds later, at the abscissa $M_{ds}$. Since $dm/dt = \epsilon K$, we have

$$M_{ds} - M_s = \epsilon D \tag{3-50}$$

If a search is started on a response curve $L^-$ (Fig. 3-27), the switching point will not be $S^-$, but $U^-$. Since $U^-$ always belongs to $L^-$, the corresponding equation (3-39) must be satisfied with $M_{ds}$

$$H_{ds} = \frac{M_{ds}^2}{2} - \epsilon M_{ds} + 1 + \lambda e^{-\epsilon M_{ds}} \tag{3-51}$$

The elimination of $\lambda$ and $M_s$ between Eqs. (3-49), (3-50), and (3-51) gives the switching response curve with $\tau$ and $\tau_d$.

$$H_{ds} = \frac{M_{ds}^2}{2} - \epsilon M_{ds} + 1 - (1 - \Delta)e^{-D} \qquad (3\text{-}52)$$

If the dead time was absent, $D = 0$, the switching response curve would be

$$H_s = \frac{M_s^2}{2} - \epsilon M_s + \Delta \qquad (3\text{-}53)$$

If the switching threshold was absent

$$H = \frac{M_s^2}{2} - \epsilon M \qquad (3\text{-}54)$$

The three switching response curves $H^-$, $H_s^-$, and $H_{ds}^-$, for $\epsilon = -1$ and corresponding to the three equations (3-52), (3-53), and (3-54), are represented in Fig. 3-27. The curves $H_{ds}$ are also parabolas; they can be obtained from the curves $H$ by a vertical shift of the quantity $[1 - (1 - \Delta)e^{-D}]$.

Once the operating point, during the course of the search, arrives at $U^-$, the switching occurs. The operating point jumps from $U^-$ to $U^+$ on $H_{ds}^+$, similar to the case shown in Fig. 3-26. For the search to be adequate and convergent to the minimum point $Q_m$, a number of conditions must be satisfied.

1. If $\Delta = 1$, the curves $H_{ds}$ and $H_s$ tend to the asymptote parabola [Eq. (3-46)]; there are on more points $S$ or $U$, the switching cannot occur, and the system diverges. Therefore, one must have $\Delta < 1$.

2. The switching can only happen if the waiting period $W$ is greater than the dead time $D$. Then, another condition $W > D$.

3. Equations (3-52) and (3-54) show that the dead time $D$ can be compensated by a choice of the threshold $\Delta$ if:

$$1 - (1 - \Delta)e^{-D} = 0 \qquad (3\text{-}55)$$

or

$$(1 - \Delta)e^{-D} = 1 \qquad \therefore \quad \Delta = 1 - e^D \qquad (3\text{-}56)$$

### 3.5.6 Search Results

What happens after the operating point $Q$ arrives at $S^+$ (Fig. 3-26) or at $U^+$, (Fig. 3-27)? Normally, with the correct sign of $\epsilon$, and an adequate variation of $M$, $Q$ must go towards $Q_m$ along $S^+$ or $U^+$.

However, after the switching instant, the condition $G = 0$ is no longer satisfied. As a consequence, $Q$ leaves $H^+$ or $U^+$, and moves along some $L^+$, initiating then a number of oscillations. Because of the rate control, the value

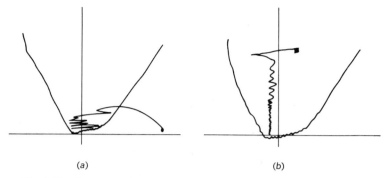

(a)                                    (b)

***Fig. 3-28***   Some experimental results of the dynamical switching search.

of $M$ is required to pass by zero, the oscillations therefore become cyclic and always smaller in magnitude. Convergent limit cycles, however, can only occur if adequate conditions on $D$, $\Delta$, and $W$ are satisfied.

This dynamical switching search method has been experimented on an alternator connected to a main supply with the objective of minimizing the reactive power produced. The principle time constant is 5.0 sec due to the measurement of the alternator current. Dead time is negligible. Typical search trajectories are shown in Fig. 3-28*a* and *b*. The values of the search parameters are, in both cases,

$$K\tau_{\min} = 1$$
$$K\tau_{\max} = 6$$

### 3.5.7   Extension to Multidimensional Problems

The question arises whether the techniques described in this section are extendible to multidimensional dynamical extremum seeking problems? When discussing steepest ascent search techniques, we have seen that the multidimensionality can be parametrically reduced to one dimension. However, the difficulties that will be encountered are (a) the contour distortions or the high-degree nonlinearities; and (b) the problem of finding the absolute extremum. When discussing the peak-holding and divider methods, we have seen that the difficulties lie in the way of coordinating the time settling and the rates of the control signals, when they are more than two.

All these difficulties will be met in the case of multidimensional dynamic systems. The parabola assumption of the performance function may not be valid. The multimodality may exist. The difficulties in coordinating phases and magnitudes of the control signals are even increased because time constants and dead times may have different values for each control input.

A number of studies have been devoted to this subject of extremum seeking in multidimensional dynamical systems, the system equations being unknown. The obtained results are rather theoretical; we shall not discuss this question further.

# REFERENCES

[3-1] Courant, R., "Variational Methods for the Solution of Problems of Equilibrium and Vibrations," *Bull. Am. Math. Soc.* **49**, (1943) 1–23.

[3-2] Stakhovskii, R. I., "Twin Channel Automatic Optimizer," *Automation and Remote Control* (USSR), August 1958.

[3-3] Feldbaum, A. A., "Automatic Optimalizer," *Automation and Remote Control* (USSR), August 1958.

[3-4] Krasovskiy, A. A., "Optimal Methods of Search in Continuous and Pulsed Extremum Control Systems," *Proc. 1st Int. Symp. on Optimizing and Adaptive Control*, I.S.A., 1962.

[3-5] Van Nice, R. I. and D. A. Burt, "Optimizing Control Systems for the Process Industries," Westinghouse Reprint n° 5467.

[3-6] Archer, D. H., "An Optimizing Control for the Chemical Process Industries," *Brit. Chem. Eng.*, February 1960, 88–94.

[3-7] Macmillan, R. H. and N. W. Rees, "Automatic Control Systems," *Process Control and Automation*, November–December 1960.

[3-8] Bell, D. A., "Intelligent Machines," Blaisdell Ed., (1962), 62–63.

[3-9] Draper, C. S., and Y. T. Li, "Principles of Optimalizing Control Systems and an Application to an Internal Combustion Engine," *ASME Publ.*, New York, 1951.

[3-10] Li, Y. T., "Optimalizing System for Process Control," *Instruments*, **25** (1952).

[3-11] Eckman, D. P. and I. Lefkowitz, "Report on Optimizing Control of a Chemical Process," *Control Eng.*, **4** (1957), n 9, 97.

[3-12] Lefkowitz, I. and D. P. Eckman, "A Review of Optimizing Computer Control," *Proc. Self-Adaptive Control Systems Symp.*, WADC-Tr 59, Wright Air Development Center, Dayton, Ohio, March 1959.

[3-13] Vasu, G., "Experiments with Optimizing Controls Applied to Rapid Control of Engine Pressures with High-Amplitude Noise Signals," *Trans. ASME*, **79** (1957), 481–489.

[3-14] Lefkowitz, I., "Computer Control," Ch. 13, **3** of [G-42].

[3-15] Frait, J. S., "An Investigation of Optimizing Circuits using the Divider-Optimizer Concept," M.S. thesis, Case Institute of Technology, 1960.

[3-16]   Fujii, S. and N. Kanda, "An Optimizing Control of Boiler Efficiency," Second IFAC Congress, Basle, June 1963.

[3-17]   Perret, R. and R. Rouxel, "Principle and Application of an Extremal Computer," Second IFAC Congress, Basle, June 1963.

[3-18]   Perret, R. and R. Rouxel, "Study of an Algorithm for Dynamic Optimization," in [G-37].

[3-19]   Chestnut, H., R. R. Duersch, and W. M. Gaines, "Automatic Optimizing of Poorly Defined Process, Part I," *Trans. IEEE, Appl. & Ind.*, March 1963, 32–41. Part II, JACC 1963.

[3-20]   Schweitzer, P. R., C. Volz, and F. De Luca, "Control System to Optimize Engine Power," *SAE Automotive Eng. Congr.*, Detroit, January 1966.

[3-21]   Hooke, R. and T. A. Jeeves, " "Direct Search" Solution of Numerical and Statistical Problems," *J. Ass. Comp. Mach.*, **8, 2** (April, 1961), 212–29.

[3-22]   Hague, D. S. and C. R. Glatt, "An Introduction to Multivariable Search Techniques for Parameter Optimization," *NASA* CR-173200 (1968), 91.

[3-23]   Hague, D. S. and C. R. Glatt, "A Guide to the Automated Engineering and Scientific Optimization Program AESOP," *NASA* CR-73201 (1967), 178.

[3-24]   Narendra, K. S. and Th. S. Baker, "Simultaneous Multiple Parameter Adjustment in Adaptive Systems using a Single Perturbation Signal," AD-653474 (April 1967), 24.

[3-25]   Forsythe, G. E., "On the Asymptotic Directions of the *s*-dimensional Optimum Gradient Method." AD-650699 (April 1967), 47.

# 4

---

# DYNAMICAL OPTIMIZATION
# TECHNIQUES

## 4.1 INTRODUCTION

THE TECHNIQUES DISCUSSED in this chapter have three characteristics:
(a) They are intended to solve optimization problems of dynamical systems,
as opposed to those techniques discussed in Chapter 2 where static systems
were considered. (b) The mathematical model of the optimized system is
assumed known, contrary to the hypothesis of Chapter 3, where it is assumed
unknown. (c) The methods discussed in this chapter lead, in principle, to
exact and complete solutions. However, numerical values of the solutions
cannot always be obtained because of computational difficulties. Methods for
overcoming these difficulties are discussed in Chapter 5.

Historically, dynamical optimization problems can be approximately
classified as follows:

1. Error minimization in the presence of noise, in the well-known Wiener-
Hopf sense. The true objective is information smoothing and filtering rather
than maximization of some concrete profit. These stochastic problems will
not be discussed here.

2. Optimum bang-bang feedback control systems with minimum-time or
minimum-error performance indices. Substantial results have been obtained
in this field [G-34; G-49; G-54; G-58] notably in linear and time-varying
linear systems of various orders. Except for a brief mention in Appendix 4D,
these problems will not be discussed here explicitly.

3. Optimization of rocket trajectories and of various space missions. The
performance index is some concrete physical parameter: a flight time, a

range, a fuel consumption, a payload, etc. Works in this field have contributed extensively to the systematic formulation and treatment of optimization problems. A number of examples are used in this text for illustration [G-29; G-44].

4. Optimization of industrial plants such as power plants, chemical reactors, steel and iron plants, etc. [G-35; G-36; G-46; G-47]. The performance index is some concrete parameter: an operating time, a yield, an efficiency, a consumption, etc. Developments in this field are strongly pushed by the progress in process control computers and in peripheral equipment. Practical implementations are, however, still at an early stage. Both experimental and theoretical efforts are required for their advancement.

Three categories of techniques will be discussed in this chapter.

A. The variational methods in Section 4.2, comprising both the classical variational method and the Pontryagin's method. The basic philosophy of these methods is the examination of the variation of the optimal trajectory (Fig. 4-1a). Conditions for nulling this variation are then derived. The method

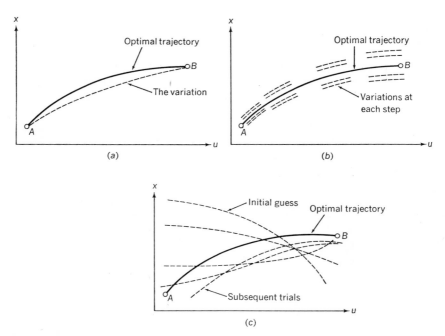

**Fig. 4-1**   Basic philosophy showing how the optimal trajectory is obtained in various techniques. (*a*) Calculus of variations and maximum principle. (*b*) Dynamic programming. (*c*) Method of gradients.

is indirect, in that it results in a set of differential equations, the solution of which is required for obtaining numerical values.

B. The method of dynamic programming, in Section 4.3, which proceeds by using a multistage decision process (Fig. 4-1b). The basic philosophy is to divide the optimal trajectory into a number of portions. Variations of each portion are prospected. The next step is taken into consideration after the optimal trajectory of the last step is found. The great advantage of this method lies in the prospection which is, in effect, a numerical computation, thus avoiding analytical difficulties of nonlinearities, boundaries, and discontinuities.

C. The method of gradients, in Section 4.4, which is, in effect, a trial-and-error process (Fig. 4-1c). The difference between this and the steepest ascent method of Chapter 3 is that the initial guess, as well as the subsequent trials, consist of a complete solution extending over the whole period $t_0 - t_f$, rather than points.

## 4.2  VARIATIONAL METHODS

In this section, we shall discuss the classical variational technique and the Pontryagin's technique. The emphasis is mainly on the optimization procedure and illustrative examples. The theories underlying these techniques are discussed in Appendices 4A and 4B. After discussing the basic extremalizing procedure, the important two-point-boundary-value problem will be shown. Finally, two special application examples, one of them concerning state-variable discontinuities, and the other, boundary inequality constraints.

### 4.2.1  Classical Variational Method

**4.2.1.1  Problem Formulation.**  In the classical Calculus of Variations, the optimization problem is formulated as follows.

$$\operatorname*{Min}_{y_k(x)} J = [G(x, y_k)]_0^f + \int_{x_0}^{x_f} F(x, y_k, y_k') \, dx \tag{4-1}$$

by choosing the functions $y_k(x)$, $k = 1, 2, \ldots, n$ satisfying:

$$\varphi_j(x, y_k, y_k') = 0 \qquad j = 1, 2, \ldots, p < n \tag{4-2}$$

$$w_r(x_0, y_{k0}) = 0 \qquad r = 1, 2, \ldots, q \tag{4-3}$$

$$w_r(x_f, y_{kf}) \;\;= 0 \qquad r = q + 1, \ldots, s \le 2n + 2 \tag{4-4}$$

Such a formulation is called the Problem of Bolza. If $F \equiv 0$ in $J$, the formulation becomes the Problem of Mayer. If $G \equiv 0$ in $J$, the formulation bears the name of the Problem of Lagrange. Note that in $J$, the $G$-part is function only of the initial and the final values of the independent variable $x$, while the $F$-part depends on the whole history between $x_0$ and $x_f$. Any Mayer problem can be transformed into a Lagrange problem by adding a new dependent variable $y$ (and a new constraint equation $\varphi$). For instance

$$J = [G(x, y_k)]_0^f = y_k^2(x_f) \qquad k = 1, 2, \ldots, n \qquad (4\text{-}5)$$

We define a new dependent variable

$$y_{n+1} = \int_{x_0}^{x_f} y_k^2 \, dx \qquad (4\text{-}6)$$

or by derivation

$$y_{n+1}' = y_k^2(x_f) \quad \text{with} \quad y_{n+1}(0) = 0 \qquad (4\text{-}7)$$

Then

$$J = y_{n+1}(x_f) = \int_{x_0}^{x_f} y_k^2 \, dx \qquad (4\text{-}8)$$

As we shall see later, this procedure is used in the Pontryagin's method.

In dynamic optimization problems, some formal differences are introduced: (1) the independent variable $x$ becomes the time $t$, (2) the dependent variables $y_k(x)$, $k = 1, 2, \ldots, n$ are split into two parts, $x_i(t)$, $i = 1, 2, \ldots, n$ (state variables), $u_j(t)$, $j = 1, 2, \ldots, r$ (control variables). The formulation becomes

$$\min_{u(t)} J = [G(x_i, u_j, t)]_{t_0}^{t_f} + \int_{t_0}^{t_f} F(\dot{x}_i, x_i, u_j, t) dt \qquad (4\text{-}9)$$

$u_j(t)$, and $x_i(t)$ are subject to the following constraints:

$$\varphi_i = \dot{x}_i - g_i(x, u, t) = 0, \qquad i = 1, 2, \ldots, n \qquad (4\text{-}10)$$

$$\psi_r[x_i(t_0), u_j(t_0), t_0] = 0, \qquad r = 1, 2, \ldots, q \qquad (4\text{-}11)$$

$$\psi_r[x_i(t_f), u_j(t_f), t_f] = 0, \qquad r = q + 1, \ldots, s \leq 2n + 2 \qquad (4\text{-}12)$$

Note that in the classical formulation, all the $y_k$-values are to be determined, and $\varphi_j$ is a system of general differential equations, while in the new formulation, only the $u_j$-values are to be determined and $\varphi_i$ is a system of first-order differential equations.

**4.2.1.2 Application of the Euler Equation.** The first step towards the optimal solution is to form the Euler equation. Considering the preceding formulation, we form the modified function.[1]

$$F^* = F + \sum_{i=1}^{n} \lambda_i \varphi_i \qquad (4\text{-}13)$$

---

[1] The part $G$ of $J$, concerning the end values of the variables, will only be implied in the transversality condition when the end points are not fixed (see Appendix 4A).

The number of variables amounts to $x: x_1, x_2, \ldots, x_n;\ u: u_1, u_2, \ldots, u_r;$ $\lambda: \lambda_1, \lambda_2, \ldots, \lambda_n$, a total of $2n + r$, all variables function of time. The application of the Euler equation yields

(A)
$$\frac{\partial F^*}{\partial x_i} - \frac{d}{dt}\left(\frac{\partial F^*}{\partial \dot{x}_i}\right) = 0 \qquad i = 1, 2, \ldots, n$$

(B)
$$\frac{\partial F^*}{\partial u_j} - \frac{d}{dt}\left(\frac{\partial F^*}{\partial \dot{u}_j}\right) = 0 \qquad j = 1, 2, \ldots, n$$

which degenerates to $(\partial F^*)/(\partial u_j) = 0$ because $F^*$ contains no $\dot{u}_j$

(C)
$$\frac{\partial F^*}{\partial \lambda_i} - \frac{d}{dt}\frac{\partial F^*}{\partial \dot{\lambda}_i} = 0 \qquad i = 1, 2, \ldots, n$$

which reverts to $\varphi_i = 0$, the original system dynamics or constraints. There are thus $2n + r$ equations in $2n + r$ unknowns. (A) and (C) are first-order differential equations, while (B) produces a set of algebraic equations.

***Example 1.*** Minimum effort problem (Fig. 4-2) [4-1].
*System.* One integrator input $\dot{x}$, output $x$. One control signal $u$, put in the integrator as $\dot{x}$ resulting to one condition

$$\varphi = \dot{x} - u = 0 \tag{a}$$

*Object.* Minimize the total cost $\int_0^T (u^2 + x^2)\, dt$ by choosing $u$, and hence the trajectory from

$$x(0) \quad \text{at} \quad t = 0 \quad \text{to} \quad x(T) \quad \text{at} \quad t = T$$

Augmented function
$$F^* = u^2 + x^2 + \lambda(\dot{x} - u) \tag{b}$$

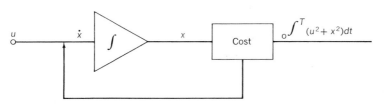

***Fig. 4-2*** A simple integrator feedback system.

Applying the Euler equation yields

$$2x - \dot{\lambda} = 0 \tag{c}$$

$$2u - \lambda = 0 \tag{d}$$

$$\dot{x} - u = 0 \tag{e}$$

Note that (c) and (e) are differential, (d) is algebraic, and (e) equivalent to (a). From (d) differentiation with respect to time gives $\dot{\lambda} = 2\dot{u}$. From (e) differentiation gives $\ddot{x} = \dot{u}$. Then $\dot{\lambda} = 2\ddot{x}$. Substituting this in (c) gives $2x - 2\ddot{x} = 0$ which leads to $x = Ae^{-t} + Be^{t}$. (A) and (B) will be chosen according to the boundary conditions $x(0)$ and $x(T)$. Finally the control $u$ is the same as $\dot{x}$.

**Example 2.** Minimum cost problem[1] [4-2].

*System.* A plant consisting of one hydro and one thermal-electric power generation unit. Let

$P_R$    Received power (imposed by the load).

$P_S$    Thermal power.

$P_H$    Hydroelectric power.

$P_L$    Transmission loss, quadratic function of $P_S$ and $P_H$.

$dV/dt$    Rate of change of the storage $V$.

$r$    Inflow of the reservoir.

$q$    Sum of the outflows $q_i$.

$S$    Nonpower generating release.

$F(P_s)$    Cost of fuel for generating $P_S$.

System conditions are:

$$\text{Static} \quad P_R = P_S + P_H - P_L \tag{a}$$
$$\text{Dynamical} \quad (dV)/(dt) = r - q - S \tag{b}$$

*Object.* Search for the discharge function $q^*(t)$ which minimizes the integral of fuel costs over the optimizing period

$$E = \int_0^T F(P_s)\, dt \tag{c}$$

---

[1] This example, together with those given in Chapter 2, Section 2.2 (on Lagrange-multiplier method) and Section 2.4 (Kuhn-Tucker Convex Programming) shows the progression in treating power-plant economic dispatching problems.

The augmented functional is

$$J = \int_0^T \left[ F(P_s) + \lambda(P_R + P_L - P_B - P_H) + \gamma\left(\frac{dV}{dt} - r + q + s\right) \right] dt \quad \text{(d)}$$

The application of the Euler equation with respect to the two state variables $P_S(t)$ and $V(t)$ and to the control variable $q(t)$ yields:

$$\frac{\partial F}{\partial P_s} - \lambda\left(1 - \frac{\partial P_i}{\partial P_s}\right) = 0 \quad \text{(e)}$$

$$-\lambda\left(1 - \frac{\partial P_L}{\partial P_H}\right) \cdot \frac{\partial P_H}{\partial q} + \gamma = 0 \quad \text{(f)}$$

$$-\lambda\left(1 - \frac{\partial P_L}{\partial P_H}\right) \cdot \frac{\partial P_H}{\partial V} - \frac{d\gamma}{dt} + \frac{\partial S}{\partial V} \cdot \gamma = 0 \quad \text{(g)}$$

We have five variables $P_s(t)$, $q(t)$, $V(t)$, $\lambda(t)$, and $\gamma(t)$ and five equations (a), (b), (e), (f), (g). Two of them, (b) and (g) are differential equations, while the three others are algebraic. All the partial derivatives of the form $\partial y/\partial x$ can be looked upon as slopes of the linear or nonlinear functions $y = y(x)$ and must be known experimentally at the operating points. The solution of the five equations must be simultaneous, thus implying the knowledge of two initial or two final conditions for solving the two first-order differential equations. In practice, $V(0)$ and $V(T)$ (endpoints of the reservoir storage) are known, but $\gamma(0)$ and $\gamma(T)$ are not. The solution of the system is therefore not easily obtained. This is the $2\,PBVP$ which will be discussed later in this section.

**Example 3.** Minimum operating time problem [4-3, 4-4].

*System.* Consider a chemical process characterized by the reaction equations

$$X + \text{Hydrogen} \overset{K_1}{=} Y \quad \text{(a)}$$

$$Y + \text{Hydrogen} \overset{K_2}{=} Z \quad \text{(b)}$$

The reactions are described by first-order equations

$$-\frac{dx}{dt} = K_1 x \quad \text{(c)}$$

$$\frac{dy}{dt} = K_2 x - K_1 y \quad \text{(d)}$$

The notations are:

$XYZ$: Chemical components.
$xyz$: Mol fractions of $XYZ$: $x + y + z = 1$.
$K_1 = A_1 p^{N_1}$.      $K_2 = A_2 p^{N_2}$.
$A_1 A_2 N_1 N_2$ are constants.
$p$: Adjustable pressure.
   By using new variables $u$, $v$, and $m$ and by letting: ($u$, $v$: state variables, $m$: control variable).
$u = y/x$.
$v = \log x_n/x$.
$m = 1 - K_2/K_1$ ($m$ adjustable via $p$).
$x_n = $ value of $x$ existing at the start of the $n$th computing interval.

Equations (c) and (d) are transformed into a single equation

$$\frac{du}{dv} = mu + 1 \qquad \text{(e)}$$

*Object.* Choose the program $m(v)$ to minimize the cost, or the total processing time which is the predominant factor in the cost equation. The total processing time $t_N$ is derived from Eqs. (c) and (d).

$$t_N = A \int_0^{v_N} (1 - m)^{1-B} \, dv \qquad \text{(f)}$$

with

$$A = \frac{A_1^{B N_1/N_2}}{A_2^B}$$

$$B = \frac{N_1}{N_2 - N_1}$$

The boundary conditions $(u_n, 0)$ $(u_N, v_N)$ are assumed given. This is therefore a fixed-boundary problem. Applying the Euler equation to the integrand of equation (f) gives

$$\frac{\partial f(v)}{\partial u} - \frac{d}{dv} \frac{\partial f(v)}{\partial u'(v)} = 0 \qquad \text{(g)}$$

where

$$f(v) = (1 - m)^{1-B} \qquad \text{(h)}$$

Extracting $m$ from the constraint equation (e), to be substituted in (h), and applying the new $f(v)$ in Eq. (g) gives

$$\frac{du}{dv} = \frac{1 - Bm \cdot}{Bu} \tag{i}$$

It is important to note that the state variable $v$ plays the role of independent variable (like $t$ in ordinary dynamical problems). There are therefore two unknowns $u(v)$ and $m(v)$, for the two first-order differential equations (e) and (i), the simultaneous solution of which will yield the curves $u(v)$ and $m(v)$. Note here again that boundary conditions are only given for $u$.

$$u(v_0) = u(0) = u_n, \qquad u(v_N) = u_N$$

These *two* conditions are theoretically sufficient for the *two* differential equations. However, they are situated at two different boundaries, and the 2 *PBVP* arises again.

**4.2.1.3 Complete Optimization Procedure.** Here we shall expose the complete procedure of using classical variational techniques for solving dynamical optimization problems. The procedure will be exposed parallel to an application example, and will be divided into five parts:

(a) Problem formulation including boundary limitations on the control vector.
(b) Derivation of the Euler equations.
(c) Terminal and corner conditions.
(d) First part of the solution and conditions related to the second variations.
(e) Second part of the solution.

The application example concerns optimal utilizations of a space rocket. [G-29, Chs. 4 and 5; 4-5 to 4-12]. The following assumptions will be adopted.

1. The rocket is considered a particle of variable mass, i.e., the rocket's moments of inertia are negligibly small, but the mass is a function of time $m = m(t)$.
2. The thrust magnitude is taken as a linear function of the mass flow rate: $T = -c\dot{m}$.
3. The external forces acting on the rocket give rise to accelerations which are functions of position and time only

$$X = X(x, y, t) \qquad Y = Y(x, y, t)$$

(Gravity accelerations are taken into account, while atmospheric resistances are neglected.)

4. The thrust direction can be changed instantaneously, i.e., any time delays in controlling the gimbal angle of the meter are neglected.

The state variables are $u$, $v$, $x$, $y$, $m$ (velocities, positions and mass, motion on a plane). The control variables are $c$, $\beta$, and $\psi$ (characteristic speed, thrust magnitude, and thrust direction).

| (a) General Formulation | Example |
|---|---|

**1. State equation (dynamic)**

Equations of motion:

$$\dot{x}_i = g_i(x_i, u_j, t)$$

or

$$\varphi_i = \dot{x}_i - g_i(x_i, u_j, t) = 0$$

state vector:     $x = \{x_1, x_2, \ldots, x_n\}$
control vector: $u = \{u_1, u_2, \ldots, u_r\}$

$$\dot{u} = X + \frac{c\beta}{m} \cos \psi \qquad (1)$$

$$\dot{v} = Y + \frac{c\beta}{m} \sin \psi \qquad (2)$$

$$\dot{x} = u \qquad (3)$$

$$\dot{y} = v \qquad (4)$$

$$\dot{m} = -\beta \qquad (5)$$

state-vector: $\{u, v, x, y, m\}$, $n = 5$
control-vector: $\{c, \beta, \psi\}$, $r = 3$
At the total, 8 dependent variables and one independent variable, $t$.

**2. Performance index**
(To be extremalized)

$$J = [G(x, u, t)]_i^f + \int_{t_i}^{t_f} F(x, u, t)\, dt$$

Minimize (fuel consumption, negative value of the range):

$$J = G(q) \big|_i^f$$

$$q = \{x, y, v, u, m, t\}$$

$$F \equiv 0 \quad \text{(Mayer Problem)}$$

**3. Limitations**

$$a \leq x \leq b$$
$$c \leq u \leq d$$

In order to characterize these limitations, new variables are introduced

$$(x - a)(b - x) = z^2$$
$$(u - c)(d - u) = \gamma^2$$

This procedure is attributed to Valentine

No limitation on $q$. Limitations on $c$, $\beta$ and $\psi$.

$$c(t) = \text{constant} \qquad (6)$$

$\psi(t)$ without limitations

$$\beta_{\min} < \beta < \beta_{\max} \qquad (7)$$

($\beta_{\min}$ and $\beta_{\max}$ are given values)
The new variable $\gamma$ is introduced:

$$(\beta - \beta_{\min})(\beta_{\max} - \beta) = \gamma^2 \qquad (8)$$

*Comments.* There are five equations of motion, Eq. (1) to (5), but 8 variables $x$, $y$, $u$, $v$, $m$, $c$, $\psi$, and $\beta$. The values of three of them: $c$, $\psi$, and $\beta$ can be arbitrarily chosen to minimize $G(q)|_i^f$. The system has three degrees of freedom.

| (b) *Euler Equations* | *Minimizing Equations* |
|---|---|

Form the so-called "augmented function":

$$F^* = F + \sum_1^n \lambda_i \varphi_i$$

The necessary extremum conditions are given by Euler equations

$$\frac{d}{dt} \frac{\partial F^*}{\partial \dot{w}_k} - \frac{\partial F^*}{\partial w_k} = 0$$

$$w_k = \{x_1, x_2, \ldots, x_n, u_1, u_2, \ldots, u_r\}$$

*Case a*: No limitations neither on $x$, nor on $u$:

in $F^*$:　$i = 1, \ldots, n$

in $w_k$　$k = 1, \ldots, (n + r)$

*Case b*: There are $n$ limitations on $x_i$, $r$ limitations on $u_j$ then:

in $F^*$: $n$ equations in $x_i$
　　　　$n$ equations in $z_i$
　　　　$r$ equations in $\gamma_j$
Total: $(2n + r)$ equations
in $w_k$: $n$ variables $z$
　　　　$r$ variables $\gamma$

　　　$k = 1, \ldots, (n + r)$

The $z$ and $\gamma$ replace the $x$ and $u$.

---

We are in the Case b:
　5 eq. (1) to (5) plus eq. (8), 6 eq. at the total. Eq. (6) does not enter in $F$, since $c$ is not variable with respect to time.
　in the Euler equations, $\gamma$ replaces $\beta$.

Augmented functions:

$$F^* = \lambda_u \left( \dot{u} - X - \frac{c\beta}{m} \sin \psi \right)$$

$$+ \lambda_v \left( \dot{v} - Y - \frac{c\beta}{m} \cos \psi \right) + \lambda_x (\dot{x} - u)$$

$$+ \lambda_y (\dot{y} - v) + \lambda_m (\dot{m} + \beta)$$

$$+ \lambda_\gamma [\gamma^2 - (\beta - \beta_{\min})(\beta_{\max} - \beta)]$$

(There are 6 Lagrange multipliers $\lambda_x \lambda_y \lambda_u \lambda_v \lambda_m \lambda_\gamma$ all function of $t$)

Euler equations:

$$\dot{\lambda}_u + \lambda_x = 0 \tag{9}$$

$$\dot{\lambda}_v + \lambda_y = 0 \tag{10}$$

$$\dot{\lambda}_x + \lambda_u \frac{\partial X}{\partial x} + \lambda_v \frac{\partial Y}{\partial y} = 0 \tag{11}$$

$$\dot{\lambda}_y + \lambda_u \frac{\partial X}{\partial y} + \lambda_v \frac{\partial Y}{\partial x} = 0 \tag{12}$$

$$\dot{\lambda}_m - \frac{c\beta}{m^2}(\lambda_u \cos \psi + \lambda_v \sin \psi) = 0 \tag{13}$$

$$\frac{c\beta}{m}(\lambda_u \sin \psi - \lambda_v \cos \psi) = 0 \tag{14}$$

$$\frac{c}{m}(\lambda_u \cos \psi + \lambda_v \sin \psi) - \lambda_m$$
$$- \lambda_\gamma (\beta_{\max} + \beta_{\min} - 2\beta) = 0 \tag{15}$$

$$\lambda_\gamma \cdot \gamma = 0 \tag{16}$$

---

*Comment 1.* The limitations on $\beta$ introduce a new equation of motion (Eq. 8). However, when we derive Euler equations, only the multiplier $\lambda_\gamma$ is introduced (instead of $\lambda_\beta$), because there is only one $\varphi$-equation more (Eq. $8 \to \varphi_\beta$). The derivative of $F$ is then taken with respect to $\gamma$. Regarding the whole problem, we have one supplementary variable $\gamma$ characterizing $\beta$-limitations. In practical applications, these delicate points must be examined carefully.

*Comment 2.* The equations (1) to (5), (6), (8), (9) to (16) form a system of 15 equations with 15 variables $x$, $y$, $u$, $v$, $m$, $c$, $\beta$, $\psi$, $\gamma$, $\lambda_x$, $\lambda_y$, $\lambda_u$, $\lambda_v$, $\lambda_m$, and $\lambda_\gamma$, 10 of these 15 equations are first-order differential equations. The system is therefore of the 10th order, and can be solved if 10 initials *or* final values are known.

| (c) *Terminal and Corner Conditions* | *Example* |
|---|---|

**(c1) *Given data***

A part of the end values are given by end constraints:

$$\psi_l(x_i, t)_i = 0 \quad l = 1, 2, \ldots, q$$
$$\psi_l(x_i, t)_f = 0 \quad l = (q + 1), \ldots, s$$
$$s = 2n + 2$$

**(c2) *Transversality***

The other $(2n + 2 - s)$ end values are supplied by the transversality condition:

$$\left[ dG + \left( F^* - \sum_{1}^{n} \frac{\partial F^*}{\partial \dot{x}_i} \dot{x}_i \right) dt \right.$$
$$\left. + \sum \frac{\partial F^*}{\partial \dot{x}_i} dx_i \right]_{t_i}^{t_f} = 0$$

*Comment.* The transversality condition signifies that the extremal (extremalizing trajectory) in the phase space must have the same slope as the trajectories $\psi_i$ and $\psi_f$ at times $t_i$ and $t_f$.

**(c1)** Given data are for instance:

$$x(t_i) = 0 \quad y(t_i) = 0 \quad v(t_f) = 0$$

**(c2)** Transversality condition:

$$dG + (\lambda_u \, du + \lambda_v \, dv + \lambda_x \, dx$$
$$+ \lambda_y \, dy + \lambda_m \, dm - C \, dt)_{t_i}^{t_f} = 0 \quad (17)$$
$$C = \lambda_u \dot{u} + \lambda_v \dot{v} + \lambda_x \dot{x} + \lambda_y \dot{y} + \lambda_m \dot{m} \quad (18)$$

If $X$ and $Y$ are not explicit functions of $t$, $C$ is a first integral

$$dC = 0$$

and

$$C = \text{constant} \quad (19)$$

Equation (19) can be used to replace one of the equations (9) to (16)

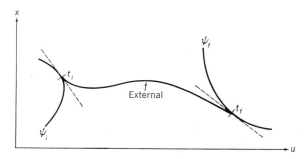

**(c3) *Corner conditions***
(Weierstrass-Erdmann conditions)

In cases where the extremalizing trajectories [e.g. $\beta(t)$] present or might present discontinuities, they consist of subarcs joined at some corners. At these corners partial derivatives of the 2

In the case of rockets, the extremalizing trajectory is often discontinuous, consisting of for example:

1 subarc maximum thrust,
1 subarc zero thrust,
Again maximum thrust, etc.

| (c3)  *Corner Conditions* | *Example* |
| --- | --- |

subarcs must be equal.

$$\left(\frac{\partial F^*}{\partial \dot{x}_i}\right)_- = \left(\frac{\partial F^*}{\partial \dot{x}_i}\right)_+$$

$$i = 1, 2, \ldots, n$$

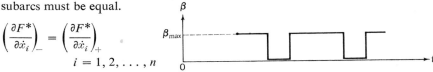

$$\left(-F^* + \sum \frac{\partial E^*}{\partial \dot{x}_i} \dot{x}_i\right)_-$$

$$= \left(-F^* + \sum \frac{\partial F^*}{\partial \dot{x}_i} \dot{x}_i\right)_+$$

Actual corner conditions:

$$\lambda_{q^-} = \lambda_{q^+}, q = \{xyuvm\} \tag{20}$$

$$C_- = C_+ \tag{21}$$

### (d)  *First Part of the Solution*

The examination of the Euler equations (9) to (16) gives a part of the solution.

(d1)  Equation (14) gives

$$\tan \psi = \lambda_u / \lambda_v \tag{22}$$

Hence,

$$\sin \psi = \pm \lambda_v / (\lambda_u^2 + \lambda_v^2)^{\frac{1}{2}} \tag{23}$$

$$\cos \psi = \pm \lambda_u / (\lambda_u^2 + \lambda_v^2)^{\frac{1}{2}} \tag{24}$$

These equations show that one can choose

$$\psi = \psi(t)$$

as well as

$$\psi = \psi(t) + \pi$$

Therefore, there is one ambiguity.

(d2)  Solving equations (8), (15) and (16), we have:

$$\text{if } \gamma \neq 0 \quad \lambda_\gamma = 0 \quad \text{then} \quad \beta_{\min} < \beta < \beta_{\max}$$

$$\text{if } \gamma = 0 \quad \left.\begin{array}{l} \lambda_\gamma \neq 0 \\ \lambda_\gamma = 0 \end{array}\right\} \quad \text{then} \quad \left\{\begin{array}{l} \beta = \beta_{\max} \\ \beta = \beta_{\min} \end{array}\right.$$

It is not clear whether we have to take extremum values of $\beta$, or some intermediate values. This is another ambiguity. The two ambiguities can be clarified by introducing the conditions related to the second variation of the augmented function $F^*$. This procedure is similar to that used in function extremum seeking: the zeroing of the first derivatives defines the extremum conditions; the sign of the second derivatives defines whether this is a maximum or a minimum.

| (d3) *Conditions Related to the Second Variation* | *Example* |
|---|---|

**(d3.1)** *Weierstrass E-function*

The necessary condition for $J$ to be a minimum is that at all points of the extremal are:

$$E = \Delta F^* - \sum_1^n \frac{\partial F^*}{\partial \dot{x}_i} \Delta \dot{x}_i \geq 0$$

for all strong variations $\Delta \dot{x}_i$ compatible with the constraints $\varphi_i$.

**(d3.2)** *Legendre-Clebsch condition* relative to all weak variations $\delta \dot{x}_i$:

$$\sum_1^n \sum_1^n \frac{\partial^2 F^*}{\partial \dot{x}_i \, \partial \dot{x}_j} \, \partial \dot{x}_i \, \partial \dot{x}_j \geq 0$$

These two inequalities change sign if we want the necessary condition for $J$ to be a maximum.

For $J = G(q_{ji}, q_{jf})$ to be a minimum, the necessary condition is that the corresponding Weierstrass $E$-function to be nonnegative. It turns out after calculations that this consists of maximalizing the function $L$:

$$L \equiv \beta \left[ \frac{c}{m} (\lambda_u \cos \psi - \lambda_v \sin \psi) - \lambda_m \right] \tag{26}$$

The function $L$ is maximum with respect to $\psi$ if:

$$\frac{\partial L}{\partial \psi} = 0 \qquad \frac{\partial^2 L}{\partial \psi^2} \leq 0 \tag{27}$$

$$\frac{\partial L}{\partial \psi} = \beta \frac{c}{m} (-\lambda_u \sin \psi + \lambda_v \cos \psi)$$

[this eq. is identical to Eq. (14)]

$$\frac{\partial^2 L}{\partial \psi^2} = \beta \frac{c}{m} (\lambda_u \cos \psi - \lambda_v \sin \psi) \leq 0$$

This inequality is satisfied if the $+$ sign is chosen in Eq. (23). Hence

$$\lambda_u \cos \psi + \lambda_v \sin \psi = (\lambda_u^2 + \lambda_v^2)^{\frac{1}{2}}$$

$\psi$ jumps of $\pi$ if $\lambda_v$ and $\lambda_u$ change sign simultaneously.

*Comment 1.* In practical applications, is it necessary to avoid confusing the Weierstrass-Edermann condition on the discontinuities with the Weierstrass $E$-function condition on the second variations.

*Comment 2.* Maximalizing $L$ with respect to $\psi$, is a function extremum-seeking problem. It is only necessary to compute derivatives, and not variations. A similar fact is also met in the methods of maximum principle and dynamic programming.

**(e)** *Second Part of the Solution*

The ambiguity on $\psi$ is clarified by the condition (27). Regarding the ambiguity on $\beta$, let us write Eq. (26) as

$$L = k\beta \tag{29}$$

This shows that $L$ is a linear function of $\beta$, the slope of which is

$$k = \frac{c}{m} (\lambda_u^e + \lambda_v^2)^{\frac{1}{2}} - \lambda_m \tag{30}$$

The figure on p. 153 shows the function $L = K\beta$ for various values of $K$. It is

(e)  *Second Part of the Solution*

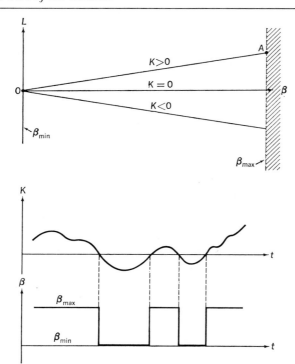

necessary, for maximalizing $L$, to take:

$$\beta = \beta_{max} \quad \text{if} \quad K > 0 \quad \text{(point A)}$$
$$\beta = \beta_{min} \quad \text{if} \quad K < 0 \quad \text{(point O)}$$

The function $K$ is called "switching function." The sign of $K$ determines the value of the control variable $\beta$. Since $\beta(t)$ can only have extremal values, we have a bang-bang control.

---

### 4.2.2  Pontryagin's Method

Under the name of Pontryagin, there is a principle and an optimization technique. The principle is the so-called maximum principle, the theory of which is discussed in Appendix 4B. The optimization technique solves certain classes of dynamical optimization problems according to a formulation that we shall call Pontryagin's formulation. In this section, we shall first discuss the Pontryagin's formulation and compare it with the classical variational formulation. Several examples are then given for illustrating the utilization of the Pontryagin's method.

**4.2.2.1   Optimization Procedure.**   The problem formulation proceeds in two steps. The first step is a Mayer problem:

$$\min_{u(t)\in U} S = \sum_{i=1}^{n} c_i x_i(t_f) \tag{4-14}$$

subject to:

$$\dot{x}_i = g_i(x_i, u_j, t) \qquad i = 1, 2, \ldots, n \qquad j = 1, 2, \ldots, r \tag{4-15}$$
$$x_i(0) = x_i^0 \qquad \text{(given)}$$

As the second step, we introduce an additional state variable $x_{n+1}$ to include nonlinear and integral relationships in the performance function $S$. For instance, if the integral-square value of one of the state variables $x_s$ is to be minimized, then we define:

$$x_{n+1} = \int_{t_0}^{t_f} x_s^2 \, dt \tag{4-16}$$

from which we have

$$\dot{x}_{n+1} = x_s^2 \qquad x_{n+1}(0) = 0 \tag{4-17}$$

and then we have the Pontryagin's formulation

$$\min_{u(t)\in U} S = x_{n+1}(t_f) = \sum_{1}^{n} c_i x_i(t_f) \tag{4-18}$$

subject to:

$$\dot{x}_i = g_i(x_i, u_j, t)$$
$$x_i(0) = x_i^0 \qquad \text{(given)}$$
$$x_{n+1}(0) = 0 \tag{4-19}$$
$$i = 1, 2, \ldots, n+1 \qquad j = 1, 2, \ldots, r$$

The derivation of the extremalizing equations proceeds as follows. We form the Hamiltonian function, by using, the momentarily unknown or co-state variables $p_i(t)$

$$H = \sum_{1}^{n+1} p_i g_i = H(x_i, u_j, p_i, t) \tag{4-20}$$

$$p_i(t_f) = -c_i \quad \text{(known)}$$

The function $H$ is a multivariable functional depending on $(n + 1)$ state variables, $r$ control variables, and $(n + 1)$ co-state variables. The $2(n + 1) + r$ unknown variables are solutions of the following set of extremalizing equations

$$\dot{x}_i = \frac{\partial H}{\partial p_i} = g_i \tag{4-21}$$

$$\dot{p}_i = -\frac{\partial H}{\partial x_i} \tag{4-22}$$

$$H(x_i, u_j, p_i, t) \text{ maximum with respect to } u_j(t) \tag{4-23}$$

Equations (4-21) and (4-22) yield $2(n + 1)$ first-order differential equations; Eq. (4-23) will yield $r$ algebraic equations if $g_i$ are linear functions in $x_i$ and $u_j$. We have thus $2(n + 1) + r$ equations for $2(n + 1) + r$ unknowns. We have written Eq. (4-23) in the indicated form to recall that it is not sufficient to write $\partial H/\partial u_j = 0$; second derivatives must be computed to avoid the stationary-point ambiguity. In the usual and more mathematical statement of the maximum principle, we say: $u$ is constrained to a closed bounded (compact) set and $u$ is a global maximum of $H$. The reader who is interested by the exact mathematical definition of the closed bounded (compact) set may consult [G-54, Ch. 3].

If we compare this procedure with the classical variational procedure, we find the following similarities:

1. The Hamiltonian function is a kind of augmented function.
2. The role of co-state variables $p_i(t)$ is similar to that of the Lagrange multipliers.
3. Equations (4-21) to (4-23) are similar to Euler equations:
   a. Completely, if in the classical procedure, conditions on $u_j$ are $\partial F/\partial u_j = 0$.
   b. Partially, if conditions on $u_j$ are derived from the Weierstrass E-function condition.

They also have the following differences:
   a. In the augmented function $F^*$, $\varphi_i = \dot{x}_i - g_i$ are used, while in the Hamiltonian function, only $g_i$ are used.
   b. No end-conditions on $\lambda_i(t)$ are known, while $p_i(t_f) = -c_i$ are known.

Table 4.1 summarizes the results of the comparison between these two procedures.

The Pontryagin's equations (4-21) to (4-23) secure, in principle, necessary and sufficient conditions for local maxima or minima with respect to the considered class of admissible controls. An interesting question then arises as to why the Euler equations secure only necessary conditions for extremals (the stationary-point ambiguity not clarified), while the Pontryagin's equations secure necessary and sufficient conditions for local maxima or minima, even though both equations are apparently similar. This can be explained shortly in the following way.

1. In the Pontryagin's derivation, the control variables $u$ are explicitly considered as causing variations in the performance index via the state variables.
2. The sign of the Hamiltonian function (see Appendix 4B) is proven representative of the sign of the variation of the performance index, i.e.,

$$H \geq 0 \Rightarrow \Delta S \geq 0 \Rightarrow S \text{ minimum}$$
$$H \leq 0 \Rightarrow \Delta S \leq 0 \Rightarrow S \text{ maximum}$$

*Table 4.1*

| Classical Formulation | Pontryagin's Formulation |
|---|---|
| *State equation* | *State equation* |
| $\dot{x}_i = g_i(x_i, u_j, t) \quad i = 1, 2, \ldots, n$ | $\dot{x}_i = g_i(x_i, u_j, t) \quad i = 1, 2, \ldots, n$ |
| $j = 1, 2, \ldots, r$ | $x_i \in X \quad u_j \in U \quad j = 1, 2, \ldots, r$ |
| $a \le x \le b$ | $X, U$ prescribed domains. |
| $c \le u \le d$ | |
| $\varphi_i = \dot{x}_i - g_i(x_i, u_j, t) = 0$ | |
| *Performance indices* | *Performance indices* |
| $$J = [G(x, u, t)]_{t_i}^{t_f} + \int_{t_i}^{t_f} F(x, u, t)\, dt$$ | e.g. $$S = \sum_1^n c_i x_i(t_f)$$ $$S = x_{n+1}(t_f)$$ $$x_{n+1} = \int_{t_0}^{t_f} x_s^2\, dt$$ |
| *Extremalizing equations* | *Extremalizing equations* |
| (a) Form the augmented function | (a) Define a new equation $g$ such as: |
| $F^* = F + \Sigma \lambda_i \varphi_i$ | $\dot{x}_{n+1} = x_s^2 = g_{n+1}$ |
| $F^* = F + \Sigma \lambda_i(\dot{x}_i - g_i)$ | $x_{n+1}(0) = 0$ |
| (b) Form the Euler equations | Define the variable co-vector $p_k(t)$ such as: |
| $$\frac{d}{dt}\frac{\partial F^*}{\partial \dot{w}_k} - \frac{\partial F^*}{\partial w_k} = 0$$ | $$\dot{p}_i = -\sum_{k=1}^{n+1} p_k \frac{\partial g_k}{\partial x_i}$$ |
| $w_k =$ | $p_i(t_f) = -c_i \quad i = 1, \ldots, (n+1)$ |
| $\{x_1, \ldots, x_n, u_1, \ldots, u_r, \lambda_1, \ldots, \lambda_n\}$ | Form the Hamiltonian equation |
| The variables $x_i(t)$ and $\lambda_i(t)$ are solutions of the differential equation $\varphi_i$ and the Euler equations. | $$H = \sum_1^{n+1} p_i g_i$$ |
| | (b) The variables $x_i(t)$ and $p_i(t)$ are solutions of the system of differential equations: |
| | $$\dot{p}_i = -\frac{\partial H}{\partial x_i} \qquad \dot{x}_i = \frac{\partial H}{\partial p_i}$$ |
| *Weierstrass E-function condition* | *Maximum principle* |
| The control vector $u$ is obtained by maximalizing | The control vector $u$ is obtained by maximalizing |
| $$L = \sum_1^n \lambda_i g_i$$ | $$H = \sum_1^{n+1} p_i g_i = H(x, u, p, t)$$ |
| with respect to $u_j$. | with respect to $u_j$. |

3. The condition (4-23) states

$$H(x, \bar{u}, p, t) \geq H(x, u, p, t)$$

where $\bar{u}$ is the optimal control vector. If this relation is solved with respect to $u$, then the sign of $H$ is secured.

**4.2.2.2   Application Examples.**

*Example 1.*   We use the same problem as in Example 1 of Subsection 4-2.3:
minimize $\int_0^T (u^2 + x_2^2) \, dt$ given $\dot{x} = u$ (values of $x_2(0)$ and $x_2(T)$ are specified) by choosing an appropriate control signal $u$.

*Solution.*   Put $\dot{x}_1 = u^2 + x_2^2$ and the problem is now to minimize $x_1(T)$. Set up

$$H = p_1 \dot{x}_1 + p_2 \dot{x}_2$$
$$= p_2 u + p_1(u^2 + x_2^2)$$

Maximize $H$ with respect to $u$ ($p_2$, $x_2$ in $H$ then considered as constants):

$$\partial H/\partial u = p_2 + 2up_1 = 0$$

yielding

$$u = -p_2/2p_1$$

Substituting the optimal value of $u$ in $H$

$$H = p_2 \left( \frac{-p_2}{2p_1} \right) + p_1 \frac{p_2^2}{4p_1^2} + p_1 x_2^2$$

The equations $\dot{x}_i = \partial H/\partial p_i$ and $\dot{p}_i = -\partial H/\partial x_i$ are, in this case,

$$\dot{x}_2 = -p_2/2p_1$$

$$\dot{x}_1 = + \frac{p_2^2}{4p_1^2} + x_2^2$$

which are slightly disguised versions of

$$\dot{x}_2 = u$$
$$\dot{x}_1 = (u^2 + x_2^2)$$

The other two equations are

$$\dot{p}_2 = -2p_1 x_2$$
$$\dot{p}_1 = 0$$

because $x_1$ does not appear owing to the formulation of the problem.

For these four equations, we have two given boundary conditions $x_2(T)$ and $x_2(0)$; two others arise from the problem $p_2(T) = 0$ and $p_1(T) = 1$

which are the coefficients $-C_2$ and $-C_1$. Since $\dot{p}_1 = 0$ and $p_1(T) = 1$, $p_1 = 1$ for all $t$. This makes $\dot{x}_2 = -p_2/2$ and $\dot{p}_2 = -2x_2$ by substituting $p_1$. The other equations are of no interest at the moment. Differentiating $\dot{x}_2 = -p_2/2$ with respect to time gives $\ddot{x} = -\dot{p}_2/2$ and substituting $\dot{p}_2 = -2x_2$ in this results in $\ddot{x} = x$ as for the previous solution in Subsection 4-2.1.1, example 1.

**Example 2.** We take the same problem used for illustrating the complete procedure of the calculus of variations, i.e., in Subsection 4-2.1.3. The differential equations describing the rocket system are:

$$\dot{x}_1 = \dot{u} = \frac{c\beta}{m} \cos \psi \qquad\qquad x_1(0) = x_1^0$$

$$\dot{x}_2 = \dot{V} = \frac{c\beta}{m} \sin \psi - g \qquad x_2(0) = x_2^0$$

$$\dot{x}_3 = \dot{x} = u \qquad\qquad\qquad x_3(0) = x_3^0 \qquad\qquad \text{(a)}$$

$$\dot{x}_4 = \dot{y} = V \qquad\qquad\qquad x_4(0) = x_4^0$$

$$\dot{x}_5 = \dot{m} = -\beta \qquad\qquad\quad x_5(0) = x_5^0$$

$$\dot{x}_6 = 1 \qquad\qquad\qquad\qquad x_6(0) = 0$$

We have five state variables $u$, $v$, $x$, $y$, $m$ and two control variables $\beta$, $\psi$. The state variable $x_6$ is added to include the possibility of time appearing in the performance function, since

$$S = x_6(t_f) = \int_{t_0}^{t_f} dt = t_f \rightarrow \dot{x}_6 = 1, \qquad x_6(0) = 0$$

The mass flow rate $\beta$ is bounded by

$$\beta_1 \leq \beta \leq \beta_2 \qquad\qquad\qquad\qquad\qquad\qquad \text{(b)}$$

The Hamiltonian function is

$$H = p_1 \frac{c\beta}{m} \cos \psi + p_2 \left( \frac{c\beta}{m} \sin \psi - g \right) + p_3 u + p_4 v - p_5 \beta + p_6 \qquad \text{(c)}$$

The differential equations defining the co-state variables are

$$\dot{p}_1 = -p_3 \qquad \dot{p}_2 = -p_4 \qquad \dot{p}_3 = 0 \qquad \dot{p}_4 = 0$$

$$\dot{p}_5 = p_1 \frac{c\beta}{m^2} \cos \psi + p_2 \frac{c\beta}{m^2} \sin \psi \qquad\qquad \dot{p}_6 = 0 \qquad\qquad \text{(d)}$$

Suppose that the performance $S$ is a linear function of the final values of the state variables. To minimize $S$, an application of the maximum principle gives for the $\beta$

$$\beta = \beta_2 \quad \text{when} \quad K > 0$$
$$\beta = \beta_1 \quad \text{when} \quad K < 0 \tag{e}$$

where

$$K = \frac{c}{m} (p_1 \cos \psi + p_2 \sin \psi) - p_5 \tag{f}$$

and

$$\tan \psi = p_2/p_1$$
$$p_1 \cos \psi + p_2 \sin \psi \geq 0 \tag{g}$$

for $\psi$. These are the same relationships obtained in Subsection 4-2.4 when the Weierstrass and Clebsch conditions were applied. Combining Eqs. (d) and (f) to obtain the new form of the switching function

$$\dot{K} = - \frac{c}{m} (p_3 \cos \psi + p_4 \sin \psi) \tag{h}$$

which indicates that there are, at most, two zeros of $K$. This leads to the conclusion that there are no more than three subarcs to the optimum path. If $K = 0$ over some period of time, the performance $S$ does not depend on $\beta$ during this time; $S$ is said to be insensitive to $\beta$ and the solution is not unique.

*Example 3.* We shall take the problem of minimizing the transfer time of a low-thrust ion rocket between the orbits of Earth and Mars, in a quite simplified version. We shall assume that the thrust direction $\psi$ is the only control variable. The orbits of Earth and Mars are circular and coplanar, and the gravitational attractions neglected.

The following nomenclature will be used.

$r$   Radial distance.
$uv$   Radial and circumferential velocities.
$m$   Mass of the rocket.
$\psi$   Thrust direction.
$T$   Thrust impulse magnitude.
$a(t) = T/(m_0 + \dot{m}t)$

The given terminal conditions are

$$\text{at} \quad t = t_0 \quad m_0 r_0 u_0 v_0$$
$$\text{at} \quad t = t_f \quad r_f u_f v_f$$

Table 4-2 shows, in the classical way and in the Pontryagin's way, how this problem is formulated and how the extremalizing equations are obtained.

**Table 4.2**

| Classical Formulation | Pontryagin's Formulation |
|---|---|

*State equation*

$$\dot{r} = u$$

$$\dot{u} = \frac{v^2}{r} - \frac{K}{r^2} + a \sin \psi$$

$$\dot{v} = -\frac{uv}{r} + a \cos \psi$$

$$\dot{x}_1 = \dot{r} = u$$

$$\dot{x}_2 = \dot{u} = \frac{v^2}{r} - \frac{K}{r^2} + a \sin \psi$$

$$\dot{x}_3 = \dot{v} = -\frac{uv}{r} + a \cos \psi$$

$$\dot{x}_4 = 1$$

*Performance index*

$$J = [G(x, u, t)]_{t_i}^{t_f} = t_f$$

$$S = \int_{t_0}^{t_f} \dot{x}_4 \, dt = t_f$$

*Extremalizing equations*

Augmented function

$$F = \sum \lambda_i \varphi_i = \lambda_r (\dot{r} - u)$$
$$+ \lambda_u \left( \dot{u} - \frac{v^2}{r} + \frac{K}{r^2} - a \sin \psi \right)$$
$$+ \lambda_v \left( \dot{v} - \frac{uv}{r} - a \cos \psi \right)$$

Hamiltonian equation:

$$H = \sum_1^{n+1} p_i g_i = p_1 u + p_2 \left( \frac{v^2}{r} - \frac{k}{r^2} \right.$$
$$\left. + a \sin \psi \right) + p_3 \left( -\frac{uv}{r} + a \cos \psi \right)$$

Differential equations

Euler-Lagrange equations

$$\dot{\lambda}_r = \lambda_u \left( \frac{v^2}{r^2} - 2\frac{K}{r^3} \right) - \lambda_v \frac{uv}{r^2}$$

$$\dot{\lambda}_u = -\lambda_r + \frac{v}{r} \lambda_v$$

$$\dot{\lambda}_v = -2\lambda_u \frac{v}{r} + \lambda_r \frac{u}{r}$$

$$\dot{p}_1 = -\frac{\partial H}{\partial x_1} = p_2 \left( \frac{v^2}{r^2} - 2\frac{K}{r^3} \right) - p_3 \frac{uv}{r^2}$$

$$\dot{p}_2 = -\frac{\partial H}{\partial x_2} = -p_1 + p_3 \frac{v}{r}$$

$$\dot{p}_3 = -\frac{\partial H}{\partial x_3} = -2p_2 \frac{v}{r} + p_3 \frac{u}{r}$$

$$\dot{p}_4 = -\frac{\partial H}{\partial x_4} = 0 \Rightarrow p_4(t_f) - p_4(t_i) = 0$$

Weierstrass $E$-function condition

$$\max_{\psi} L \rightarrow \tan \psi = \frac{\lambda_u}{\lambda_v}$$

since $p_4(t_f) = -c_4 = -1$ $\therefore$ $p_4(t_0) = -1$

Maximum principle:

$$\max_{\psi} H \Rightarrow \tan \psi = p_2/p_3$$

Rewrite the state equations:

$$\dot{r} = u$$

$$\dot{u} = \frac{v^2}{r} - \frac{K}{r^2} + a \frac{\lambda_u}{(\lambda_u^2 + \lambda_v^2)^{\frac{1}{2}}}$$

$$\dot{v} = -\frac{uv}{r} + a \frac{\lambda_v}{(\lambda_u^2 + \lambda_v^2)^{\frac{1}{2}}}$$

Then: $\dot{x}_1 = u$

$$\dot{x}_2 = \frac{v^2}{r} - \frac{K}{r^2} + a \frac{p_2}{(p_2^2 + p_3^2)^{\frac{1}{2}}}$$

$$\dot{x}_3 = -\frac{uv}{r} + a \frac{p_3}{(p_2^2 + p_3^2)^{\frac{1}{2}}}$$

$$\dot{x}_4 = 1 \Rightarrow x_4(t_f) - x_4(t_0) = t_f$$

In either formulation, we have a system of six differential equations of first order in $(r u v \lambda_u \lambda_v \lambda_r)$ or in $(x_1 x_2 x_3 p_1 p_2 p_3)$ with unknown end-conditions of $\lambda(t)$ or $p(t)$. Two-point boundary problem.

### 4.2.3  State Variable Discontinuities

Much of the work, as shown in the preceding sections, is based on certain continuity assumptions concerning the state variables and, in some cases, even the derivatives of the state variables. In a great many practical problems of system optimization, however, such continuity assumptions are not valid. One very important problem in which the assumption is not valid is that of optimizing the trajectory of a multistage rocket vehicle. At staging points one of the state variables, mass, is discontinuous. Orbital transfer problems involving velocity impulses pose similar difficulties. The effects of such state variable discontinuities in the solution of variational problems will be discussed here [4-13].

There are two important types of state variable discontinuities.

1. Defined discontinuities which occur when certain given conditions are encountered, and the magnitudes of the discontinuities are determined by given relationships. It is impossible to exert any direct control over either the timing or the magnitudes of the discontinuities.

2. Only-magnitude defined discontinuities characterized by the fact that only the magnitudes of the discontinuities are determined by given relationships: the timing is directly controllable and can be used to optimize the system performance.

Here we shall consider only type 1 discontinuities, with one possible problem formulation as follows.

*System.* a.  System dynamics

$$\dot{x}^{(a)} = f^{(a)}(x, u, t) \quad \text{for} \quad t_{a-1} < t < t_a \tag{4-24}$$

$$a = 1, 2, \ldots, A$$

$$x(t_0) \text{ assumed given.}$$

b.  Terminal constraints

$$\Omega[x(t_A), t_A] = 0 \tag{4-25}$$

c.  Corner-time conditions

$$\theta_a(x, t) = 0 \qquad a = 1, 2, \ldots, A - 1 \tag{4-26}$$

where $\theta_a$ are known functions of $x$ and $t$.

d.  Discontinuity definition at each corner time $t_a$

$$\lim_{\epsilon \to 0} x(t_a + \epsilon) = \lim_{\epsilon \to 0} \{x(t_a - \epsilon) + \xi_a[x(t_a - \epsilon), t_a - \epsilon]\}$$

$$a = 1, \ldots, A - 1 \tag{4-27}$$

*Object.* Determine $u(t)$ so that Eqs. (4-24) through (4-27) are satisfied, and so to minimize

$$\varphi[x(t_A), t_A]$$

a performance index function of the final state $x(t_A)$ and the final time $t_A$.

The difference between such a problem formulation and the basic Mayer problem formulation lies in the presence of conditions (4-26) and (4-27) which characterize mathematically the discontinuities. The meanings of the symbols are:

$x$   $n$-dimensional vector of state variables.

$u$   $r$-dimensional vector of control variables.

$f$   $n$-dimensional vector of known functions.

$\Omega$   $q$-dimensional vector of known functions.

$\xi_a$   $n$-dimensional vector determining the magnitude of the discontinuity in $x$ at $t = t_a$.

The superscript $(a)$ notation in Eq. (4-24) with $\dot{x}$ is used because the derivative expressions are not necessarily continuous through the entire solution from $t_0$ to $t_A$. Discontinuities can result from changes in the algebraic form of the derivatives, from a change in some parameter in the system, or from a discontinuity in $u$ or in $x$. Because of the existence of these discontinuities, a complete solution to Eq. (4-24) consists of a number of segments (Fig. 4-3). Each segment is referred to as a subarc, and the juncture of two subarcs is called a corner. The length and position of a subarc are defined by the corner times: in particular, the $a$th subarc begins at $t_{a-1}$ and ends at $t_a$. Because discontinuities in $x$ are permitted at the corner times, the derivative expressions are defined over open intervals. It is assumed, however, that $f^{(a)}(x, u, t)$ is sectionally continuous. In Eq. (4-27), $\xi_a$ must be a sectionally continuous and differentiable function of $x$ and $t$.

The presence of the discontinuities introduces complications in the establishment of the extremalizing conditions. At this time, there is no known work treating their influence on the second variation. The investigation presented

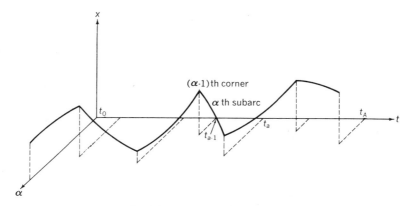

***Fig. 4-3***   A discontinuous solution.

here concerns only their influence on the first-variation necessary conditions. The mathematical procedure of this investigation is as follows:

(1) Form an augmented function $J$.
(2) Calculate analytically the variation $\Delta J$.
(3) Derive conditions making $\Delta J = 0$.

The augmented function is obtained by adjoining to the performance index $\varphi$ all the constraints by means of Lagrange multipliers:

$$J = \varphi + v^T \Omega + \sum_{a=1}^{A-1} \mu_a \theta_a + \lim_{\epsilon \to 0} \sum_{a=1}^{A} \int_{t_{a-1}+\epsilon}^{t_a-\epsilon} \lambda^T (\dot{x}^{(a)} - f^{(a)}) \, dt \quad (4\text{-}28)$$

Where:

$v$   is a $q$-dimensional vector of Lagrange multipliers associated with the $\Omega$ constraints.

$\mu_a$   multipliers associated with the $\theta_a$ constraints.

$\lambda$   an $n$-dimensional vector of multiplier functions associated with the differential constraints.

$T$   transpose.

The variation of $J$ is

$$\Delta J = \left(\frac{\partial \varphi}{\partial x}\right)_{t_A} (\Delta x)_{t_A} + \left(\frac{\partial \varphi}{\partial t}\right)_{t_A} \Delta t_A + v^T \left[\left(\frac{\partial \Omega}{\partial x}\right)_{t_A} (\Delta x)_{t_A} + \left(\frac{\partial \Omega}{\partial t}\right)_{t_A} \Delta t_A\right]$$

$$+ \sum_{a=1}^{A-1} \mu_a \left[\left(\frac{\partial \theta_a}{\partial x}\right)_{t_a-\epsilon} (\Delta x)_{t_a-\epsilon} + \left(\frac{\partial \theta_a}{\partial t}\right)_{t_a-\epsilon} \Delta t_a\right]$$

$$+ \lim_{\epsilon \to 0} \sum_{a=1}^{A} \int_{t_{a-1}+\epsilon+\Delta t_{a-1}}^{t_a-\epsilon+\Delta t_a} \lambda^T \left[\delta \dot{x} - \frac{\partial f^{(a)}}{\partial x} \delta x - \frac{\partial f^{(a)}}{\partial u} \delta u\right] dt \quad (4\text{-}29)$$

where:

$\Delta(\ )$   indicates a total variation including the variation of the independent variable, $t$.

$\delta(\ )$   indicates a variation at a fixed time point.

The relation between $\Delta(\ )$ and $\delta(\ )$ at a fixed time point is

$$(\delta x)_{t_a \pm \epsilon} = (\Delta x)_{t_a \pm \epsilon} - (\dot{x})_{t_a \pm \epsilon} \Delta t_a \quad (4\text{-}30)$$

By using Eqs. (4-24) through (4-27), neglecting higher ordered terms, making $\Delta J = 0$, omitting the intermediate derivations the following necessary conditions are obtained:

$$\lambda^T(t_A) = -\left(\frac{\partial \varphi}{\partial x} + v^T \frac{\partial W}{\partial x}\right)_{t_A} \tag{4-31}$$

$$[\lambda^T f^{(A)}]_{t_A} = \left(\frac{\partial \varphi}{\partial t} + v^T \frac{\partial \Omega}{\partial u}\right)_{t_A} \tag{4-32}$$

$$\lambda^T \frac{\partial f^{(a)}}{\partial u} = 0 \tag{4-33}$$

$$\dot{\lambda} + \left(\frac{\partial f^{(a)}}{\partial x}\right)^T \lambda = 0 \tag{4-34}$$

$$\lambda^T(t_a - \epsilon) = \lambda^T(t_a + \epsilon)\left[I + \left(\frac{\partial \xi_a}{\partial x}\right)_{t_a - \epsilon}\right] - \mu_a\left(\frac{\partial \theta_a}{\partial x}\right)_{t_a - \epsilon} \tag{4-35}$$

$$\lambda^T(t_a - \epsilon)f^{(a)}(t_a - \epsilon) = \lambda^T(t_a + \epsilon)\left[f^{(a+1)}(t_a + \epsilon) - \left(\frac{\partial \xi_a}{\partial t}\right)_{t_0 - \epsilon}\right] + \mu_a\left(\frac{\partial \theta_a}{\partial t}\right)_{t_a - \epsilon} \tag{4-36}$$

where $I$ is the identity matrix.

Equations (4-31) and (4-32) are the familiar transversality conditions, and Eqs. (4-33) and (4-34) are the Euler equations. Equation (4-35) and (4-36) are analogous to the Erdmann-Weierstrass corner conditions, and they are the necessary conditions resulting from the state variable discontinuities.

As an analytical example, consider the problem of determining the minimum time path between two points for a particle moving at constant velocity.

*System.* a. System dynamics

$$\dot{z} = V \cos \gamma \tag{a}$$

$$\dot{y} = V \sin \gamma \tag{b}$$

State variables : $z$, $y$ position coordinates.
Control variable : $\gamma$ direction of the velocity vector.
Parameter: $V$ given constant velocity magnitude.
Given boundary conditions

$$z(0) = y(0) = 0 \tag{c}$$

$$z(T) = Z \qquad y(T) = Y \tag{d}$$

$$T = \text{final time}$$

b. Discontinuities
Condition:

$$\theta = z - aZ = 0 \tag{e}$$

Magnitude:

$$\xi = \begin{Bmatrix} 0 \\ by \end{Bmatrix}_{t=t_1-\epsilon}$$

*Object.* Determine the control program $\gamma(t)$ in order to minimize the final time

$$\varphi = T \tag{g}$$

*Solution.* a. Euler equations

$$\dot{\lambda}_z = 0 \qquad \dot{\lambda}_y = 0 \tag{h}$$

$$V(-\lambda_z \sin \gamma + \lambda_y \cos \gamma) = 0 \tag{i}$$

Equation (h) states that $\lambda_z$ and $\lambda_y$ are constant over each subarc. Equation (i) states that $\gamma$ is also constant over each subarc and is equal to

$$\gamma = \arctan (\lambda_y/\lambda_z) \tag{j}$$

b. Transversality conditions

$$V(\lambda_z \cos \gamma + \lambda_y \sin \gamma)_{t=T} = 1 \tag{k}$$

It can be seen from Eq. (4-36) that Eq. (k) holds true throughout the entire solution. This fact can be used to eliminate $\lambda_z$ from the remaining equations.

c. Discontinuity conditions

The discontinuity in $\lambda_y$ at $t = t_1$ is

$$\lambda_y^{(-)} = \lambda_y(t_1 - \epsilon) = (1 + b)\lambda_y(t + \epsilon) = (1 + b)\lambda_y^{(+)} \tag{m}$$

Equations (a) through (m) can be combined to obtain the following equation which must be solved for $\lambda_y^{(+)}$

$$Y = (1 + b)^2 aZV\lambda_y^{(+)}/[1 - V^2(1 + b)^2(\lambda_y^{(+)})^2]^{\frac{1}{2}}$$
$$+ (1 - a)ZV\lambda_y^{(+)}/[1 - V^2(\lambda_y^{(+)})^2]^{\frac{1}{2}} \tag{n}$$

Equation (n) can be solved by trial and error. The value of $\lambda_y^{(-)}$ can then be computed from Eq. (m), and the control program $\gamma(t)$ can be determined as

$$\gamma = \arcsin (V\lambda_y) \tag{p}$$

The minimum final time for the solution is

$$T = \{aZ/V[1 - V^2(1 + b)^2(\lambda_y^{(+)})^2]^{\frac{1}{2}}\} + \{(1 - a)Z/V[1 - V^2(\lambda_y^{(+)})^2]^{\frac{1}{2}}\} \tag{q}$$

### 4.2.4 Optimization Problems with Boundary Inequality Constraints

In Chapter 1, Section 1.2.3, we have enumerated various types of constraints encountered in dynamic optimization problems. In the preceding sections of this chapter, we have discussed the influences of (a) magnitude constraints $u_0 \le u \le u_1$, (b) equality constraints $\dot{x} - g_i = 0$, (c) end-point constraints $[w(x, u, t)]_0^f = 0$, and (d) state discontinuities. In this section, we shall discuss the influence of inequality constraints, thus completing the *tour d'horizon* of all the constraints.

We consider the Mayer formulation of the Bolza problem. Determine $u(t)$ in the interval $t_0 \le t \le t_f$ so as to maximize [4-14, 4-15, 4-16]

$$J = \phi[x(t_f), t_f] \tag{4-37}$$

subject to the usual constraints

system $\qquad \dot{x} = f[x(t), u(t), t], \qquad x_0 \text{ and } t_0 \text{ given} \tag{4-38}$

terminal $\qquad M = M[x(t_f), t_f] = 0 \tag{4-39}$

and the boundary inequality constraints

$$C(x, u, t) \le 0 \tag{4-40}$$

or

$$S(x, t) \le 0 \tag{4-41}$$

where $x$ is the state $n$-vector, $u(t)$ a scalar variable freely chosen, $f$ an $n$-vector of known functions, $M$ a $p$-vector of known terminal constraint functions, $C$ a scalar function of $x$, $u$, and $t$; $S$ a scalar function of $x$ and $t$.

The presence of the boundary inequality constraints will introduce additional conditions to the necessary conditions of the Euler equations, the derivation of which will follow a similar procedure as in Subsection 4-1.3. Consider Fig. 4-4 representing the schematic behavior of the state variable $x$; for those periods of an extremal solution on the constraints boundary, $u(t)$ is determined in terms of $x(t)$ and $t$ by the relation

$$C[x(t), u(t), t] = 0 \tag{4-42}$$

Thus, the neighboring solution for those periods must satisfy

$$(\partial C/\partial x)\, \delta x + (\partial C/\partial u)\, \delta u = 0 \tag{4-43}$$

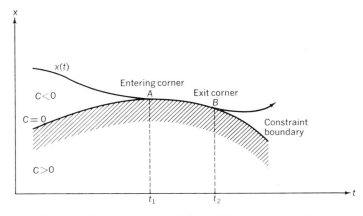

***Fig. 4-4***   Schematic representation of the state vector history.

where $\partial C/\partial x = (\partial C/\partial x_1, \ldots, \partial C/\partial x_n]$ and $\partial C/\partial u$ are evaluated along the extremal solution. On the other hand, neighboring solutions must also satisfy the perturbation differential equations

$$\frac{d}{dt} \delta x = \frac{\partial f}{\partial x} \delta x + \frac{\partial f}{\partial u} \delta u \tag{4-44}$$

obtained from Eq. (4-38), and where:

$$\frac{\partial f}{\partial x} = \begin{bmatrix} \dfrac{\partial f_1}{\partial x_1} & \cdots & \dfrac{\partial f_1}{\partial x_n} \\ & & \\ \cdot & & \cdot \\ \cdot & & \cdot \\ \cdot & & \cdot \\ \dfrac{\partial f_n}{\partial x_1} & \cdots & \dfrac{\partial f_n}{\partial x_n} \end{bmatrix} \qquad \frac{\partial f}{\partial u} = \begin{bmatrix} \dfrac{\partial f_1}{\partial u} \\ \cdot \\ \cdot \\ \cdot \\ \dfrac{\partial f_n}{\partial u} \end{bmatrix}$$

are also evaluated along the extremal solution. Substituting Eq. (4-43) into Eq. (4-44) yields

$$\frac{d}{dt}(\delta x) = \left[\frac{\partial f}{\partial x} + \frac{\partial f}{\partial u}\left(\frac{\partial C}{\partial u}\right)^{-1}\frac{\partial C}{\partial x}\right]\delta x \tag{4-45}$$

It follows that the Euler equations for determining an extremal solution are

$$\frac{d\lambda}{dt} = \begin{cases} -\left(\dfrac{\partial f}{\partial x}\right)^T \lambda & \text{when } \quad C < 0 \\[2mm] -\left[\dfrac{\partial f}{\partial x} - \dfrac{\partial f}{\partial u}\left(\dfrac{\partial C}{\partial u}\right)^{-1}\dfrac{\partial C}{\partial x}\right]^T \lambda & \text{when } \quad C = 0 \end{cases} \tag{4-46}$$

$$\lambda^T \frac{\partial f}{\partial u} = 0 \text{ determines } u(t) \text{ when } C < 0 \tag{4-47}$$

$$C(x, u, t) = 0 \text{ determines } u(t) \text{ when } C = 0 \tag{4-48}$$

where $\lambda(t)$ is an $n$-vector of *influence functions* on the Lagrange-multiplier functions. In an interval of $C = 0$, the inequality

$$\lambda^T \left(\frac{\partial f/\partial u}{\partial C/\partial u}\right) > 0 \tag{4-49}$$

must be satisfied. This is equivalent to the requirement that the variational Hamiltonian be minimized, according to the maximum principle, with respect to $u$, by the $u$ from Eq. (4-48). Equations (4-46) through (4-49) represent additional conditions to the Euler equations because of the presence of $C$. The boundary conditions for the influence functions are

$$\lambda^T(t_f) = \left[\frac{\partial\phi}{\partial x} + \nu^T\frac{\partial M}{\partial x}\right]_{t=t_f} \tag{4-50}$$

$$(\lambda^T \dot{x})_{t=t_f} = -\left[\frac{\partial\phi}{\partial t} + \nu^T\frac{\partial M}{\partial t}\right]_{t=t_f} \tag{4-51}$$

where $\nu$ is a $p$-vector of Lagrange-multiplier constraints. The $(\lambda)_{t=t_f}$, $t_f$, and $\nu$ constitute $n + 1 + p$ quantities, which are determined to satisfy Eqs. (4-50), (4-51), and (4-39). We note that in Eqs. (4-46) and (4-49), there are derivatives of $C$ taken with respect to $u$. Now, what happens if the boundary constraints are of the type $S(x, t)$ which does not contain $u$ explicitly? This is a surprising aspect of the optimization problems; if $S$ does not contain $u$ explicitly, its time derivatives may contain $u$, and the additional conditions still hold. To see this, we reconsider those periods of an extremal solution on the constraints boundary, for which the state variables are interrelated by

$$S[x(t), t] = 0 \tag{4-52}$$

Since in this case, $S$ must vanish identically, it follows that its time derivatives also must vanish

$$d^k S/dt^k = 0 \tag{4.53}$$
$$k = 0, 1, 2, \ldots$$

on the constraint boundary. Now

$$dS/dt = (\partial S/\partial t) + (\partial S/\partial x)\dot{x}$$
$$= (\partial S/\partial t) + (\partial S/\partial x)f \tag{4-54}$$

as $\dot{x} = f$ by Eq. (4-38), and since $f$ is function of $u$, $dS/dt$ is also. Besides, if $ds/dt$ is not explicitly a function of $u(t)$, one may consider the second derivative, third derivative, etc., until some $q$th time derivative which does explicitly involves $u(t)$. The $q$th order state variable inequality constraint $S^{(q)}(x, u, t) = 0$ then plays the same role as $C(x, u, t) = 0$, and the differential equations for $\lambda(t)$ are the same as in Eq. (4-46) with $C$ replaced by $S^{(q)}$.

However, in addition to the fact that $S^{(q)}(x, u, t) = 0$ on the constraint boundary, it must also be stipulated that at entering corners (like points $A$ on Fig. 4-4), the following conditions be met:

$$S[x(t_1), t_1] = 0$$
$$S^{(1)}[x(t_1), t_1] = 0$$
$$\vdots \tag{4-55}$$
$$S^{(q-1)}[x(t_1), t_1] = 0$$

These conditions play the role of terminal constraints (with respect to time, just like $M$) for the unconstrained arc. It follows that $\lambda(t)$ must satisfy other relations similar to those in Eqs. (4-50) and (4-51) at $t = t_1$. These relations are

$$\lambda^T(t_1^-) = \lambda^T(t_1^+) + \left[ \mu_0 \frac{\partial S}{\partial x} + \mu_1 \frac{\partial S^{(1)}}{\partial x} + \cdots + \mu_{q-1} \frac{\partial S^{(q-1)}}{\partial x} \right]_{t=t_1} \quad (4\text{-}56)$$

$$[\lambda^T \dot{x}]_{t=t_1^-} = [\lambda^T \dot{x}]_{t=t_1^+} - \left[ \mu_0 \frac{\partial S}{\partial t} + \mu_1 \frac{\partial S}{\partial t} + \cdots + \mu_{q-1} \frac{\partial S^{(q-1)}}{\partial t} \right]_{t=t_1} \quad (4\text{-}57)$$

where $\mu_0, \ldots, \mu_{q-1}$ are $q$ Lagrange-multiplier constraints. The $(\lambda)_{t=t_1}$, $t_1$, $\mu_0, \ldots, \mu_{q-1}$ constitute $n + 1 + q$ quantities that are determined to satisfy Eqs. (4-56), (4-57), and (4-55).

***Example.*** A brachistochrone problem.
Given

$$\dot{x} = (2gy)^{\frac{1}{2}} \cos \gamma$$

$$\dot{y} = (2gy)^{\frac{1}{2}} \sin \gamma$$

$$x(0) = y(0) = 0$$

where $x$ is horizontal distance, $y$ is vertical distance (positive downward), $g$ is the gravity acceleration, and $\gamma$ is path angle to the horizontal (Fig. 4-5).

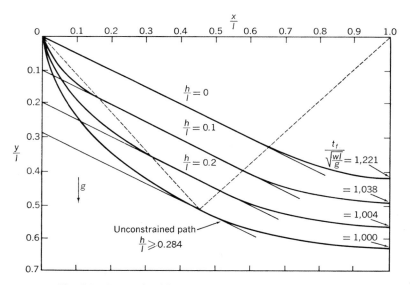

***Fig. 4-5*** Constrained brachistochrone problem with $\tan \theta = 1/2$.

Find $\gamma(t)$ to minimize the time to reach $x = 1$ with the constraint that $y \leq x \tan \theta + h$, with $\theta$ and $h$ constant. We have

$$S = y - x \tan \theta - h \leq 0$$
$$\dot{S} = (2gy)^{\frac{1}{2}} \sec \theta \sin (\gamma - \theta)$$

$S$ does not contain the control variable $\gamma$; $\dot{S}$ does. On $S = 0$, $\dot{S} = 0$ implies that $\gamma = \theta$.

The solution to the unconstrained problem,

$$h/l \geq (2/\pi)\{1 - [(\pi/2) - \theta] \tan \theta\}$$

is as follows:

$$\gamma(t) = \frac{\pi}{2} - \omega t \quad \text{where} \quad \omega = \left(\frac{\pi}{4}\frac{g}{l}\right)^{\frac{1}{2}}$$

$$\frac{x}{l} = \frac{2}{\pi}\left(\omega t - \frac{\sin 2\omega t}{2}\right)$$

$$\frac{y}{l} = \frac{2}{\pi} \sin^2 \omega t$$

$$t_f = \left(\frac{rl}{g}\right)^{\frac{1}{2}}$$

minimum final time

$$\left.\begin{array}{l} \lambda_x = -\omega/g \\ \lambda_y = -\dfrac{\omega}{g} \operatorname{ctn} \omega t \end{array}\right\} \quad \text{where} \quad dt_f = (\lambda_x \, \delta x + \lambda_y \, \delta y)_{t \leq t_f}$$

$$H = \lambda_x \dot{x} + \lambda_y \dot{y} = -1$$

variational Hamiltonian

The solution to the constrained problem,

$$h/l < (2/\pi)\{1 - [(\pi/2) - \theta] \tan \theta\}$$

is

$$\gamma(t) = \begin{cases} \dfrac{\pi}{2} - \omega_1 t & 0 \leq t \leq t_1 \\ \theta & t_1 \leq t \leq t_2 \\ \omega_2(t_f - t) & t_2 \leq t \leq t_f \end{cases}$$

where

$$\omega_1 = \left[\frac{g}{2}\frac{\theta - (\pi/2) + \operatorname{ctn}\theta}{h\operatorname{ctn}\theta}\right]^{\frac{1}{2}}$$

$$\omega_2 = \left(\frac{g}{2}\frac{\theta + \operatorname{ctn}\theta}{l + h\operatorname{ctn}\theta}\right)^{\frac{1}{2}}$$

$$t_1 = \frac{(\pi/2) - \theta}{\omega_1}$$

$$t_f - t_2 = \theta/2\omega_2$$

$$t_f = \left[\frac{2}{g}(l + h\operatorname{ctn}\theta)(\theta + \operatorname{ctn}\theta)\right]^{\frac{1}{2}} - \left[\frac{2h}{g}\operatorname{ctn}\theta\left(\theta - \frac{\pi}{2} + \operatorname{ctn}\theta\right)\right]^{\frac{1}{2}}$$

= minimum final time

$$\lambda_x(t_1^-) - \lambda_x(t_1^+) = -\mu_0\tan\theta$$

$$\lambda_y(t_1^-) - \lambda_y(t_1^+) = \mu_0$$

where $\mu_0 = (\operatorname{ctn}\theta/g)(\omega_2 - \omega_1)$. Note that $\mu_0 \to 0$ and $t_1 \to t_2$ as $h/l \to (2/\pi)\{1 - [(\pi/2) - \theta]\tan\theta\}$

$$H = \lambda_x\dot{x} + \lambda_y\dot{y} = -1 \qquad 0 < t < t_f$$

Figure 4-5 shows the solution for $\tan\theta = 1/2$ for several values of $h/l$.

### 4.2.5   The Two-Point-Boundary-Value Problem

The great drawback of the variational methods, is the so-called two-point-boundary-value problem. We have previously seen that, at the end of the optimization procedure, when the control $u(t)$ is obtained as a function of $x$ and $p$ by solving $\partial H/\partial u = 0$, and substituted into the Hamiltonian system, we obtain a set of two systems of first-order differential equations.

$$\dot{x} = \dot{x}(x, p, t) \tag{4-58}$$

$$\dot{p} = \dot{p}(x, p, t) \tag{4-59}$$

$x$ and $p$ both being $n$-vectors.

Two points are to be noted here.

1. Except in the cases where the functional $F$ is linear or quadratic, and where $g_i$ are linear or time-varying linear equations, the systems (4-58) and (4-59) are generally difficult to solve.

2. For the two systems of $n$ first-order differential equations, there is need of knowing $2n$ terminal values for identifying the $2n$ integration constants.

Since it is difficult to solve (4-58) and (4-59) analytically, we can have recourse to numerical integration methods. This will be successful if the $2n$ terminal values are known at the same side, i.e., either $x(t_0) = x_0$ and $p(t_0) = p_0$, or $x(t_f) = x_f$ and $p(t_f) = p_f$. To see this, let us consider the following examples:

(a) $$\dot{x} = x^2 - 3x - 1 \qquad x(t_0) = 5$$

We write first, assuming $\Delta t = 1$

$$\Delta x = x(t_{n+1}) - x(t_n) = [x^2(t_n) - 3x(t_n) - 1]\,\Delta t$$

or

$$x(t_{n+1}) = x^2(t_n) - 2x(t_n) - 1$$

Starting with $x(t_0) = 5$, we compute successively

$$x(t_1) = 25 - 10 - 1 = 14$$

$$x(t_2) = 196 - 28 - 1 = 167 \cdots$$

so that we construct the whole solution $x(t)$ from $t = t_0$ to $t = t_f$

(b) $$\dot{x} = x^2 - 7xp - 6 \qquad x(t_0) = 1$$
$$\dot{p} = p^2 - 6xp - 7 \qquad p(t_0) = 2$$

Assuming $\Delta t = 1$, we write

$$x(t_{n+1}) = x^2(t_n) - x(t_n)[7p(t_n) + 1] - 6$$

$$p(t_{n+1}) = p^2(t_n) - p(t_n)[6x(t_n) + 1] - 7$$

Starting with $x(t_0) = 1$ and $p(t_0) = 2$, we compute successively

$$x(t_1) = 1 - (14 + 1) - 6 = -20$$

$$p(t_1) = 4 - 2(6 + 1) - 7 = -17$$

$$x(t_2) = 400 + 20(-119 + 1) - 6 = -1966$$

$$p(t_2) = 289 + 17(-120 + 1) - 7 = -1741 \cdots$$

so that we construct the two solutions $x(t)$ and $p(t)$ from $t = t_0$ to $t = t_f$.

If the values $x_f$ and $p_f$ are known instead of $x_0$ and $p_0$, a similar procedure can be used by computing in the reverse $t$-direction.

Unfortunately, in the variational methods, the terminal values are known at separate ends, i.e., $x(t_0)$ at the $t_0$-side, while $p(t_f)$ at the $t_f$-side. Since in computing $x(t_1)$ and $p(t_1)$, we need both $x(t_0)$ and $p(t_0)$, the procedure cannot even be started.

One method for solving the $2PBVP$ is to use a trial-and-error process on the basis of the numerical integration method. The values of $x_0$ and $p_f$ are assumed known. We start by choosing an arbitrary $p(t_0) = p_0'$ (Fig. 4-6). Using $x_0$ and $p_0'$ we compute successively $x_1, p_1', x_2, p_2', \ldots$ until the value of $p_f'$ is

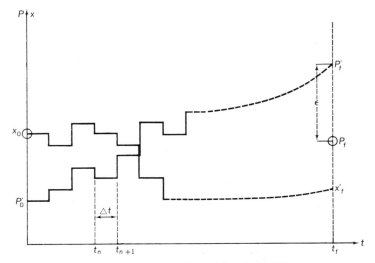

***Fig. 4-6***   Trial process for solving the 2*PBVP*.

obtained. We compare the calculated $p_f'$ to the given $p_f$. If they are not identical, we restart the process by using another value $p_0^*$, etc., till the calculated $p_f^*$ becomes identical to the known $p_f$.

Such a trial process suffers from a great disadvantage. There is no formal way of using the last results to improve the new guess. The process has no inherent convergence property and can last very long even when a fast computer is used. The situation is clearly aggravated if the set of differential equations contains more than two variables. Methods for overcoming the 2 *PBVP* will be discussed in Chapter 5.

## 4.3   METHOD OF DYNAMIC PROGRAMMING

The method of "dynamic programming," contains a principle, called the principle of optimality, and an optimization procedure, based on a recurrent function derived from the principle of optimality. In this section, we shall discuss successively [G-18; G-24; G-25; G-26; G-36; G-56]:

1. Principle of optimality.
2. Optimization procedure by the dynamic programming.
3. Problem of dimensionality.
4. Effects of the constraints.
5. Successive optimality.

### 4.3.1   Principle of Optimality—Concepts

The maximum principle is a form of mathematical reasoning, and its exposition necessitates a mathematical background. The principle of optimality is much more intuitive or conceptual; it has some mathematical formulation, but it can be explained by readily accessible and plausible reasoning. We shall do this by using an example. Before starting, we must make the following comments:

1. The basic concept underlying the principle of optimality is not difficult to understand.

2. The vocabulary, or the formalism of its application, which is also not difficult to understand, is, however, complicated. The reason is that a number of meanings are covered by some apparently simple subscripts and symbols. The reader is asked to pay special attention to this problem.

Let us consider a problem of highway construction. (Fig. 4-7). The highway must go from the $A$ town to the $Z$ town (or vice-versa), in six steps. At each step, several towns or points can be chosen, and a specific cost of construction (known value) is related to each choice. The question is how the itinerary must be chosen so that the total construction cost is minimized.[1] Clearly, for problems with a small number of steps, a simple enumeration suffices to solve the problem. However, the number of combinations rapidly becomes very high if there are many steps. A systematical procedure, such as the dynamic programming procedure, must be employed. This procedure will be explained here first without mathematics.

1. We consider how to reach $Z$ town in two steps [4-17, 4-18]. We have to answer two questions before calculating.
   a. From which of the towns $O, P, Q, R$ are we supposed to start? This question of initial conditions belong to the formulation problem.
   b. Once the starting town is fixed, through which of the towns $S, T, U$ do we have to pass? This question of decision, or choice of control variable, belongs to the extremalizing problem. Note that the true control variable is not the town, but the path (including its cost), because to one town, several paths (arriving and leaving) are related. Here the decision is made on one path between 0th step and 2nd step.

Since all the points $O, P, Q, R$ are intermediate points, we calculate the costs of all the possible itineraries. For the point $O$, we have 12 for the path

---

[1] A simplist understanding of the dynamic programming is: start from the end point, choose step-by-step the minimum path, and eliminate all the others. Several times, the author has made the mistake of starting by considering only one step. In our case, the minimal trajectory thus obtained would be $AIEKPSZ$ which is not the true minimum.

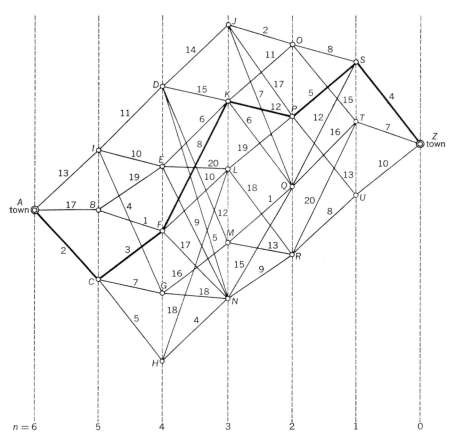

**Fig. 4-7**  A minimum-cost highway project.

*OSZ* and 22 for *OTZ*. We note by a dash (see Fig. 4-8, which summarizes the results of the calculations) the minimum 12 of these two values, and write the corresponding itinerary *OSZ* between parenthesis after *O*. For the point *P*, we have 9 for *PSZ*, and 23 for *PUZ*, the minimum itinerary between *P* and *Z* is *PSZ* with a cost of 9. Continuing the same way, we find the minimum itineraries *QSZ* (16) and *RUZ* (18). Our results of optimization up to this stage are.

  (1) If we fix the initial condition, say, *Q*, then the minimum-cost itinerary is *QSZ* with 16

  (2) If the initial condition is left to our decision, then the minimum minimorum is *PSZ* (underlined) which is the minimum-cost itinerary of the free-initial-end 2-step optimization problem.

| Two steps $n = 2$ | Three steps $n = 3$ | Four steps $n = 4$ | Five steps $n = 5$ | Six steps $n = 6$ |
|---|---|---|---|---|
| Initial points Q, P, Q, R | Initial points J, K, L, M, N | Initial points D, E, F, G, H | Initial points I, B, C | Initial points A |
| O (OSZ) | J (JOSZ) | D (DJOSZ) | I (IEKPSZ or IGMQSZ) | A (ACFKPSZ) |
| 12 — | 14 — | 28 — | 39 | 50 |
| 22 | 26 | 36 | 37 — | 47 |
|  | 23 | 29 | 37 — | 34 — |
| P (PSZ) |  | 32 |  |  |
| 9 — | K (KPSZ) |  | B (BFKPSZ) |  |
| 23 | 23 | E (EKPSZ) | 46 |  |
|  | 21 — | 27 — | 30 — |  |
| Q (QSZ) | 22 | 46 |  |  |
| 16 — |  | 36 | C (CFKPSZ) |  |
| 23 | L (LRUZ) |  | 32 — |  |
|  | 28 | F (FKPSZ) | 40 |  |
| R (RUZ) | 26 — | 29 — | 36 |  |
| 27 |  | 36 |  |  |
| 18 — | M (MQSZ) | 44 |  |  |
|  | 17 — |  |  |  |
|  | 31 | G (GMQSZ) |  |  |
|  |  | 33 — |  |  |
|  | N (NRUZ) | 45 |  |  |
|  | 31 |  |  |  |
|  | 27 — | H (HNRUZ) |  |  |
|  |  | 44 |  |  |
|  |  | 31 — |  |  |

*Fig. 4-8*

2. We consider now the three-step problems. Possible initial points are *JKLMN*. First we take *J*, with possible paths of reaching *Z*: *JOSZ*, *JOTZ*, *JPSZ*, *JPLZ*, *JQSZ*, *JQTZ*. Since we know already the minimum costs at points *O*, *P*, and *Q*, we only have to consider three possibilities *JO*, *JP*, and *JQ*. For *JOXZ* for instance, the minimum is necessarily 2 (*JO*) + 12 (*OSZ*) = 14 (*JOSZ*), *X* = *S*, because the minimum from *O* to *Z* has been determined previously. Similarly, from *J* to *Z* by *Q*, the minimum cost is 7 (*JQ*) + 16 (*QS*) = 23, and from *J* to *Z* by *P* 17 (*JP*) + 9 (*PSZ*) = 26. This reduction of six itineraries to be calculated to three is one of the basic properties of the dynamic programming. Now, between *J* and *Z*, the true minimum is *JOSZ* with a cost of 14. In a similar way, we calculate the minimum itineraries from *KLMN* to *Z*. The results are also indicated in Fig. 4-8, from which we can see that the optimum itinerary of the free-initial-end 3-step problem is *JOSZ* (underlined). Our results of optimization now are:

    (1) For calculating new minimum, (for instance *JOSZ*), we use the old minimum (*OSZ*) and the new one-step cost (*JO*).

(2) The portion $OSZ$ of the optimal itinerary between $J$ and $Z$, i.e., $JOSZ$, is itself optimal between $O$ and $Z$.

3. By proceeding in a similar way with the 4-step, 5-step, and finally the 6-step case, we recover the original problem formulation, i.e., starting from $A$ to reach $Z$. The 6-step minimum is $ACFKPSZ$, which is also the minimum of the complete problem, since when $n = 6$, one town only can be chosen. The value of the minimum cost is 34.

Based on this example, we can state a general form of the principle of optimality:

A policy is optimal if and only if, at a given period or phase, regardless of the preceding decisions made, the decisions which remain to be made constitute an optimal policy with respect to the result of the preceding decisions.

This principle can easily be verified by using the itineraries of the highway example. We shall give another simple proof by using our more familiar state-space language. We consider an optimal trajectory $ADB$ unique in the state space $(x, u)$. At a certain phase $t_1$, the result of the preceding decisions is that our state arrives at the point $D$ (Fig. 4-9). We affirm that the portion $DCB$ belonging to $ADB$ is also optimal. This affirmation (which is another way of expressing the principle of optimality) can be proved by contradiction. Supposing that $DCB$ was not optimal, then there would exist another path $DEB$ optimal. Then the path $ADEB$ would also be optimal, which contradicts our basic assumption of unicity.

The dynamic programming procedure is the exploitation of the principle of optimality resulting in a recurrent algorithm. The very general form of the

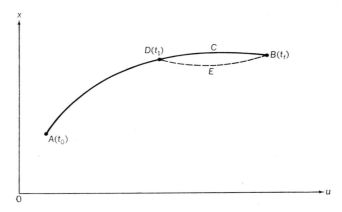

**Fig. 4-9**  Simple proof of the principle of optimality.

algorithm can be written as

$$F_{n+1} = \operatorname*{Min}_{u} (\Delta V + F_n) \qquad n = 2, \ldots, N$$

$$F_1 = \operatorname*{Min}_{u} [V(c, u)] \qquad x(0) = c \text{ given}$$

(4-60)

where $V$ is some value function $V = V(x, u)$.

$F = \operatorname{Min} V$ ($F$ is the minimum value among all possible values of $V$).

$F_n$, $F_{n+1}$ are values of $F$ at steps $n$ and $n + 1$.

$u$ is the control variable.

The recurrent function (4-60) means that the minimum value of $V$ at step $n + 1$ ($F_{n+1}$, unknown) is obtained by determining or searching the value of $u$ which minimizes the expression ($\Delta V + F_n$), in which the minimum value of $V$ at step $n(F_n)$ is known from the last search and the increase ($\Delta V$) (or modification) of the value function is a function of $u$. By using this recurrent function and starting from a known end, we can construct the whole optimal trajectory portion by portion. As it is formulated, the recurrent function applies to all optimization problems, static or dynamical. The essential requirement is the possibility of formulating the problem under the process of multistage decision process.

### 4.3.2    Mathematical Formulation of the Optimization Procedure

The formulation comprises various parts: (1) the problem, including the variables, the constraints, and the performance function ; (2) the recurrent relation, the exact form of which depends upon that of (1); (3) the intermediate notations needed notably for differentiating various intermediate minima relative to various step-lengths and initial conditions, and here again, this depends on forms of (1) and (2). In this subsection, we shall expose the formulation process by using two examples.

*Example 1.*    The first example is the one used in the preceding subsection. We define the following:

$n$    The order of the steps, $n$ is an index, the values of $n$ are from 0 to 6.

$N$    The length of the considered itinerary, $N$ is a variable; here, the values of $N$ are from 0 to 6.

$x$    The state variable, here towns $A, B, C, \ldots, Z$.

$u$    The control variable, here portions like $AI, AB, \ldots, TZ, UZ$ and their related cost.

For a better understanding, we can compare this highway terminology with our familiar state-space terminology as shown in Fig. 4-10. We have a four-dimensional problem where we have one independent variable $t$ or $n$, and

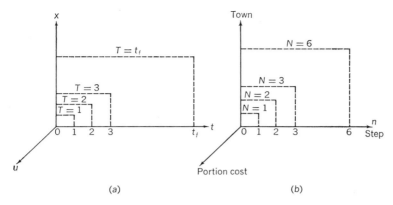

**Fig. 4-10** Comparison of terminology. (*a*) State-space terminology. (*b*) Highway terminology.

three dependent variables $x$ or town, $u$ or portion cost, $T$ or the itinerary length $N$. The particularity is that $N$ has the same value than $n$. A common source of confusion is the letter $N$ which designates here a *variable* length, while in general practice this letter indicates the ending constant of the index $n = 1, 2, \ldots, N$. Now we call

$v_0(u_0, x_1)$  Cost of the portion 0–1 (which depends upon $x_1$ and $u_0$).

$v_1(u_1, x_2)$  Cost of the portion 1–2 (which depends upon $x_2$ and $u_1$).

$v_2(u_2, x_3)$  Cost of the portion 2–3 (which depends upon $x_3$ and $u_2$).

$v_3(u_3, x_4)$  Cost of the portion 3–4 (which depends upon $x_4$ and $u_3$).

$v_4(u_4, x_5)$  Cost of the portion 4–5 (which depends upon $x_5$ and $u_4$).

$v_5(u_5, x_6)$  Cost of the portion 5–6 (which depends upon $x_6$ and $u_5$.

The total cost of the highway will depend upon these variables, and will be expressed as:

$$V(u_0, \ldots, u_5, x_1, \ldots, x_6) = v_0(u_0, x_1) + v_1(u_1, x_2) + v_2(u_2, x_3)$$
$$+ v_3(u_3, x_4) + v_4(u_4, x_5) + v_5(u_5, x_6)$$

The objective is to find the minimum value $f$ of $V$ by a suitable choice of the control variables $u_0, \ldots, u_5$.

$$\tilde{f} = \operatorname*{Min}_{\substack{u_i \\ i=0,\ldots,5}} V(u_0, \ldots, u_5, x_1, \ldots, x_6) \tag{4-61}$$

Note that we do not have an explicit system equation or constraint. The literal relation is: when two towns ($x_1$, $x_2$, for instance) are determined, the corresponding portion cost ($u$) is known automatically.

Following the sequence of Subsection 4-3.1:

1. We consider the two-step problem $N = 2$, and calculate the corresponding minimum costs designated as

$$f_2(x_2) = \text{Min} \ [v_1(u_1, x_2) + v_0(u_0, x_1)] \qquad (4\text{-}62)$$
$$\quad{}_{u_0, u_1}$$

The notation $f_2(x_2)$ means: minimum value of the two-step problem, function of the initial state $x_2$. The second member means: find the minimum of the expression $(v_1 + v_0)$ by varying $u_0$ and $u_1$. From Fig. 4-7, we can see that the possible values of $u_0$ are 4 $(SZ)$, 7 $(TZ)$, 10 $(UZ)$, those of $u_1$ are 8 $(OS)$, 5 $(PS)$, 12 $(QS)$, 15 $(OT)$, 16 $(QT)$, 20 $(RT)$, 13 $(PU)$, 8 $(RU)$. From Fig. 4-8, we see that the values of $f_2(x_2)$ are:

$$f_2(O) = 12 \qquad f_2(P) = 9 \qquad f_2(Q) = 16 \qquad f_2(R) = 18$$

from which we deduce the minimal minimum:

$$\bar{f}_2 = 9 = \text{Min} \ f_2(x_2)$$
$$\quad{}_{x_2}$$

Note that $\bar{f}_2$ is no longer a function of the initial conditions $x_2$, since it has become independent of them. Note also that Eq. (4-62) can be expressed as a recurrent form by replacing $v_0(u_0, x_1)$. Here we have $v_0(u_0, x_1) = f_1(u_0, x_1)$ because the choice of $u_1$ will fix automatically $x_1$ and $u_0$.

2. Consider now the three-step problem, $N = 3$. The corresponding minimum costs are calculated by the relation:

$$f_3(x_3) = \text{Min} \ [v_2(u_2, x_3) + f_2(u_2)]$$
$$\quad{}_{u_2}$$

For each defined initial state $x_3$, the decision $u_2$ determines automatically the state $x_2$, and also the value of $f_2(x_2)$ which has been computed previously. For example,

$$f_3(J) = \text{Min} \begin{bmatrix} 2 + f_2(O) \\ 17 + f_2(P) \\ 7 + f_2(Q) \end{bmatrix} = \begin{bmatrix} 2 + 12 \\ 17 + 9 \\ 7 + 16 \end{bmatrix} = \begin{bmatrix} 14 \\ 26 \\ 23 \end{bmatrix} = 14$$

The others are

$$f_3(K) = 21 \qquad f_3(L) = 26 \qquad f_3(M) = 17 \qquad f_3(N) = 27$$

and

$$\bar{f}_3 = \text{Min} \ f_3(x_3) = f_3(J) = 14$$
$$\quad{}_{x_3}$$

3. The minimum costs of the four-step case can be calculated by:

$$f_4(x_4) = \text{Min} \ [v(u_3, x_4) + f_3(u_3)] \qquad (4\text{-}63)$$
$$\quad{}_{u_3}$$

$$\bar{f}_4 = \text{Min} \ f_4(x_4) \qquad (4\text{-}64)$$
$$\quad{}_{x_4}$$

The general expressions of the minimum costs of the $N$-step problem is therefore:

$$f_N(x_n) = \underset{u_{n-1}}{\text{Min}} \, [v(u_{n-1}, x_n) + f_{N-1}(u_{n-1})] \qquad (4\text{-}65)$$

$$\bar{f}_N = \underset{x_n}{\text{Min}} \, f_N(x_n) \qquad (4\text{-}66)$$

Note in these expressions that

(1) $f_N(x_n)$ is searched by varying only one unknown parameter, $u_{n-1}$; the values of $x_n$, and $f_{N-1}$ are known.

(2) Distinct subscripts are used for $f$ on one hand, and for $x$ and $u$ on the other; $n$ designates the rank in the step-dimension (or $t$-dimension in dynamical problems), while $N$ designates the length of the considered itinerary, (or the total duration in dynamical problems), although they have the same numerical values at a given step.

(3) The minimal minimum $\bar{f}$ conceptually is a function of both the length of the itinerary $N$ and the initial state $x_n$: $\bar{f} = \bar{f}(N, x_n)$. In other words, the fixed-initial-end problem is solved by considering the whole category of variable-initial-end problems. This feature, called "imbedding process," is one of the essential characteristics of the dynamic programming. Furthermore, this imbedding process is extended to the length of the itinerary $N$: a short-itinerary problem $(N = k, k = 1, 2, \ldots, N - 1)$ is solved by considering a longer-itinerary problem. (Fig. 4-11).

(4) Since the problem is reversible (the costs are the same in either direction; one can start the computations either from $A$ town or from $Z$ town) the initial state can be designated as $x_0 = c$: $\bar{f} = \bar{f}(N, c)$, this is the common designation found in the literature.

***Example 2.*** The second example chosen belongs to the control field [G-31]. For ease of comparison, we shall expose the general formulation procedure parallel with the example. One of the differences in this example with the preceding one is the explicit existence of the system equation

$$\varphi_i = \dot{x}_i - g_i(x_i, u_j, t) = 0 \qquad (4\text{-}67)$$

To be handled by the dynamic programming, this equation is approximated by the following discrete form (considered as a multistage process):

$$x_i(t + \Delta t) = x_i(t) + \Delta t \cdot g_i[x_i(t), u_j(t), t] \qquad (4\text{-}68)$$

Letting

$$t_f - t_0 = N \, \Delta t = N\tau, \qquad n = 1, 2, \ldots, N$$

this equation can be written as

$$x_{i(n+1)} = x_{in} + \tau \cdot g_i(x_{in}, u_{jn}, n\tau) \qquad (4\text{-}69)$$

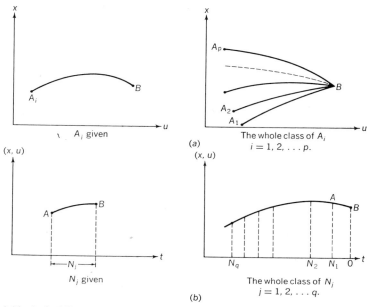

**Fig. 4-11** Imbedding process. (*a*) Initial-condition imbedding. (*b*) Itinerary-length imbedding.

We note that in the preceding recurrent equations [Eq. (4-65)] the second member contains the state variable at two different instances, i.e., $x_{n-1}$, $x_n$ because $f_{N-1}(u_{n-1})$ contains in reality $x_{n-1}$. With the help of Eq. (4-68), an expression in $(x_{n-1})$ can be converted into an expression in $(x_n)$ and $(u_{n-1})$. When $x_n$ is defined, the minimization is simply effectuated by an ordinary differentiation with respect to one single variable $u_{n-1}$. The system equation (4-67) is no longer a constraint as it is in the calculus of variations, but an aid. This point is another of the essential features of the dynamic programming.

| *Procedure* | *Example* |
|---|---|
| (a) Given | (a) Given |

$$x_{n+1} = g(x_n, u_n)$$

the state equation,

$$J_n(x_n, u_n)$$

the performance index,

$$t_f - t_0 = N \cdot \Delta t = N \cdot \tau$$

the duration,

$$x(t_0) = x_0$$

the state initial value, and

$$x(t_f) = x(N\tau) = 0$$

$$x_{n+1} = a \cdot x_n + Kbu_n \qquad (a)$$

$$a = \exp\left(-\frac{\tau}{T}\right)$$

$$b = T(1 - a)$$

$$T = \text{time constant}$$

$$\tau = \Delta t$$

$$K = \text{system parameter}$$

$$x(t_f) = x(N \Delta t) = 0$$

for stability requirements.

### Procedure

(b) Find $u_n^*$ $(n = 1, \ldots, N)$ or $u^*(t)$ which minimizes

$$\sum_{n=0}^{N} J_n(u_n, x_n)\tau$$

(c) Define the optimum performance index $f_N(x_0)$ function of $x_0$ and covering $N$ steps as:

$$f_N(x_0) = \min_{u_n} \sum_{n=0}^{N} J_n(x_n, u_n)\tau$$

The principle of optimality gives the recurrent algorithm for the minimum search:

$$f_N(x_0) = \min_{u_n} [J_0(x_0, u_0) + \sum_{n=1}^{N} J_n(x_n, u_n)]\tau$$

(d) Choice of the initial control according to the total number $N$ of steps. Search procedure.

(d1) *One-step problem*

0 $\Delta t$        1 $\Delta t$

at $N = 0$, decision on $u_0$ to be taken $x(t_0) = x_0$ gives $x(t_i) = x_1 = 0$ nonexisting or will be the result

$$f_1(x_0) = \min_{u_0} J_0(x_0, u_0)$$

Since $x_1 = 0$, $g(x_0, u_0) = 0$ gives $u_0^*$ $(N = 1)$

(d2) *Two-step problem*

at $N = 0$, decision on $u_0$ to be taken $x(t_0) = x_0$ given $x(t_1) = x_1$ exist $x(t_2) = x_2$ nonexisting or will be the final result

$$f_2(x_0) = \min [J_0(x_0, u_0) + J_1(x_1, u_1)]$$

### Example

(b) Find the sequence $u_0, \ldots, u_N$ minimizing

$$\sum_{n=0}^{N} (x_n^2 + \lambda u_n^2)\tau, \quad \lambda = \text{weighting factor}$$

(c) Define:

$$f_N(x_0) = \min_{u_n} \sum_{0}^{N} (x_n^2 + \lambda u_n^2)\tau \quad \text{(b)}$$

The recurrent algorithm for the minimum search is:

$$f_N(x_0) = \min_{u_n} [(x_0^2 + \lambda u_0^2) + \sum_{1}^{N} (x_n^2 + \lambda u_n^2)]\tau \quad \text{(c)}$$

$N = 1$

$$f_1(x_0) = \min_{u_0} (x_0^2 + \lambda u_0^2)\tau \quad \text{(d)}$$

$$g(x_0, u_0) = 0 \Rightarrow ax_0 + bku_0 = 0$$

and

$$u_0^*(N = 1) = -\frac{ax_0}{kb} \quad \text{(e)}$$

then

$$f_1(x_0) = \left(1 + \frac{a^2\lambda}{k^2 b^2}\right) x_0^2 \tau \quad \text{(f)}$$

$N = 2$

$$f_2(x_0) = \min_{u_0} [(x_0^2 + \lambda u_0^2) + (x_1^2 + \lambda u_1^2)]\tau$$

<table>
<tr><td>

*Procedure*

two unknowns $x_1$, $u_1$ to be determined with respect to $u_0$

$$x_1 = g(x_0, u_0)$$

$u_1$ is first determined with respect to $x_1$ by considering $(1 \Delta t - 2 \Delta t)$ as a one-step problem and $x_1$ as the initial condition.

The bracket is then only a function of $x_0$ and $u_0$, such as:

$$J(x_0, u_0)$$

</td><td>

*Example*

Considering $(1 \Delta t - 2 \Delta t)$ as a one-step problem, $u_1$ is determined by Eq. (e)

$$u_1 = -\frac{ax_1}{kb}$$

then

$$f_2(x_0) = \min_{u_0} \left[ (x_0^2 + \lambda u_0^2) + \left(1 + \frac{a^2\lambda}{b^2k^2}\right)x_1^2 \right]\tau \quad (g)$$

$x_1$ is given by Eq. (a)

$$x_1 = ax_0 + bku_0$$

then

$$f_2(x_0) = \min_{u_0} \left[ (x_0^2 + \lambda u_0^2) + \left(1 + \frac{a^2\lambda}{b^2k^2}\right)(ax_0 + bku_0)^2 \right]\tau = \min_{u_0} J(x_0, u_0) \quad (h)$$

</td></tr>
<tr><td>

Since $x_0$ is known, the optimum $u_0^*(N = 2)$ is obtained by:

$$\frac{dJ}{du_0} = 0$$

(a function extremum search problem).

</td><td>

By setting $dJ/du_0 = 0$ we have:

$$u_0^*(N = 2) = -\frac{kab[1 + (a^2\lambda/b^2k^2)]x_0}{\lambda + k^2b^2[1 + (a^2\lambda/b^2k^2)]} = qx_0 \quad (i)$$

With this value, we compute

$$x_1^* = ax_0 + bkqx_0 = (a + bkq)x_0 = rx_0$$

$$u_1^* = -\frac{ax_1^*}{kb} = -\frac{ar}{kb}x_0$$

$$f_2(x_0) = \ldots, \text{etc., by Eq. (h) or (g).}$$

</td></tr>
</table>

**(d3)** *N-step problem*

In a similar way, we compute $u_0^*$ $(N = N)$

In the course of the computation, we obtain

$U_0^*(N = N)$ and $u_1^*, u_2^*, u_3^*, \ldots, u_{N-1}^*$

$x_1^*, x_2^*, x_3^*, \ldots, x_{N-1}^*$

$f_1(x_0), f_2(x_0), f_3(x_0), \ldots, f_{N-1}(x_0), f_N(x_0)$

$$\underline{N = N}$$

Exposed as such, the algorithm for the extremum search is quite simple, but the practical procedure of using this algorithm is not so simple. The $N$-step extremum search passes necessarily by:

(a) A one-step extremum search, or more generally, by a sequence of one-step, two-step, three-step, . . . , $(N - 1)$-step searches.

(b) Recomputations of sequences $(u_0^* u_1^*)$, $(u_0^* u_1^* u_2^*)$, $(u_0^* u_1^* u_2^* u_3^*)$, . . . , $(u_0^*, u_1^*, u_2^*, \ldots, u_{N-1}^*)$ for each of the new search: two-step search, three-step search, . . . , until $N$-step search. For a given value of $N$, these sequences are even different if the initial condition is different.

The initial functional $J(x, u)$ minimization problem is reduced to a sequence of functions $J(u_0)$ minimization problems, ordinary methods (derivative Fibonacci, Newton-Raphson) can then be used. In the practical point of view, this situation will help to a great extent in handling problems where $J(x, u)$ and $g(x, u)$ have nonlinearities, boundary conditions, and limitations, and $dJ/du$ is either discontinuous or laborious to find.

### 4.3.3   Effects of the Constraints

The optimization procedure by dynamic programming, as previously given, is effective for a normal problem without the kinds of constraints discussed in the section on variational calculus. Here we shall examine the influence of these constraints on the dynamic programming.

(a) TERMINAL CONSTRAINTS.   The final-time constraints or conditions of the type $\psi_1[x(t_f), u(t_f), t_f] = 0$ will help the determination of the control vector $u$ at the 2-step starting optimizing stage. For example, in the highway problem (Fig. 4-7), for a chosen $u_1$ (e.g. $OS$), or $x_1(S)$, the final step constraint (only one way to go from $S$ to $Z$) automatically fixes the value of $u_0(4)$.

The initial-time constraints or conditions of the type $\psi_2[x(t_0), t_0] = 0$ restrains freedom in the choice of the initial state in the imbedding process (Fig. 4-11a). Instead of being arbitrary, the points $A_1, A_2, \ldots, A_p$ must be related to each other by $\psi_2$. For instance, if $\psi_2 = x_1 + x_2 - 4 = 0$ we can choose $x_1$ arbitrarily, but $x_2$ must be equal to $4 - x_1$.

(b) BOUNDARY LIMITATIONS ON $u$ AND $x$.   Boundary limitations of the type $c \leq u \leq d$ are taken into account automatically in the $v$- or $J$-minimization process; this point is similar to one in the maximum principle with respect to magnitude and stationarity problems. There are, however, some differences, as shown in Fig. 4-12. In the maximum principle procedure, the derivation $dJ/du$ is operated with respect to an entire function of time: $u(t)$. The resulting optimal control $u^*(x, p, t)$ is, furthermore, a function of two unknown variables $x(t)$ and $p(t)$, the solutions of which can only be obtained later by solving the extremalizing differential equations. In the

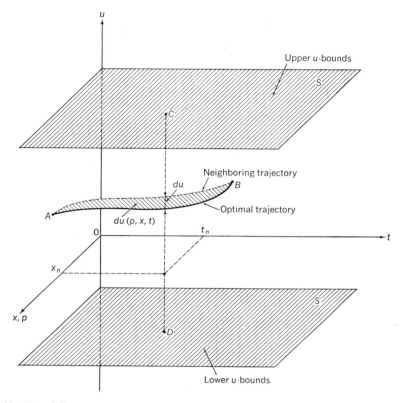

**Fig. 4-12**  Two different ways of operating $J$-minimization. Maximum principle: $du(x, p, t)$ between two surfaces $S$ and $S'$. Dynamic programming: $du$ on line $CD$.

dynamic-programming procedure, the derivation $dJ/du$ is operated at a given time $t_n$; the resulting optimal value $u^*$ might be a function of $x_n$ but $x_n$ is known at the step.

Boundary limitations of the type $a < x_n < b$ are taken into account in two places:

1. When effectuating $\bar{f}_N = \underset{x_n}{\text{Min}} f_N(x_n)$ [Eq. (4-66)].

2. When converting an expression like [see Eq. (4-65)]

$$v_N = v(u_{n-1}, x_n) + f_{N-1}(x_{n-1}, u_{n-1})$$

which is a function of the arguments $u_{n-1}, x_n, x_{n-1}$ into an expression

$$v_N = v_N(u_{n-1}, x_n)$$

which is a function of the arguments $u_{n-1}$, $x_n$ only by using the discrete state equation (4-69), one takes care not to violate the constraints.

(c) INEQUALITY CONSTRAINTS. The type $S(x, t) \leq 0$ imposes the nonviolation condition to the initial-condition choice.

The type $C(x, u, t) \leq 0$

1. Facilitates the determination of $u_0$ as in (a).
2. Imposes a nonviolation condition to the initial-condition choice.
3. Has the same effects as in (b).

(d) NONLINEARITIES AND DISCONTINUITIES. We shall discuss non-linearities and discontinuities in the performance function $J(x, u, t)$ and the system function $g(x, u, t)$. They are not exactly constraints. However, in the variational calculus, they form a source of analytical difficulties when the differentiations are concerned. Such difficulties do not exist in the dynamic programming. Instead of differentiations, one can effectuate a direct computation of $J$ or $g$, by systematically varying $u$ and $x$ in their permissible zone and according to their nonlinear and discontinuous behavior. The $J$-minimization is then reduced to a comparison of the computed values.

Some general conclusions may be deduced from the preceding discussions. The variational calculus is essentially an analytical procedure. As an advantage, an analytical solution, in principle, results from the optimization process. The same solution is valid for all special cases. As a disadvantage, analytical difficulties arise in the presence of particular constraints, non-linearities, and discontinuities. The dynamic programming is practically a direct-computation procedure. Another disadvantage is that the resulted solution is numerical and is valid for the set of particular parameters considered. However, since the direct computation is used, particular constraints, nonlinearities, and discontinuities cause no special difficulties.

### 4.3.4  Application Obstacle—The Dimensionality Problem

The method of dynamic programming has been applied for solving problems in a number of fields [G-26; G-36; 4-19 to 4-22]. The application is, however, limited by the high requirement of computer memory, especially in multidimensional problems. This problem, now classical and called dimensionality problem, is inherent to the dynamic-programming procedure. The main concern is the storage necessity of the intermediate results $\{f_N(c)\}$, $\{u_N(c)\}$, $N$ being the length of the itinerary, and $c$ the initial state variable. In Subsection 4.3.2, the initial state variable has been designated $x_n$ for a better understanding of the context. Here we come back to the letter $c$ commonly used in the literature. But the reader must remember that $c$ is indexed, and changes value from step to step. To explain the dimensionality problem, more details on the computing process are needed.

Let the problem be minimizing

$$J(u) = \int_0^I v(x, u)\, dt \tag{4-70}$$

subject to:

$$\dot{x} = g(x, u) \qquad x(0) = c \tag{4-71}$$

By using the discrete formulations

$$J(u_n) = \sum_1^{N_c} v(x_n, u_n)\tau \tag{4-72}$$

$$x_{n+1} = x_n + \tau g(x_n, u_n)$$

we have seen that the minimum functions can be found by the recurrence relation of the form

$$f_1(c) = \operatorname*{Min}_u\, [v(c, u)\tau] \tag{4-73}$$

$$f_N(c) = \operatorname*{Min}_u\, [v(c, u)\tau + f_{N-1}(c + g(c, u)\tau)] \tag{4-74}$$

$$N = 2, 3, \ldots$$

Concrete results desired are the numerical values of the functions $f_N(c)$, $N = 1, 2, \ldots$, and the corresponding sequences of the optimal control $u^* = \{u_1^*, u_2^*, \ldots, u_n^*\}$, $n = 0, 1, \ldots, N$.

In an open-initial-end problem, the value of the initial state is determined by choice. Thinking strictly in numerical lines in terms of computer operations, we first define the grid of $c$-values and the set of allowable $u$-values. We start with Eq. (4-73), and choose for instance $c = c_1$ leading to

$$f_1(c_1) = \operatorname*{Min}_u\, [v(c_1, u)\tau] \tag{4-75}$$

Since $\tau$ and the function $v$ are known, we start with $u = u^0$, calculate and store $v(c_1, u^0)\tau$. We change $u = u^1$, compute $v(c_1, u^1)\tau$, compare it with $v(c_1, u^0)\tau$, and store the smallest of the two values. In a similar manner by increasing $u$ till $u_{\max}$, we finally obtain the smallest of the $v(c_1, u)\tau$-values. The result is $f_1(c_1)$. At the same time, we obtain the value of $u$ making $v(c_1, u)\tau$ minimum; let us call it $u(c_1)$. By varying $c_1$ from $c_{\min}$ to $c_{\max}$, we obtain $f_1(c_{\min}), \ldots, f_1(c_{\max})$ or, in other words, the function $f_1(c)$. At the same time, we obtain $u(c_{\min}), \ldots, u(c_{\max})$ or, in other words, the function $u_1(c)$.

Once $f_1(c)$ has been determined, we are ready to determine $f_2(c)$ by Eq. (4-74)

$$f_2(c) = \operatorname*{Min}_u\, [v(c, u)\tau + f_1(c + g(c, u)\tau)] \tag{4-76}$$

The argument of $f_1$ is now $c' = c + g(c, u)\tau$. Since $g$ is given, for fixed $u$, $c$, and $\tau$, we can compute $c'$ easily. On the other hand, since we know the whole function (that we have stored in memory) $f_1(c)$ for any known value of the argument $c$, $f_1(c')$ is also known. The repetition of the previous operations will therefore determine the functions $f_2(c)$ and $u_2(c)$.

Proceeding in a similar way, by increasing $N$, we obtain a set of functions $\{f_k(c)\}$, $\{u_k(c)\}$ with $k = 1, \ldots, N$ which can be shown in the form of a chart (Table 4.3). This chart, which has been established for all possible values of itinerary-lengths $N$, and all possible initial-state values $c$, forms a general reservoir of solutions.

**Table 4.3**

| $c$ | 1-Step | | 2-Step | | | N-Step | |
|---|---|---|---|---|---|---|---|
| | $f_1(c)$ | $u_1(c)$ | $f_2(c)$ | $u_2(c)$ | $\cdots$ | $f_N(c)$ | $u_N(c)$ |
| $c_{\min}$ | $\cdots$ | $\cdots$ | $\cdots$ | $\cdots$ | | $\cdots$ | $\cdots$ |
| $c_1$ | $\cdots$ | $\cdots$ | $\cdots$ | $\cdots$ | | $\cdots$ | $\cdots$ |
| $c_2$ | $\cdots$ | $\cdots$ | $\cdots$ | $\cdots$ | | $\cdots$ | $\cdots$ |
| $c_3$ | $\cdots$ | $\cdots$ | $\cdots$ | $\cdots$ | $\cdots$ | $\cdots$ | $\cdots$ |
| $\vdots$ | $\vdots$ | $\vdots$ | $\vdots$ | $\vdots$ | | $\vdots$ | $\vdots$ |
| $c_{\max}$ | $\cdots$ | $\cdots$ | $\cdots$ | $\cdots$ | | $\cdots$ | $\cdots$ |

As numerical application, we take

$$v(x, u) = x^2 + 4xu + u^2$$

$$g(x, u) = 2x + u$$

grid $x$ or $c$      $1, 2, \ldots, 10,$      $\tau = 1$

grid $u$      $(-20), (-19), \cdots (+20)$

$$f_1(c) = \underset{u}{\text{Min}} \, (c^2 + 4cu + u^2)$$

Operating as previously, we obtain Table 4.4

$$f_2(c) = \underset{u}{\text{Min}} \, \{c^2 + 4cu + u^2 + f_1[c + g(c, u)]\}$$

$$= \underset{u}{\text{Min}} \, [c^2 + 4cu + u^2 + f_1(c + 2c + u)]$$

$$= \underset{u}{\text{Min}} \, [c^2 + 4cu + u^2 + f_1(3c + u)]$$

**Table 4.4**

| $c$ | $f_1(c)$ | $u_1(c)$ |
|---|---|---|
| 1 | $-3$ | $-2$ |
| 2 | $-12$ | $-4$ |
| 3 | $-27$ | $-6$ |
| 4 | $-48$ | $-8$ |
| 5 | $-75$ | $-10$ |
| 6 | $-108$ | $-12$ |
| 7 | $-147$ | $-14$ |
| 8 | $-192$ | $-16$ |
| 9 | $-243$ | $-18$ |
| 10 | $-300$ | $-20$ |

Compute all values of $f_1$ for all possible combinations of $(c, u)$

$$u = -1 \qquad c = 1 \qquad f_1 = -12$$
$$c = 2 \qquad f_1 = -75, \ldots, \text{until } c = 10$$
$$u = -2 \qquad c = 1 \qquad f_1 = -3$$
$$c = 2 \qquad f_1 = -48, \ldots, \text{until } c = 10$$
$$u = 20 \qquad \cdots \qquad c = 10$$

Then search for minimum value of

$$v_2 = c^2 + 4cu + u^2 + f_1$$

where $f_1$ is now a known numerical value, for all combinations of $(c, u)$, leading to those shown in Table 4.5.

**Table 4.5**

| $c$ | $f_2(c)$ | $u_2(c)$ |
|---|---|---|
| 1 | $-6$ | $-2$ |
| 2 | $-24$ | $-4$ |
| 3 | $-54$ | $-6$ |
| 4 | $-96$ | $-8$ |
| 5 | $-150$ | $-10$ |
| 6 | $-216$ | $-12$ |
| 7 | $-294$ | $-14$ |
| 8 | $-384$ | $-16$ |
| 9 | $-486$ | $-18$ |
| 10 | $-600$ | $-20$ |

Operating now successively according to the expressions

$$f_3(c) = \min_u [c^2 + 4cu + u^2 + f_2(3c + u)]$$

.

.

.

$$f_N(c) = \min_u [c^2 + 4cu + u^2 + f_{N-1}(3c + u)]$$

we finally obtain the complete results shown in Table 4.6. The exploitation of this reservoir for any specific imbedded problem with given numerical values of $c$ and $N$ is as follows.

*Table 4.6*

| $c$ | $N = 1$ $f_1(c)$ | $u_1(c)$ | $N = 2$ $f_2(c)$ | $u_2(c)$ | $N = 3$ $f_3(c)$ | $u_3(c)$ | $N = 4$ $f_4(c)$ | $u_4(c)$ |
|---|---|---|---|---|---|---|---|---|
| 1 | $-3$ | $-2$ | $-6$ | $-2$ | $-9$ | $-2$ | $-12$ | $-2$ |
| 2 | $-12$ | $-4$ | $-24$ | $-4$ | $-36$ | $-4$ | $-48$ | $-4$ |
| 3 | $-27$ | $-6$ | $-54$ | $-6$ | $-81$ | $-6$ | $-108$ | $-6$ |
| 4 | $-48$ | $-8$ | $-96$ | $-8$ | $-144$ | $-8$ | $-192$ | $-8$ |
| 5 | $-75$ | $-10$ | $-150$ | $-10$ | $-225$ | $-10$ | $-300$ | $-10$ |
| 6 | $-108$ | $-12$ | $-216$ | $-12$ | $-324$ | $-12$ | $-432$ | $-12$ |
| 7 | $-147$ | $-14$ | $-294$ | $-14$ | $-441$ | $-14$ | $-588$ | $-14$ |
| 8 | $-192$ | $-16$ | $-384$ | $-16$ | $-576$ | $-16$ | $-768$ | $-16$ |
| 9 | $-243$ | $-18$ | $-486$ | $-18$ | $-729$ | $-18$ | $-972$ | $-18$ |
| 10 | $-300$ | $-20$ | $-600$ | $-20$ | $-900$ | $-20$ | $-1200$ | $-20$ |

1. We start an $N$-stage process in state $c$. Referring to Table 4.6 above: we see that the optimal choice of $u$ is given by the value $u_N(c)$. Consequently, the new $c$-value will be given by

$$c_1 = c + g[c, u_N(c)]$$

*Example:*  $N = 4 \qquad c = 3$

then

$$u_4(c) = -6 \qquad c_1 = c + 2c + u(c) = 3$$

2. Having made the initial choice, we face an $(N - 1)$-stage process in state $c_1$. Referring to Table 4-6, we see that the optimal choice of $u$ at the second stage is given by $u_{N-1}(c_1)$. The new state is then given by

$$c_2 = c_1 + g[c_1, u_{N-1}(c_1)]\tau$$

In our example,

$$u_3(c_1) = -6 \qquad c_2 = c_1 + 2c_1 + u_3(c_1) = 3$$

3. Continuing in this way, we obtain the succession of states $\{c_i\}$, and the succession of $u$-values for the optimal control. In our example, we have:

State: 3   3   3   3

Control: $-6$   $-6$   $-6$   $-6$

$$f_N(c) = f_4(c) = -108$$

Check: $f_4(c) = v_1 + v_2 + v_3 + v_4$

$$= 4(9 - 72 + 36) = -108$$

We have seen in the preceding computing process that it was necessary:

1. To have a complete table of $f_{N-1}(c)$ for consultation in order to find the numerical value of $f_{N-1}$ when $c$ is given at the $f_N$-minimization step.

2. To know the function $g$ for computing $c_N$ at each stage.

3. To store all the values of $u_k(c)$ and $f_k(c)$, $k = 1, 2, \ldots, N$, at the end of the compilation of the reservoir for later exploitation.

Let us take a problem having a six-dimensional state $x$, three-dimensional control $u$ to see what happens. The computation of $c_n$ causes no special problem: since we know the functions $g$, we only have to store a limited number of coefficients. The table of consultation $f_{N-1}(c)$ is indispensable for the iterative process. For a $c$-grid of 100 values, there are $100^6 = 10^{12}$ possible pairs of $c$-values and $f_{N-1}$-values. Even the greatest computer today does not have such a big memory capacity. The memory capacity for storing $u_k(c)$ and $f_k(c)$ will depend upon the value of $N$ and the dimensions of $c$. With $k = 50$, a $c$-grid of 100, and six state variables, we need $2 \times 6 \times 100 \times 50 = 60,000$ words, truly a negligible number with respect to $10^{12}$. Such is the dimensionality problem.

### 4.3.5  Successive Optimality

The storage of the sequences $f_k(c)$ and $u_k(c)$, $k = 1, 2, \ldots, N$, as we have seen in the preceding subsection, constitutes a stringent requirement for the memory capacity of the computer. However, the possession of this information is in some cases a great advantage; this is best explained by an example.

Consider a space vehicle including its payload, rocket, optimizing computer, and sensors. Supposing that the optimal control program $u_0$ has been computed previously by using initial conditions $c_0$ at time $t_0$, resulting to an optimal trajectory $\Gamma_0(x, u_0)$ (Fig. 4-13). At a certain instant $t_1$ of the flight, because of the various perturbations, the state $x$ takes the value $c_1$ different from $x_1$: the theoretical value of $x$ on $\Gamma_0$ at $t = t_1$. This value of $c_1$ forms the initial state of reduced itinerary $(t_1 - t_f)$ (Fig. 4-14).

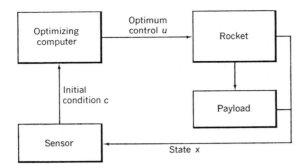

**Fig. 4-13**  Space vehicle including an optimizing computer.

We have seen previously (see the highway example of subsection 4.3.1) that with respect to the shorter itinerary $(t_1 - t_f)$, the optimal trajectory is no longer a portion of the original optimal trajectory if the state takes a value different from the one it must take. In our case, therefore, the optimal trajectory at $t = t_1$ is no longer $CB$, portion of $ACB$, but $C_1B = \Gamma_1(x, u_1)$ determined by $c_1$, resulting in the control law $u_1(t)$.

If the on-board optimizing computer is mechanized so that the sequences $u_k(c)$ and $f_k(c)$ are all stored in memory, the implementation of the new control law $u_1(t)$ becomes very simple. Based on the value of the state sensed at $t = t_1$, a simple consultation of the table $u_k(c)$, $k = 1$, $c = c_1$, immediately permits the derivation of numerical values of $u_1(t)$ and the modification of the control signal $u$ accordingly.

At subsequent times $t = t_2, \ldots, t = t_f - \Delta t$, similar operations permit us to determine trajectories $\Gamma_2(x, u_2)$, $\Gamma_3(x, u_3)$, $\ldots$ and $u_2(t)$, $u_3(t)$, $\ldots$. These trajectories $\Gamma_2$, $\Gamma_3$, $\ldots$ are all optimal with respect to the actual system state

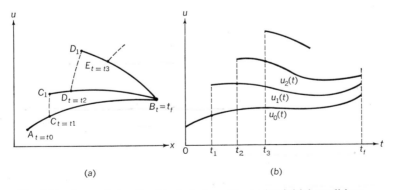

(a)    (b)

**Fig. 4-14**  Successive optimal trajectories with varying initial conditions.

and perturbations. This successive optimality is one of the important features of the dynamic-programming method.

It is interesting to see what we could do in the same circumstances if the variational method was used in the optimizing computer. There could be two alternatives.

1. We could feed back the $x$-signal and adjust the $u$-signal so that the differences $(c_1 - x_1)$, $(c_2 - x_2)$, . . . become zero; then the $u$-values would be different from the ones they must take on $\Gamma_0(x, u_0)$, and the new trajectory would not be optimal.

2. We could calculate the same new optimal trajectories $\Gamma_1$, $\Gamma_2$, . . . , and $u_1(t)$, $u_2(t)$, . . . , each time we have to solve the complete set of the optimizing differential equations. The computing time would be a question of minutes instead of question of microseconds needed for a simple consultation.

The successive optimality of the dynamic programming is, of course, valid for other optimization problems using the scheme of Fig. 4-13.

## 4.4 METHOD OF GRADIENTS

From the preceding sections, we have seen that the variational methods (classical and Pontryagin) are mainly analytical methods, yielding a set of extremalizing differential equations, valid for all problems of the same category. Concrete numerical results are only obtained when the specific set of parametric values is given and when the equations are solved. On the other hand, the method of dynamic programming is a computing method. By using the recurrent computing algorithm, and the specific set of parametric values, the numerical solutions are readily obtained. In this respect, the method of gradient is similar to the method of dynamic programming. It also consists of a recurrent computing algorithm yielding the numerical solutions at the end of the iteration. There are, however, several differences:

1. The dynamic-programming method operates by steps, each one being an element of time; once the optimal control $u^*(t_1)$ is obtained at that time $t_1$, we proceed to find the optimal control $u^*(t_1 + \Delta t)$ of the next time $t_1 + \Delta t$. The method of gradients operates on entire functions of time, starting from a wrong function $u_a(t)$, $t_0 < t < t_f$; by iteration, we obtain $u_b(t)$, $u_c(t)$, . . . , until the optimal function $u^*(t)$.

2. The recurrent algorithm in the method of gradients explicitly uses expressions such as: the Lagrange multipliers $\lambda$, partial derivatives, etc., belonging to the variational calculus, while this is not the case with the dynamic programming.

3. Where optimizing and computing algorithms are concerned, and this is just like search techniques discussed in Chapter 3, there are two extremalizations. One concerns the performance function, and the other the search or the computing speed. The method of dynamic programming has a computing speed which depends upon the dimensionality of the treated problem, while the algorithm of the method of gradients ensures the quickest search from any initial state to an extremum state (at least local).

In this section, we shall discuss successively:

(a) Basic concepts,
(b) Computing procedure, and
(c) Effects of other constraints.

### 4.4.1 Basic Concepts

To introduce the basic concepts, we shall first consider a simple dynamical optimization problem: find the control program $u(t)$, $0 < t < T$ (one single variable) to minimize the performance functional $J = J[x(T)]$ subject to the constraint:

$$\dot{x}_i = f_i(x_i, u, t) \qquad i = 1, 2, \ldots, n \tag{4-77}$$

Then we compare the eventual solution of this problem by the method of gradients with the search techniques by steepest ascent (descent) applied to the static multidimensional optimization problem (Subsection 3.3.3) as shown in Table 4.7.

The various questions arising from this comparison indicate the points to be emphasized. Answers to Questions 1 to 3 will be developed in this subsection, while the answer to Question 4 will be given in the next one. Given $x_i(0)$, $u(0)$, and $T$, the admissible trajectories have a fixed initial point $A$. However, the final point $B$ can be situated anywhere along $t = T$ (Fig. 4-15). Any variation of the control $u(t)$ will cause some movement to the trajectory because it modifies the state $x$ by virtue of the constraints $\dot{x}_i = f_i(x_i, u, t)$. The situation can be compared to that of shaking a cord from $A$. Distinct values of $x_i(T)$, and hence $J[x_i(T)]$ will result. Therefore, although there is only *one* control variable, the problem is multidimensional. This is the answer to Question 1. The preceding considerations also establish the following cause-and-effect relationship:

Variation of $u(t) \rightarrow$ variation of $x_i(t)$ through constraints

$$\dot{x}_i = f_i(x_i, u, t) \rightarrow \text{variation of } x_i(T) \rightarrow \text{variation of } J[x_i(T)].$$

This means that the study of the variation of $J[x_i(T)]$ which is the object of the optimization, must pass through that of $x_i(t)$ with respect to $u(t)$, by

*Table 4.7*

| Static Multidimensional Optimization Problem | Dynamical Constrained Multidimensional Optimization Problem |
| --- | --- |
| *Variables* | *Variables* |
| $$x_1, x_2$$ | $$x_i(t) = i = 1, 2, \ldots, n$$ |
| (All controllable variables; no distinction between state and control variables) | State variables $u(t)$: one control variable $x_i$ and $u$ functions of $t$ |
| *Performance* | $$J = J[x_i(T)]$$ |
| $$y = y(x_1, x_2)$$ | A functional; analytical form known. |
| A function; analytical form unknown, $y$-values measurable. | |
| *System equations* | *System equations* |
| None | $$\dot{x}_i = f_i(x_i, u, t)$$ |
| | called constraints |
| *Initial slope tests* | *Initial slope tests* |
| $$\Delta x_1 = x_{1a} - x_{1b} = k y_{x1}$$ | Question 1: There is only one adjustable variable: $u(t)$. Is this still a multi-dimensional problem? |
| $$\Delta x_2 = x_{2a} - x_{2b} = k y_{x2}$$ | Question 2: What is the role played by the constraints $\dot{x}_i = f_i$? |
| $$\Delta y \cong y_{x1} \Delta x_1 + y_{x2} \Delta x_2$$ | |
| $$\cong k[y_{x_1}^2 + y_{x_2}^2]$$ | |
| where: | |
| $$y_{x1} = \frac{\Delta y_1}{\Delta x_1} \quad y_{x2} = \frac{\Delta y_2}{\Delta x_2}$$ | |
| $\Delta y_1$, $\Delta y_2$, $\Delta y$ are measured values of $y$-increments when $(\Delta x_1)$, $(\Delta x_2)$, $(\Delta x_1$ and $\Delta x_2)$ increments are applied $K$ a constant | |
| *Gradient determination* | *Gradient determination* |
| $$K = \Delta y / [y_{x_1}^2 + y_{x_2}^2]$$ | Question 3: Is it anything corresponding to $K$? |
| *Search steps* | *Search steps* |
| Continue with | Question 4: What is the extremalizing procedure? |
| $$x_{1 \text{ new}} = x_{1 \text{ old}} + K y_{x1 \text{ old}}$$ | |
| $$x_{2 \text{ new}} = x_{2 \text{ old}} + K y_{x2 \text{ old}}$$ | |
| until the top where $[y_{x_1}^2 + y_{x_2}^2]$ is very small as compared to $[y_{x1a}^2 + y_{x2a}^2]$ | |

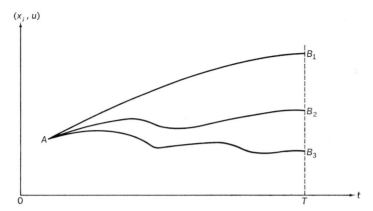

**Fig. 4-15**  Influence of variations of $u(t)$.

using the constraints $\dot{x}_i = f_i(x_i, u, t)$. This is the answer to Question 2. To study the variations of $x_i(t)$, we shall consider the equations describing small perturbations around the nominal trajectory [4-22 to 4-26]:

$$\frac{d}{dt}(\delta x_i) = \sum_{j=1}^{n} \frac{\partial f_i}{\partial x_j}\,\delta x_j + \frac{\partial f_i}{\partial u}\,\delta u \qquad (4\text{-}78)$$

where $\delta x_i\ \delta x_j\ \delta u$ are functions of time, and the partial derivatives $\partial f_i/\partial x_j$ and $\partial f_i/\partial u$ are evaluated along the nominal trajectory. These partial derivatives are known functions of time, and do not contain either $x_i$ or $u$. Therefore, Eqs. (4-78) are linear differential equations in $\delta x$ with time-varying coefficients.

The set of equations adjoint to Eqs. (4-78) is defined as

$$\frac{d\lambda_i}{dt} = -\sum_{j=1}^{n} \frac{\partial f_j}{\partial x_i}\,\lambda_j \qquad (4\text{-}79)$$

where, it will be noted, the negative transpose of the coefficient matrix $\partial f_i/\partial x_j$ occurs.[1] $\lambda_i(t)$ is the *influence function* corresponding to $x_i(t)$. We note here the similarity between the definition of $\lambda_i$ in Eq. (4-79) and those of $\lambda_i$ [Eq. (4-3), Section 4.2.1.2] and $p_i$ [Eq. (4-22), Section 4.2.2.1].[2] Although the

---

[1] This is indicated by the inversion of the subscripts $j$ and $i$.
[2] In Eq. (4-13), Section 4.2.1.2, for a Mayer problem, $F \equiv 0$, then $F^* = \Sigma\,\lambda_i\varphi_i$, so that Eq. (A) becomes

$$\frac{\partial F^*}{\partial x_i} - \frac{d}{dt}\left(\frac{\partial F^*}{\partial \dot{x}_i}\right) = \lambda_j - \frac{d}{dt}\sum \lambda_i \frac{\partial g_i}{\partial x_i} = 0$$

or

$$\sum \dot{\lambda}_i \frac{\partial g_i}{\partial x_i} = \lambda_j$$

introduction of $\lambda_i$ and $p_i$ seems arbitrary there, here the significance and the utility of $\lambda_i$ will be shown more physically.

Multiplying Eq. (4-78) by $\lambda_i$ and Eq. (4-79) by $\delta x_i$ and summing, we obtain

$$\frac{d}{dt}\sum_{i=1}^{n}\lambda_i\,\delta x_i = \sum_{i=1}^{n}\lambda_i\frac{\partial f_i}{\partial u}\,\delta u + \sum_{i=1}^{n}\sum_{j=1}^{n}\left(\lambda_i\frac{\partial f_i}{\partial x_j}\,\delta x_j - \lambda_j\frac{\partial f_j}{\partial x_i}\,\delta x_i\right) \quad (4\text{-}80)$$

The term involving the double summation is easily seen to vanish by interchanging indices on one set of terms. Obviously the definition of Eq. (4-79) was contrived to make this happen. Integrating Eq. (4-80) over the interval of the trajectory, $0 \leq t \leq T$, we obtain:

$$\left[\sum_{i=1}^{n}\lambda_i\,\delta x_i\right]_0^T = \int_0^T \lambda_u(t)\,\delta u(t)\,dt$$

where
$$\quad (4\text{-}81)$$

$$\lambda_u = \sum_{i=1}^{n}\lambda_i\frac{\partial f_i}{\partial u}$$

The functions $\lambda_i(t)$ are determined by Eq. (4-79) except the terminal conditions at time $t = 0$ or $t = T$, so we can let[1]

$$\lambda_i(T) = [\partial J/\partial x_i]_{t=T} \quad (4\text{-}82)$$

Since

$$\delta J = \frac{\partial J}{\partial x_i}\,\delta x_i \quad (4\text{-}83)$$

Clearly

$$[\delta J]_{t=T} = \left[\sum_{i=1}^{n}\lambda_i\,\delta x_i\right]_{t=T} \quad (4\text{-}84)$$

So that Eq. (4-81) becomes

$$\delta J = \int_0^T \lambda_u(t)\,\delta u(t)\,dt + \left[\sum_{i=1}^{n}\lambda_i\,\delta x_i\right]_{t=0} \quad (4\text{-}85)$$

Since $\delta J$ means the variation of the performance functional, Eq. (4-85) supplies a more physical interpretation of $\lambda_u$ and $\lambda_i$:

---

so that

$$\dot{\lambda}_i = -\sum_{j=1}^{n}\frac{\partial g_j}{\partial x_i}\lambda_j$$

(the Jacobian matrix transposed).

From Eqs. (4-21) and (4-22), Section 4.2.2.1

$$\dot{p}_i = -\sum_{j=1}^{n+1}\frac{\partial g_j}{\partial x_i}p_j$$

[1] See the similarity between this procedure and the one used for determining $p_i(T) = -c_i$ in the Pontryagin's formulation (Section 4.2.2.1).

1. $\lambda_u$ can be called the *control influence function* which tells how small changes in the control function, $\delta u(t)$ will affect $J$. Actually, $\lambda_u(t)$ plays the role of the classical weighting function of the basic feedback control theory, it tells what a unit impulse function (Dirac function) in $\delta u$ at time $t$ would do to $J$. The value of $\lambda_u$ is known [second of the Eq. (4-81)] if those of $\lambda_i$ and $\partial f/\partial u$ are known.

2. $\lambda_i(t_0)$ the values of $\lambda_i(t)$ at time $t = 0$ can be called initial-state influence functions; they tell how a unit change in $\delta x_i(t_0)$ would affect $J$. In our case, we have assumed $x_i(t_0)$ given, then $\delta x_i(t_0) = 0$ and $[\sum \lambda_i \, \delta x_i]_{t=t_0} = 0$.

3. We see more clearly now that $\lambda_i(t)$ for $0 \leq t \leq T$ can be called the state *influence functions*, or the co-state functions, or the co-variant functions; they tell how a unit change in the state, $\delta x_i(t)$, $0 \leq t \leq T$ would affect the performance $J$. The values of $\lambda_i(t)$ are determined by the differential equations (4-79) with integration constants undetermined. In our case, Eqs. (4-82) supply necessary conditions for determining these integration constants.

According to Eq. (4-85), the greatest change $\delta J$ in $J$ for a given value of $\int_0^T (\delta u)^2 \, dt$ is obtained when [4-29] $\delta u = K\lambda_u(t)$ where $K = $ constant (4-86). This is the "steepest descent" (or ascent) direction to the minimum (or maximum) of $J$. Substituting Eq. (4-86) into Eq. (4-85), we obtain

$$\delta J = K \int_0^T [\lambda_u(t)]^2 \, dt + \left[ \sum_{i=1}^n \lambda_i \, \delta x_i \right]_{t=0} \tag{4-87}$$

We then ask for some small change of $J$ in the desired direction—i.e., a decrease or an increase. Typically, we try $\delta J$ between 5 and 10 percent of $J$, initially, by manipulating $u(t)$. This determines $K$, since, from Eq. (4-87)

$$K = \delta J \Big/ \int_0^T [\lambda_u(t)]^2 \, dt \tag{4-88}$$

with $0 < |\delta J/J| < 0.1$, the values of $\delta J$ and $\lambda_u(t)$ being calculated from their analytical expressions. Obviously, the coefficient $K$ represents the constrained gradient (the greatest change in $\delta J[x_i(I)]$ for a given $u(t)$ satisfying at the same time the constraints $\dot{x}_i = f_i(x_i, u, t)$. This is the answer to Question 3.

The new $u(t)$ program for the next iterative step is

$$u_{\text{new}}(t) = u_{\text{old}}(t) + K\lambda_u(t) \tag{4-89}$$

This answers, in principle, Question 4. Details will be given in the net subsection.

Assuming now that in our special case, the initial point $A$ is not fixed, then the influence of $\delta x_i(0)$ can be determined separately. Generally, the changes in $J$ that are possible by manipulating initial conditions are much less than those that are possible by manipulating $u(t)$. Therefore, changes in $J$ between 0 and 1 percent might be used to determine the appropriate $\delta x_i(t_0)$.

$$\delta x_i(t_0) = \frac{\delta J}{\lambda_i(t_0)} \quad \text{with} \quad 0 < \left|\frac{\delta J}{J}\right| < 0.01 \tag{4-90}$$

the values of $\delta J$ and $\lambda_i(t_0)$ being calculated from their respective analytical expressions. Calling

$$K_0 = \frac{\Delta J}{\sum\limits_{1}^{n} \lambda_i^2(t_0)} \tag{4-91}$$

The new initial conditions for the next iterative step are

$$x_i(t_0)_{\text{new}} = x_i(t_0)_{\text{old}} + K_0\lambda_i(t_0) \tag{4-92}$$

Note that in this special case, $x_i(t_0)$ are considered as manipulated variables; but they are independent from the control variables $u(t)$.

Apart from the answers to the preceding questions, other interesting conclusions may be drawn from the preceding discussions.

1. The Lagrange-multiplier type functions $\lambda(t)$, issued from the constraints, influence the effects of $\delta u$ on $\delta J$. These functions are calculated by using partial derivatives of the type $\partial f_i / \partial x_j$ (the Jacobian).

2. In the variational calculus, $\lambda(t)$ is unknown as $x(t)$ and $u(t)$, their values are obtained by simultaneous solution of the extremalizing equations. These values are immediately optimal. In the method of gradients, the values of $\lambda(t)$, $u(t)$, $x(t)$ are nonoptimal until the extremum is reached by successive search [4-27, 4-28].

### 4.4.2 The Computing Procedure

To illustrate the computing procedure briefly sketched in the preceding section, we use an application example: total heating minimization of a hypersonic vehicle.

SYSTEM EQUATIONS.

$$V' = -(D/mV) - [(g \sin \gamma)/V] \tag{a}$$
$$\gamma' = [\cos \gamma/(R + h)] + (L/mV^2) - [(g \cos \gamma)/V^2] \tag{b}$$
$$h' = \sin \gamma \tag{c}$$
$$x' = \cos \gamma/(1 + (h/R)) \tag{d}$$
$$q' = C\rho^{\frac{1}{2}}V^2 \tag{e}$$
$$t' = 1/V \tag{f}$$

Independent variable: $s$, distance along the flight path.
State variables: $V$, $\gamma$, $h$, $x$, $q$, $t$.
Control variable: $\alpha(x)$.

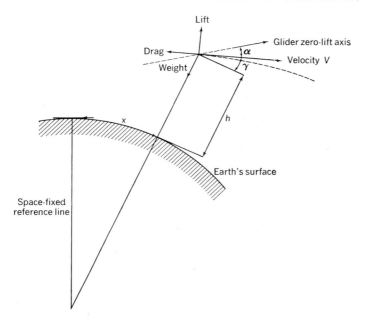

*Fig. 4-16*  Planar motion of a hypersonic vehicle on the earth's atmosphere.

Where (Fig. 4-16):

$x$ = surface range in nautical miles.

$g = g_0[R/(R + h)]^2$.

$R = 3,440$ m; $g_0 = 32.2$ ft sec$^{-2}$.

$p = C_D(\tfrac{1}{2})\rho V^2 S$ = drag force on vehicle ($S$ = reference area).

$L = C_L(\tfrac{1}{2})\rho V^2 S$ = lift force on vehicle.

$\rho = \rho(h)$ density of atmosphere (*ARDC* standard atmosphere).

$C_D^* = C_{D0} + C_{DL}|\sin^3 \alpha|$ (with $C_{D0}$, $C_{DL}$ constants) = drag coefficient.

$C_L^* = C_{L0} \sin \alpha \cos \alpha |\sin \alpha|$ (with $C_{L0}$ a constant) = lift coefficient.

$\alpha$ = angle of attack of vehicle from zero-lift direction.

$(\ )' = d(\ )/ds$, $s$ = distance along flight path.

$q$ = heat absorbed per unit area at the stagnation point.

$C = 1.98 \times 10^{-8}$ BtU slug$^{-\frac{1}{2}}$ sec$^2$ ft$^{-\frac{7}{2}}$ for a nose radius of 1 ft.

$t$ = time.

$C_{DL} = 1.82$, $C_{D0} = 0.042$, $C_{L0} = 1.46$.

$(C_D)$ max = 1.502, $(L/D)$ max = 2.00 at $\alpha = 20.5°$.

$(C_L)$ max = 0.702 at $\alpha = 54.7°$.

*Performance function.* Total heating $\displaystyle\int_0^T q\, dt$ to be minimized.

ADJOINT EQUATIONS [corresponding to Eq. (4-79)].

$$\lambda_w' + [(2g \sin \gamma)/V^2]\lambda_w + [(2g \cos \gamma)/V^2\lambda_\gamma + 2C\rho^{\frac{1}{2}} V^2\lambda_q - (1/V)\lambda_t = 0 \quad (g)$$

$$\lambda_\gamma' - [(g \cos \gamma)/V^2]\lambda_w + \sin \gamma\{(g/V^2 - [1/(R+h)]\}\lambda_\gamma$$
$$- \{\sin \gamma/[1 + (h/R)]\}\lambda_x + (\cos \gamma)\lambda_h = 0 \quad (h)$$

$$\lambda_h' + \left[\frac{2g \sin \gamma}{V^2(R+h)} - \frac{\cos \gamma}{(R+h)^2} + \frac{L}{mV^2} \frac{1}{\rho} \frac{d\rho}{dh}\right]\lambda_\gamma \quad (i)$$

$$- \frac{\cos \gamma}{R[1 + (h/R)]^2}\lambda_x + \tfrac{1}{2}C\rho^{\frac{1}{2}}V^2 \frac{1}{\rho}\frac{d\rho}{dh}\lambda_q = 0 \quad (j)$$

$$\lambda_x' = 0 \quad (k)$$

$$\lambda_q' = 0 \quad (m)$$

$$\lambda_t' = 0 \quad (n)$$

where

$$\omega = \log (V/V_0) \quad (V_0 \text{ is some reference velocity})$$

$$\lambda_w = V\lambda_v$$

In state-space notation, this is a first-order differential equation of the form

$$\dot{x} = A(t)x \qquad x: \text{an } n\text{-vector}$$

with variable coefficients $A(t)$. $A(t)$ is an $n \times n$ matrix, the Jacobian $\partial f_i/\partial x_j$ of Eq. (4-79). It is well known that the condition for the existence of a solution is that the matrix $A(t)$ is nonsingular. In our case, the system is solvable, leading to the following expression for the influence function relating small changes in $\alpha$ to change in the performance function [see Eq. (4-81), second equation]:

$$\lambda_\alpha = (\rho S/2m)[-(dC_D/d\alpha)\lambda_\omega + (dC_L/d\alpha)\lambda_\gamma] \quad (o)$$

ADDITIONAL CONSTRAINT.   Acceleration constraint for a human pilot is introduced. The acceleration expression is

$$a = (1/m)\sqrt{D^2 + L^2} = (\rho V^2 S/2m)\sqrt{C_D^2 + C_L^2} \quad (p)$$

The tolerance rate function introduced

$$d\psi/dt = 1/[\tau(a)] \quad (q)$$

means that an acceleration $a$ can be tolerated for $\tau(a)$ sec. The endurance limit requires that

$$\psi_f = \int_t^T \frac{dt}{\tau[a(t)]} \leq 1 \quad (r)$$

where

$$\psi_f = \psi[T]$$

The introduction of this constraint adds one more dependent variable $\psi$, one more system equation

$$\psi' = 1/V[\tau(a)] \tag{s}$$

Additional terms on the left-hand side of Eqs. (g) and (i)

$$\lambda_\psi((2a/V)(d/da)\{1/[\tau(a)]\} - \{1/[V\tau(a)]\}) \tag{t}$$

$$\lambda_\psi((a/V)(d/da)\{1/[\tau(a)]\}(1/\rho)(d\rho/dh)) \tag{u}$$

An additional term on the right-hand side of Eq. (n)

$$\lambda_\psi \left\{ \frac{\rho S}{2m} V \frac{d}{da} \left[ \frac{1}{\tau(a)} \right] \frac{C_L(dC_L/d\alpha) + C_D(dC_D/d\alpha)}{(C_L^2 + C_D^2)^{\frac{1}{2}}} \right\} \tag{v}$$

and another differential equation in the adjoint system:

$$\lambda'_\psi = 0 \tag{w}$$

NUMERICAL VALUES.

$$mg_0/S = 27.3 \text{ lb ft}^{-2} \qquad \gamma(0) = 0.18°$$
$$V(0) = 25{,}920 \text{ ft sec}^{-1} \qquad x(0) = q(0) = t(0) = 0$$
$$h(0) = 300{,}000 \text{ ft} \qquad x_f = 21{,}600 \text{ nm} \qquad h_f = 0$$

COMPUTATION RESULTS.    For the given numerical values, there are two values of $\alpha$ giving $x_f = 21.600$ nm. The one yielding the smaller total heat is $\alpha = 36°$ and the resulting trajectory is the dashed curve in Fig. 4-17. This constant angle-of-attack trajectory is taken as a nominal trajectory and

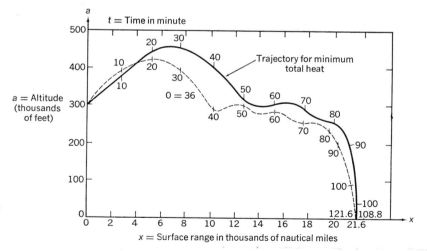

**Fig. 4-17**  Nominal and optimal trajectories. Initial conditions: $V_{(0)} = 25{,}920$ ft/sec, $\gamma = 0.18$ deg, $h_0 = 300{,}000$ ft., $mg_0/S = 27.3°/\text{ft}^2$.

the steepest-descent technique is then used to find the $\alpha(t)$ program that minimizes total heat holding final range, $x_f$, constant.

Four successive computations are sufficient to reach the extremum point. The total heat was reduced 39 percent below the value for the nominal trajectory. The optimum angle-of-attack trajectory is shown as the solid curve in Fig. 4-17.

Application of the method of gradients to dynamical optimization problems is as follows:

1. Overcomes the 2PBVP of the variational calculus. Similar difficulties, in a much slighter measure, can occur when solving the $\lambda_i$—adjoint equations. Arbitrary initial conditions [for instance, all $\lambda_i(t_0) = 0$ but 1] can be used to start the recurrent procedure.

2. Overcomes the dimensionality problem of the dynamic programming.

3. Requires differentiabilities of $J$ and $f$, since the derivatives are needed for calculating the influence function.

One great disadvantage of the method of gradients is the convergence uncertainty. Around an extremum, the individual slopes and the general gradients have small values. The approach to the extremum therefore becomes sluggish. On the other hand, for performance functionals with high sensitivity, jumps often occur over the hill summit resulting in undesirable oscillations.

### 4.4.3   Effects of Constraints and Limitations

The treatment given in Subsection 4.4.1 corresponds to the following specifications: (1) Final time $t_f = T$ given, (2) performance depending upon the final state of the state variables $J = J[x_i(T)]$, and (3) given constraints are of equality form: $\dot{x}_i - f_i = 0$.

We now ask what happens if:

(a) Performance $J$ is of a general form.

(b) Constant magnitude limitations and discontinuities are on $x$ and $u$.

(c) Final time $t_f$ and/or initial time $t_0$ are not given.

(d) There are terminal constraints either at $t = t_0$ or at $t = t_f$:

$$[\psi(x, u, t)]_t^{t_f} = 0$$

(e) Inequality constraints $C(x, u, t) \geq 0 \ S(x, t) \geq 0$.

As we have seen from preceding discussions, the method of gradients uses a mixed strategy: analytical arguments derived from the variational calculus and direct computing procedure as in the dynamic programming. The same line can then be used for treating various constraints and limitations.

(a) *General form performance index.* As we have seen in the method of Pontryagin, the Lagrange problem will be converted into a Mayer problem, and then the treatment of Subsection 4.4.1 applies [4-27].

(b) *Magnitude constraints and discontinuities on $x$ and $u$.* As in the dynamic programming, these limitations will be taken into account directly during the computing process. Values of $x$ and $u$ will be chosen so that the limitations are not violated.

(c) $t_f$ *and $t_0$ not fixed.* In the calculation of the performance variation $\delta J$, $dt_f$, and $dt_0$ will be introduced, leading to the addition of supplementary influence functions.

(d) *Terminal constraints.* In the calculation of $\delta J$, the variations of $\delta \psi$ will be taken into account. As a result, $\lambda_i(t_0)$ or $\lambda_i(t_f)$ will be defined differently, leading to a different set of $\lambda_i(t)$ and $\lambda_u(t)$ [4-24]. Consequently the gradient coefficient $K$ will be modified.

(e) *Inequality constraints.* A mixed strategy must be used here. Two situations may occur: (1) During the computing process, chosen values of $x$ and $u$ are checked so that the constraints are not violated. (2) If the constraint boundary is reached, then treat $C(x, u, t) = 0$ or $S(x, t) = 0$ just as $\dot{x} - f_i = 0$, by introducing corresponding influence functions.

# The Theory of the Classical Variational Methods

The name variational method covers actually both the Pontryagin's method and the classical variational method, because in both cases, the variation of the performance functional is used to derive extremalizing conditions. The theories of these methods, although leading to quite similar results, use different ways of demonstration. In the Pontryagin's theory, emphasis is on geometrical concepts. In the classical variational theory, analytical derivations are generally used. In the following, we shall briefly discuss these theories. In this appendix, we begin with the classical variational theory in which two categories of conditions are obtained.

1. Necessary conditions for local extremals, comprising the Euler, Transversality, Weierstrass-Erdmann Corner, and Jacobi conditions.

2. Necessary and sufficient conditions for local maxima or local minima, comprising Weierstrass E-function and Legendre-Clebsch conditions.

Conditions of group (1) only define the stationarity, i.e., there are extremals, but this could be maxima or minima. Once these conditions are satisfied, then the conditions of group (2) determine whether this is maxima or minima. Regarding terminology, tests or equations are often used instead of conditions.

***4A.1 Euler Condition*** [G-6, Ch. 1]. The basic necessary condition for local extremals is the satisfaction of the Euler equation.[1] Here we shall indicate how it is derived by using the simplest one-dimensional case. For this, we consider the problem of extremalizing the functional

$$J[y(x)] = \int_{x_0}^{x_1} F[x, y(x), y'(x)] \, dx$$

with fixed and given terminal points $y(x_0) = y_0$ and $y(x_1) = y_1$ and without any side constraints (Fig. 4A-1*a*). The mathematical problem is to find that function

---

[1] Very often called Euler-Lagrange equation. However, according to O. Bolza [G-8], the real and only author is Euler who has established it in 1744. Lagrange had contributed to its dissemination.

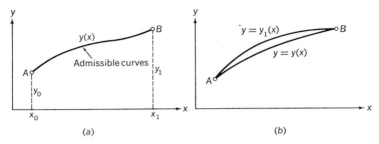

**Fig. 4A-1**  Optimal trajectory and its neighboring curves. (*a*) Admissible curves $y(x)$ with fixed boundaries $y_0$ and $y_1$. (*b*) Optimal trajectory $y(x)$ and its neighboring curve $y_1(x)$.

$y(x)$ which makes $J[y(x)]$ have an extremum. The general lines of the derivation are:

1. Although $J$ is a functional, we try to use the results of the ordinary theory of maxima and minima.

2. For this, we organize $y(x)$ and a neighboring $y_1(x)$ so that they are functions of one single parameter, say $\alpha$.

3. We then organize $J[y(x)]$ so that it becomes a function of the same parameter, leading to $J(\alpha)$.

4. Then we search conditions making $dJ(\alpha)/d\alpha = 0$.

The one-parameter family of curves is defined as follows (Fig. 4A-1*b*):

$$y(x, \alpha) = y(x) + \alpha[y_1(x) - y(x)] \tag{1}$$

which becomes $y(x)$ when $\alpha = 0$ and $y_1(x)$ when $\alpha = 1$. The difference

$$\delta y = y_1(x) - y(x) \tag{2}$$

is called by definition a variation of the function $y(x)$. Using $\delta y$, (1) becomes

$$y(x, \alpha) = y(x) + \alpha \, \delta y \tag{3}$$

Note that $\delta y$ is a function of $x$, and belongs to the same class as $y$, and we have $(\delta y)' = \delta y'$, $(\delta y)'' = \delta y''$, ..., $(\delta y)^{(k)} = \delta y^{(k)}$. We now consider the value of the functional $J[y(x)]$ taken along the curves of the family $y = y(x, \alpha)$ *only*; then we have a function of $\alpha$

$$J[y(\alpha, x)] = \varphi(\alpha) \tag{4}$$

If we take the derivative of $\varphi(\alpha)$ with respect to $\alpha$ and setting $\alpha \to 0$, we have by definition:

$$\varphi'(0) = \left\{ \frac{\partial}{\partial \alpha} J[y(x) + \alpha \, \delta y] \right\}_{\alpha=0} = \delta J \tag{5}$$

The necessary condition for an extremum $\delta J = 0$ turns out therefore to be $\varphi'(0) = 0$.

Since

$$\varphi(\alpha) = \int_{x_0}^{x_1} F[x, y(x, \alpha), y_x'(x, \alpha)] \, dx \tag{6}$$

we have

$$\varphi'(\alpha) = \int_{x_0}^{x_1} \left[ F_y \frac{\partial}{\partial \alpha} y(x, \alpha) + F_{y'} \frac{\partial}{\partial \alpha} y'(x, \alpha) \right] dx \tag{7}$$

where $F_y = \partial F / \partial y$ and $F_{y'} = \partial F / \partial y'$. Because of the relations

$$\frac{\partial}{\partial \alpha} y(x, \alpha) = \frac{\partial}{\partial \alpha} [y(x) + \alpha \delta y] = \delta y$$

$$\frac{\partial}{\partial \alpha} y'(x, \alpha) = \frac{\partial}{\partial \alpha} [y'(x) + \alpha \, \delta y'] = \delta y'$$

It follows that

$$\varphi'(\alpha) = \int_{x_0}^{x_1} \{ F_y[x, y(x, \alpha), y'(x, \alpha)] \, \delta y + F_{y'}[x, y(x, \alpha), y'(x, \alpha)] \, \delta y' \} \, dx \tag{8}$$

$$\varphi'(0) = \int_{x_0}^{x_1} \{ F_y[x, y(x), y'(x)] \, \delta y + F_{y'}[x, y(x), y'(x)] \, \delta y' \} \, dx \tag{9}$$

The condition $\varphi'(0) = 0$ becomes

$$\int_{x_0}^{x_1} (F_y \, \delta y + F_{y'} \, \delta y') \, dx = 0 \tag{10}$$

Integrating the second term by parts

$$\delta J = [F_{y'} \, \delta y]_{x_0}^{x_1} + \int_{x_0}^{x_1} \left( F_y - \frac{d}{dx} F_{y'} \right) \delta y \, dx \tag{11}$$

Since all the admissible curves pass through the fixed end-points, it follows that $\delta y(x = x_0) = 0$ and $\delta y(x = x_1) = 0$. The first term of $\delta J$ then vanishes and

$$\delta J = \int_{x_0}^{x_1} \left( F_y - \frac{d}{dx} F_{y'} \right) \delta y \, dx \tag{12}$$

FUNDAMENTAL LEMMA OF THE CALCULUS OF VARIATIONS.    If a function $A(x)$ is continuous in an interval $(x_0, x_1)$ and if

$$\int_{x_0}^{x_1} A(x) n(x) \, dx = 0$$

for an arbitrary function $n(x)$ subject to some conditions of only a general character, then $A(x) \equiv 0$ throughout the interval $(x_0, x_1)$. The conditions on $n(x)$ are, for instance, that $n(x)$ should be a first or higher-order differentiable function, that $n(x)$ should vanish at the end-points $x_0$ and $x_1$, and $|n(x)| < \epsilon$ or both $|n(x)| < \epsilon$

and $|n'(x)| < \epsilon.$[1] Applying this lemma to (12), we have $n(x) = \delta y(x)$ and

$$A(x) = F_y - \frac{d}{dx} F_{y'}$$

The condition $\delta J = 0$ becomes therefore

$$F_y - \frac{d}{dx} F_{y'} = 0 \tag{13}$$

or written explicitly

$$F_y - F_{xy'} - F_{yy'}y' - F_{y'y'}y'' = 0 \tag{14}$$

which is the *Euler equation* for the one-dimensional nonconstrained problem.

Consider now the constrained multidimensional problems

$$J = \int_{x_0}^{x_1} F(x, y_1, y_2, \ldots, y_n, y_1', y_2', \ldots, y_n') \, dx \tag{15}$$

$$\varphi_i(x, y_1, \ldots, y_n) = 0 \qquad i = 1, 2, \ldots, m < n \tag{16}$$

We form a new performance functional

$$J^* = \int_{x_0}^{x_1} \left( F + \sum_1^m \lambda_i \varphi_i \right) dx = \int_{x_0}^{x_1} F^* \, dx \tag{17}$$

where $\lambda_i = \lambda_i(x)$ are arbitrary functions of $x$. The preceding derivation, applied to $F^*$, leads to the Euler equations.

$$F_{y_j}^* - \frac{d}{dx} F_{y_j'}^* = 0 \qquad j = 1, 2, \ldots, n \tag{18}$$

The $(n + m)$ functions $y$ and $\lambda$ are to be solutions of the $(m + n)$ equations (16) and (18). The equations (16) can also be considered as Euler equations, if the $\lambda$'s are considered as arguments of $J$. The equations $\varphi_i = 0$ are assumed independent, e.g. there is a nonvanishing Jacobian of order $m$.

$$\frac{D(\varphi_1, \varphi_2, \ldots, \varphi_m)}{D(y_1, y_2, \ldots, y_m)} \neq 0 \tag{19}$$

The Euler equation (14) has known coefficients $F_y$, $F_{xy'}$, $F_{yy}$, and $F_{y'y'}$ and is a second-order differential equation in $y$. Its general solution contains, therefore, two arbitrary constants so that

$$y = y(x, \alpha, \beta) \tag{20}$$

The two constants $\alpha$ and $\beta$ are to be determined by the initial and final conditions $y_0 = y(x_0, \alpha, \beta)$ and $y_1 = y(x_1, \alpha, \beta)$ at the points $A$ and $B$. Every solution of the Euler equation (curve or function) is called, according to Kneser, an extremal; there is a double infinitude of extremals in the plane.

---

[1] The conditions imposed at the end points and on the derivatives show that, when there are terminal constraints, boundary limitations, discontinuities on either the state of the control variables, the Euler necessary condition must be completed by other conditions.

The Euler equation is reduced to a first-order differential equation if $F$ does not contain explicitly the independant variable $x$ (which is the time $t$ in dynamic optimization problems.) [G-7, p. 16]. To prove this, we write

$$\frac{dF}{dx} = \frac{\partial F}{\partial y} y' + \frac{\partial F}{\partial y'} y'' + \frac{\partial F}{\partial x} \tag{21}$$

By (13), $F_y = (d/dx)F_{y'}$, and, since $\partial F/\partial x = 0$,

$$\frac{dF}{dx} = y' \frac{d}{dx}\left(\frac{\partial F}{\partial y}\right) + \frac{\partial F}{\partial y'} y'' = y' \frac{d}{dx}\left(\frac{\partial F}{\partial y'}\right) + \frac{\partial F}{\partial y'}\frac{dy'}{dx} \tag{22}$$

which is of the form $uv' + vu'$. The first integral of this equation is

$$F = y' \frac{\partial F}{\partial y'} - K \qquad \text{or} \qquad -F + y' \frac{\partial F}{\partial y'} = K \tag{23}$$

which is a first-order differential equation in $y$.

In order to discuss the validity of the Euler condition, we must note that in the preceding development (1) only the variation of the dependent variable is taken into account, (2) the variation of the independent variable $\delta x$ is not taken into account to cause the variation of the performance function, and (3) terms of a higher order than two are neglected.

*4A.2 Transversality and Corner Conditions.* In the derivation of the necessary condition as done in the preceding section, we have assumed a fixed end-point problem; i.e., the values $y(x_0) = y_0$ and $y(x_1) = y_1$ are given. Here we shall discuss (1) the free end-point problem; i.e., $y(x_0)$ and $y(x_1)$ are unknowns, and (2) the constrained end-point problem, i.e., at the end values $x_0$, $x_1$, the dependent variable $y$ must satisfy the relations of the type $\psi[y(x_0)] = 0$ or $\psi[y(x_1)] = 0$. For simplicity, we consider first the one free end-point problem; i.e., $y(x_0) = y_0$ given, but $y(x_1)$ free. All possible extremals $y$ then form a pencil of curves such as shown in Fig. 4A-2a.

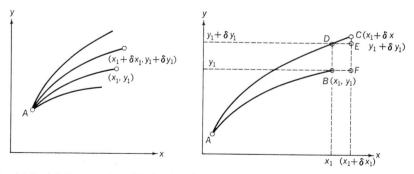

**Fig. 4A-2** (a) Curves $y(x)$ with fixed end $y_0$ and free end $y_1$. (b) Variations at the end point $(x_1, y_1)$.

The freedom of the end points imposes the so-called *transversality* conditions, which are necessary conditions and are to be satisfied as complementary conditions to the Euler condition. The derivation of the transversality condition is similar to that of the Euler equation; i.e., it is based on the statement that the variation $\delta J$ must be zero at the end points. We must therefore calculate $\delta J$ and search the conditions satisfying the relation $\delta J = 0$ [G-6, Ch. 2].

Since $y(x_0) = y_0$ is fixed, the extremals $y = y(x, \alpha, \beta)$, solutions of the Euler equation, become $y = y(x, \beta)$ functions of $x$ and of only one parameter $\beta$. Then for a given $\beta$ and a given value $x_1$, $y$ takes on definite value $y_1$. It turns out, therefore, that the functional $J$ becomes an univalent function of $y_1$ and $x_1$ at the final point. As a consequence, its variation turns into the differentiation of this function when the point $(x_1, y_1)$ moves to the point $(x_1 + \delta x_1, y_1 + \delta y_1)$ (Fig. 4A-2a). In such a case, we have the main linear part of $\Delta J$, linear with respect to $\delta x_1$ and $\delta y_1$

$$\Delta J = \int_{x_0}^{x_1 + \partial x_1} F(x, y + \delta y, y' + \delta y')\, dx - \int_{x_0}^{x_1} F(x, y, y')\, dx$$

$$= \int_{x_1}^{x_1 + \partial x_1} F(x, y + \delta y, y' + \delta y')\, dx$$

$$+ \int_{x_0}^{x_1} [F(x, y + \delta y, y' + \delta y') - F(x, y, y')]\, dx \qquad (24)$$

Using the mean value theorem, the first term of the right-hand member writes

$$\int_{x_1}^{x_1 + \delta x_1} F(x, y + \delta y, y' + \delta y')\, dx = [F]_{x = x_1 + \theta \delta x_1}\, \delta x_1 \qquad (25)$$

where $0 < \theta < 1$. Because of the continuity of $F$

$$[F]_{x = x_1 + \theta \delta x_1} = [F]_{x = x_1} + \epsilon_1 \qquad (26)$$

where $\epsilon_1 \to 0$ as $\delta x_1 \to 0$ and $\delta y_1 \to 0$. So that

$$\int_{x_1}^{x_1 + \delta x_1} F(x, y + \delta y, y' + \delta y')\, dx = [F(x, y, y')]_{x = x_1}\, \delta x_1 + \epsilon_1\, \delta x_1 \qquad (27)$$

Using the Taylor series expansion, and by neglecting terms of a higher order than $\delta y$ or $\delta y'$, the second term of the right-hand member of (24) writes

$$\int_{x_0}^{x_1} [F(x, y + \delta y, y' + \delta y') - F(x, y, y')]\, dx$$

$$= \int_{x_0}^{x_1} [F_y(x, y, y')\, \delta y + F_{y'}(x, y, y')\, \delta y']\, dx + 0(\epsilon^2) \qquad (28)$$

Integrating by parts

$$\int_{x_0}^{x_1} (F_y\, \delta y + F_{y'}\, \delta y')\, dx = [F_{y'}\, \delta y]_{x_0}^{x_1} + \int_{x_0}^{x_1} \left( F_y - \frac{d}{dx} F_{y'} \right) \delta y\, dx \qquad (29)$$

Since the Euler equation (13) is to be satisfied, $F_y - \dfrac{d}{dx} F_{y'} = 0$. Furthermore, the point $(x_0, y_0)$ being fixed, $[\delta y]_{x=x_0} = 0$, so that

$$\int_{x_0}^{x_1} (F_y \, \delta y + F_{y'} \, \delta y') \, dx = [F_{y'} \, \delta y]_{x=x_1} \tag{30}$$

Substituting (30) into (28), and then both (28) and (27) into (24) yields

$$\Delta J = [F]_{x=x_1} \, \delta x_1 + [F_{y'} \, \delta y]_{x=x_1} \tag{31}$$

Let us now consider Fig. 4A-2b where the details of the change from the point $(x_0, y_0)$ to the point $(x_1, y_1)$ are given. The following relations hold true:

$$BD = [\delta y]_{x=x_1} . \qquad FC = \delta y_1 \qquad EC \approx y'(x_1) \, \delta x_1$$

so that

$$BD = FC - EC$$

and

$$[\delta y]_{x=x_1} = \delta y_1 - y'(x_1) \, \delta x_1 \tag{32}$$

Substituting this value of $[\delta y]_{x=x_1}$ into (31) yields

$$\Delta J = [F]_{x=x_1} \, \delta x_1 + [F_{y'}]_{x=x_1}[\delta y_1 - y'(x_1) \, \delta x_1]$$
$$= [(F - y'F_{y'}) \, \delta x_1 + F_{y'} \, \delta y_1]_{x=x_1} \tag{33}$$

The complementary necessary condition for the free end point $(x_1, y_1)$ problem takes then the form

$$[(F - y'F_{y'}) \, \delta x_1 + F_{y'} \, \delta y_1]_{x=x_1} = 0 \tag{34}$$

If both the end points $(x_0, y_0)$ and $(x_1, y_1)$ are free, by using a similar development, we shall have the following condition:

$$\left[ (F - y'F_{y'}) \, \delta x + F_{y'} \, \delta y \right]_{\substack{x=x_0 \\ y=y_0}}^{\substack{x=x_1 \\ y=y_1}} = 0 \tag{35}$$

If $\delta x_1$ and $\delta y_1$ are independent, then instead of (34), we have

$$[F - y'F_{y'}]_{x=x_1} = 0 \quad \text{and} \quad [F_{y'}]_{x=x_1} = 0 \tag{36}$$

In the case of multidimensional problems, the corresponding equation to (35) writes

$$\left( F - \sum_1^n y'_j F_{y_j'} \right) \delta x + \sum_1^n F_{y_j'} \, \delta y_j = 0 \tag{37}$$

TRANSVERSALITY CONDITION. If the right end point $(x_1, y_1)$ can move along a certain curve (Note that $y_1$ is an explicit function of $x_1$)

$$y_1 = \varphi(x_1) \tag{38}$$

Then $\delta y_1 \approx \varphi'(x_1)\,\delta x_1$, and (34) writes

$$\{[F + (\varphi' - y')F_{y'}]\,\delta x_1\}_{x=x_1} = 0 \tag{39}$$

Since $\delta x_1$ varies arbitrarily

$$[F + (\varphi' - y')F_{y'}]_{x=x_1} = 0 \tag{40}$$

This condition establishes some relation between directional coefficients $\varphi'$ and $y'$ at the point $(x_1, y_1)$ and is called the *transversality* condition at the point $(x_1, y_1)$. Similarly, if the point $(x_0, y_0)$ can move along a curve $y_0 = \varphi(x_0)$, then we shall have the transversality condition at the point $(x_0, y_0)$

$$[F + (\varphi' - y')F_{y'}]_{x=x_0} = 0 \tag{41}$$

If the relations at the end points are not explicit, but implicit such as $\psi_0[x_0, y(x_0)] = 0$ and $\psi_1[x_1, y(x_1)] = 0$, then we state that these equations and (35) must be satisfied simultaneously. If the performance function also contains a term $G$ such as $J = [G(x, y)]_0^1 + \int_{x_0}^{x_1} F(x, y, y')\,dx$, then the condition (35) becomes

$$[dG + (F - y'F_{y'})\,dx + F_{y'}\,dy]_0^1 = 0 \tag{42}$$

Note that if $F$ does not contain the independent variable $x$ explicitly, then (23) holds, and (42) becomes

$$[dG - K\,dx + F_{y'}\,dy]_0^1 = 0 \tag{43}$$

CORNER CONDITIONS.   There are certain variational problems which are characterized by discontinuous solutions (Fig. 4A-3), i.e., solutions in which $y'$ experiences

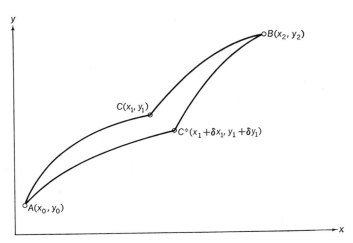

**Fig. 4A-3**   Curves $y(x)$ with corners.

a jump at a finite number of points. These points are called *corner points*. The entire solution is still called extremal arc, while each component portion is called a subarc. The derivation of complementary conditions is similar to that of the transversality conditions. We write first

$$J = \int_{x_0}^{x_2} F(x, y', y) \, dx = \int_{x_0}^{x_1} F(x, y, y') \, dx + \int_{x_1}^{x_2} F(x, y, y') \, dx \qquad (44)$$

where $(x_1, y_1)$ is the coordinates of the corner point $C$. We then look at the point $x_1$ as $x_1 - 0$ when we come from the subarc $AC$, and as $x_1 + 0$ when we come from the subarc $BC$. The variation $\delta J$ can then be calculated as before and is

$$\delta J = [(F - y'F_{y'}) \, \delta x_1 + F_{y'} \, \delta y_1]_{x=x-0} - [(F - y'F_{y'}) \, \delta x_1 + F_{y'} \, \delta y_1]_{x=x_1+0} = 0 \qquad (45)$$

The corner condition (called the Erdmann-Weierstrass Corner condition) become then

$$[(F - y'F_{y'}) \, \delta x_1 + F_{y'} \, \delta y_1]_{x=x'-0} = [(F - y'F_{y'}) \, \delta x_1 + F_{y'} \, \delta y_1]_{x_1=x+0} \qquad (46)$$

If $\delta x_1$ and $\delta y_1$ are independent, this becomes

$$[F_{y'}]_{x=x_1-0} = [F_{y'}]_{x=x_1+0} \qquad (47)$$

$$[F - y'F_{y'}]_{x=x_1-0} = [F - y'F_{y'}]_{x=x_1+0} \qquad (48)$$

If $F$ does not contain $x$ explicitly, then only the condition (47) is needed.

In the case of multidimensional problems, the corresponding equation to (46) is

$$\left[\left(F - \sum_{1}^{n} y'_j F_{y'_j}\right) \delta x_1 + \sum_{1}^{n} F_{y'_j} \, \delta y_j\right]_{x=x_1-0}$$

$$= \left[\left(F - \sum_{1}^{n} y'_j F_{y'_j}\right) \delta x_1 + \sum_{1}^{n} F_{y'_j} \, \delta y_j\right]_{x=x+0} \qquad (49)$$

In order to discuss later the validity of these conditions, the same observations as those at the end of the preceding section can be made here.

*4A.3 Jacobi Condition.* The Jacobi condition differs from the preceding ones in that it does not pertain to the neighboring curves of an extremal arc, but to the validity of the range between the two end points of this extremal arc. For illustration, we consider an example. We know [G-7, p. 28] that the smaller arc of the great circle through two points $A$ and $B$ (Fig. 4A-4) on a sphere is the shortest distance on the surface of the sphere between the two points. Let us consider two consecutive great circles $G_1$ and $G_2$ through $A$, they intersect at the point $A^*$, called the conjugate point of $A$, and which is diametrically opposite to $A$. It is easy to see that $G_1$ cannot be declared a unique minimum between $A$ and $B$: in order to have an actual minimum, the great-circle range, which begins at $A$, must not extend beyond $B$.

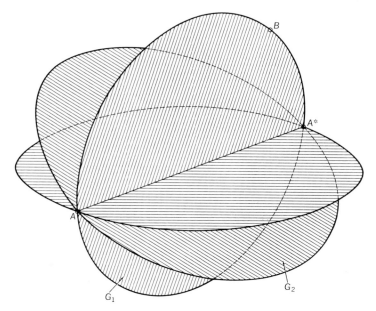

**Fig. 4A-4**   Example showing the range-limitation of an extremal arc.

We know that if the Euler equation is satisfied, the extremals are the curves of the form $y = y(x, \alpha, \beta)$. Assuming $A$ is fixed, i.e., $y_0 = y(x_0, \alpha, \beta)$ given, then the extremals are represented by a one-parameter function $y = y(x, \alpha)$. Note that as $\alpha$, we may take the slope of the extremals at $A$.

$$p = \frac{dy(x_0, \alpha)}{dx} \qquad (50)$$

Graphically, the family of extremals $y = y(x, \alpha)$ or $y = y(x, p)$ represents a pencil of curves. This pencil of curves is called a central field in the domain $D$ if the curves do not intersect each other in $D$ except at $A$. Otherwise, it is not a central field. (Fig. 4A-5a and 5b.) Further properties of the family $y = y(x, \alpha)$ will help to locate the intersecting points. It is known that in a given family of curves $\varphi(y, x, \alpha) = 0$,

1. There exists a $\alpha$-discriminant locus defined by the equations:

$$\varphi(y, x, \alpha) = 0 \qquad \partial \varphi / \partial \alpha = 0 \qquad (51)$$

The $\alpha$-discriminant locus is the envelope of the family and the locus of the multiple point of a curve of the family.

2. The point $A^*$ common to the curve passing through $A$ and to the locus is called the point conjugate to $A$.

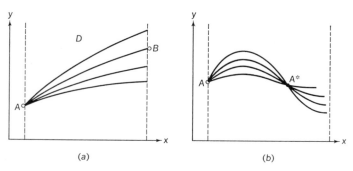

**Fig. 4A-5** Field of extremals with fixed initial point. (a) Central field in D. (b) Noncentral field in D.

3. Two neighboring curves of the family intersect on the locus or near by it.

On the basis of these properties, we see that (1) if the range $AB$ of an extremal contains the conjugate point $A^*$, then the extremals do not form a central field and (2) if this range does not contain $A^*$, then we have a central field of extremals. We have then the literal statement of the Jacobi condition.

> To form a central field of extremals with center at a point $A$ admitting an arc $AB$ of an extremal curve, it is necessary and sufficient that the point $A^*$, conjugate to the point $A$, (if $A^*$ exists) does not lie on the arc $AB$.

Note immediately that this condition does not apply if the initial point is free or constrained. In that case, we have to first satisfy conditions such as (35) or (41), resulting in a fixed-point situation, and then satisfy the Jacobi condition.

For obtaining the analytical form of the Jacobi condition, we define the function

$$u = \frac{\partial y(x, \alpha)}{\partial \alpha} = u(x) \tag{52}$$

which is a function of $x$ only if the partial derivative is taken on an extremal where $\alpha$ is fixed. The derivative of $u$ with respect to $x$ is

$$u'_x = \frac{\partial^2 y(x, \alpha)}{\partial \alpha\, \partial x} \tag{53}$$

Since $y = y(x, \alpha)$ is a solution of the Euler equation,

$$F_y[x, y(x, \alpha), y'_x(x, \alpha)] - \frac{d}{dx} F_{y'}[x, y(x, \alpha), y'_x(x, \alpha)] \equiv 0 \tag{54}$$

Differentiating this identity with respect to $\alpha$, and setting $\partial y(x, \alpha)/\partial \alpha = u$

$$\partial y(x, \alpha)/\partial \alpha = u$$

$$F_{yy}u + F_{yy'}u' - \frac{d}{dx}(F_{yy'}u + F_{y'y'}u') = 0 \tag{55}$$

or

$$F_{yy}u + F_{yy'}u' - \left( u'F_{yy'} + u\frac{d}{dx}F_{yy'} + u''F_{y'y'} + u'\frac{d}{dx}F_{y'y'} \right) = 0$$

$$\left( F_{yy} - \frac{d}{dx}F_{yy'} \right)u + \left( F_{yy'} - F_{yy'} - \frac{d}{dx}F_{y'y'} \right)u' - u''F_{y'y'} = 0$$

$$\left( F_{yy} - \frac{d}{dx}F_{yy'} \right)u - \left( \frac{d}{dx}F_{y'y'} \right)u' - (F_{y'y'})u'' = 0 \tag{56}$$

If $y = y(x, \alpha)$ is a solution of the Euler equation, $\alpha = \alpha_0$ is known; then all the partial derivatives are known functions of $x$. The linear homogeneous equations of the second order in $u$ is called the Jacobi's equation. If a solution $u$ vanishes at $x = x_0$ and *also* at some other point $x$, $x_0 < x < x_1$, then the conjugate point $A^*$ lies on the arc $AB$. If there is a solution $u$ which vanishes only at $x = x_0$, and at nowhere else, then the arc $AB$ is the unique extremal.

The Jacobi equation (56) is a necessary condition for the uniqueness of extremals. Since it is based on the Euler equation, it is only valid for small variations $\delta y$ and $\delta y'$.

*4A.4 Necessary Conditions for a Local Maximum or Minimum.* The major difficulty for practicing engineers, when using mathematical conditions in optimization techniques, is to know the validity of these conditions, notably, local or global, necessary or sufficient, weak or strong. Here, we try to clarify these questions. For this, we consider again the one-dimensional performance functional and the special type of variation

$$\delta y = y(x) + \epsilon v(x) \tag{57}$$

where $\epsilon$ is a small arbitrary quantity and $v$ an arbitrary regular function of $x$. Assuming that we consider a fixed end-point problem and that

$$F(x, y + \epsilon v, y' + \epsilon v')$$

can be expanded in a uniformly converging series, the variation of $J$ then writes:

$$\Delta J = \int [F(x, y + \epsilon v, y' + \epsilon v') - F(x, y, y')] \, dx$$

$$= \epsilon I_1 + \tfrac{1}{2}\epsilon^2 I_2 + R_3 \tag{58}$$

with

$$I_1 = \int (vF_y + v'F_{y'}) \, dx \tag{59}$$

$$I_2 = \int (v^2 F_{yy} + 2vv'F_{yy'} + v'^2 F_{y'y'}) \, dx \tag{60}$$

and where $\epsilon I_1$ is called the first variation, $(1/2)\epsilon^2 I_2$ the second variation and $R_3$ denotes the aggregate of terms involving third and higher powers of $\epsilon$. (Note here why $\epsilon$ is introduced in $\delta y$). We first note that in all cases in the classical variational calculus, we always consider variations in the neighborhood of the extremals (either for obtaining $\epsilon I_1$ or $(1/2)\epsilon^2 I_2$), the resulting conditions are therefore only valid for local maxima or minima. Second, we note that all the preceding conditions (Euler, Transversality, Corner, Jacobi) are derived on the basis of the zeroing of the first variation. They secure only a *stationary* quality of the solutions; they are necessary for extremals, but insufficient for maxima or minima. Third, we note that, just as in the ordinary theory of maxima and minima, the sufficient conditions can only be derived by examining the sign of the variation $\Delta J$. We can do this by using:

$$\Delta J = (1/2)\epsilon^2 I_2 \text{ if } I_1 = 0 \text{ and } R_3 \text{ neglected, } \Delta J = \epsilon I_1 + (1/2)\epsilon^2 I_2,$$

$$\text{or } \Delta J = \epsilon I_1 + (1/2)\epsilon^2 I_2 + R_3 \text{ if } R_3 \text{ has some importance.}$$

We know that, if $\Delta J \geq 0$ under the considered conditions, than the value of $J$ can only increase, thus securing that it has attained its minimum value. Likewise, if $\Delta J \leq 0$ its value can only decrease, securing that it has attained its maximum value. The point is, however, that the situations $\Delta J \geq 0$ (or $\Delta J \leq 0$) may occur not only under one kind of condition, but under various kinds of conditions. Sufficient conditions may be, therefore, stated valid with respect to *that* kind of condition under examinations. The various kinds of conditions under examination refer mainly to the types of variations of the arguments $y$, $y'$, and $x$ of the performance functional $J$. Two important types are generally differentiated [G-7, p.8].

*Weak variations.* Those small variations in which the derivatives of the variation are of the same order of smallness as the variation itself.

*Strong variations.* Those small variations in which the derivatives of the variation are not limited to be small, though the actual displacement is definitely small. (See also Appendix 2A, the definitions on closeness.)

The kinds of variations, used precedingly in the derivation of necessary conditions for extremals, and which are characterized by (1) $\delta y$ and $\delta y'$ small, and (2) the variation of the independent variable not taken into account in the calculation of the varied functional $[F(x, y + \delta y, y' + \delta y')$ instead of $F(x + \delta x, y + \delta y, y' + \delta y')]$ belong to the type of weak variations. Furthermore, they are named *special variations*.

WEIERSTRASS $E$-FUNCTION CONDITION.  Let us consider again the fixed end-point problem. Let $C$ be the extremal arc obtained when both the Euler and Jacobi conditions are satisfied. We now consider a very general value $p$ of the derivative $y'$, and the curve $C^*$ resulted from $p$. The increment of the performance function is

$$\Delta J = \int_{C^*} F(x, y, y') \, dx - \int_{C} F(x, y, y') \, dx \tag{61}$$

Since $C^*$ is obtained from $C$ by varying $y'$

$$\int_{C} F(x, y, y') \, dx = \int_{C^*} [F(x, y, p) + (y' - p)F_p(x, y, p)] \, dx \tag{62}$$

I.e., the right-hand member becomes the left-hand member when $p = y'$. Substituting (62) into (61) yields

$$\Delta J = \int_{C^*} F(x, y, y')\, dx - \int_{C^*} [F(x, y, p) + (y' - p)F_p(x, y, p)]\, dx \qquad (63)$$

The function

$$E(x, y, y', p) = F(x, y, y') - F(x, y, p) - (y' - p)F_p(x, y, p) \qquad (64)$$

is called the *Weierstrass E-function*, from which the following conditions are resulted

1. If $E \geq 0$ (then $\Delta J \geq 0$), we have a sufficient and necessary condition for a local minimum.

2. If $E \leq 0$ (then $\Delta J \leq 0$), we have a sufficient and necessary condition for a local maximum.

3. The minimum (maximum) is *weak* if the condition $E \geq 0$ ($E \leq 0$) holds for all small variations of $x$ and $y$, and for all small variations of $y'$, i.e., $p$ close to $y'$.

4. The minimum (maximum) is *strong* if the condition $E \geq 0$ ($E \leq 0$) holds for all small variations of $x$ and $y$ and for arbitrary variations of $y'$, i.e., $p$ may be very different from $y'$.

For multidimensional problems, Weierstrass $E$-Function condition writes

$$\Delta F - \sum_1^n \frac{\partial F}{\partial \dot{y}_j} \Delta \dot{y}_j \geq 0 \qquad (\leq 0) \qquad (65)$$

The Weierstrass $E$-Function conditions are generally uneasy to handle, because we have to assume various kinds of $p$, and for each $p$, we have to compute the $E$-function and to check its sign.

LEGENDRE-CLEBSH CONDITION. If $F(x, y, y')$ has third derivative with respect to $y'$, we have, by Taylor's formula,

$$F(x, y, y') = F(x, y, p) + (y' - p)F_p(x, y, p) + \frac{(y' - p)^2}{2!} F_{y'y'}(x, y, q) \qquad (66)$$

where $q$ is a value between $p$ and $y'$. Substituting this value in the $E$-function, yields

$$E(x, y, p, q) = \frac{(y' - p)^2}{2!} F_{y'y'}(x, y, q)$$

Since the coefficient is always positive, the sign of the $E$-function is equal to that of $F_{y'y'}(x, y, q)$, which is much simpler to compute. The Legendre-Clebsch conditions result:

1. If $F_{y'y'}(x, y, q) \geq 0$ ($\leq 0$), we have a sufficient and necessary condition for a local minimum (maximum).

2. The minimum (maximum) is *weak*, if the condition $F_{y'y'}(x, y, q) \geq 0$ ($\leq 0$) holds true for all small variations of $x$, $y$, and $y'$.

3. The minimum (maximum) is *strong*, if the condition $F_{y'y'}(x, y, q) \geq 0$ ($\leq 0$) holds true for all small variations of $x$ and $y$ and for arbitrary values of $y'$, i.e., $y' \leq q \leq p, p$ very different from $y'$.

If $F_{y'y'}(x, y, y') \neq 0$, then, because of the continuity property, it keeps its sign for all $q$, $q$ close to $y'$. The weak conditions $F_{y'y'} \geq 0$ and $F_{y'y'} \leq 0$ can then be replaced by $F_{y'y'} > 0$ and $F_{y'y'} < 0$. The conditions $F_{y'y'} > 0$ and $F_{y'y'} < 0$ are called Legendre conditions for a weak local minimum or maximum.

In multidimensional problems, the Legendre-Clebsch conditions writes

$$\sum_{k=1}^{n} \sum_{j=1}^{n} \frac{\partial^2 F}{\partial y'_k \, \partial y'_j} \, \delta_{y_k'} \cdot \delta_{y_j'} \geq 0 \qquad (\leq 0) \tag{68}$$

# The Theory of the Maximum Principle

In Appendix 4A, we have emphasized that the various conditions for maxima and minima are valid *only* with respect to given types of variations. This is, in fact, the motivation of the theory of the Maximum Principle. Apart from some formal differences in the problem formulation, attention is first brought upon a detailed definition of the dependent control variable $u(t)$. The theory stipulates that the Maximum principle is valid upon a certain class $D$ of admissible controls defined by the following conditions:

1. All the controls $u(t)$, $t_0 \leq t \leq t_1$, are measurable and bounded (controls with limiting values).

2. If $u(t)$, $t_0 \leq t \leq t_1$, is an admissible control, $v$ is an arbitrary point of $U$, $u(t) \in U$, and $t'$ and $t$ are numbers so that $t_0 \leq t' \leq t'' \leq t_1$, the control $u_1(t)$, $t_0 \leq t \leq t_1$, defined by the formula

$$u_1(t) = \begin{cases} v & \text{for } t' \leq t \leq t'' \\ u(t) & \text{for } t_0 \leq t \leq t' \text{ or } t'' \leq t \leq t_1 \end{cases}$$

is also admissible (control arc with constant portions).

3. If the interval $t_0 \leq t \leq t_1$ is broken up by means of subdivision points into a finite number of subintervals, on each of which the control $u(t)$ is admissible, then this control is also admissible on the entire interval $t_0 \leq t \leq t_1$ (discrete controls). A control obtained from an admissible control $u(t)$, $t_0 \leq t \leq t_1$, by a translation in time (i.e., the control $u_1(t) = u(t - \alpha)$, $t_0 + \alpha \leq t \leq t_1 + \alpha$) is also admissible (time-independentness from the system behavior).

Such detailed definitions of admissible controls, and hence of their variations, imply (1) definitions and notations of time instants, time intervals, and their combinations, (2) complicated derivations of the resultant state trajectory, as a consequence of the various types of controls, at various time instants and during various time intervals, (3) complex and ramiform mathematical arguments for deriving the extremalizing conditions, and (4) impossibility of showing the theory in its whole extent in a short space. Consequently, we shall only discuss the main

**221**

steps of the Pontryagin's proof. However, we shall first discuss one simple presentation in order to show some basic concepts.

**4B.1 Kopp's Presentation** [G-29, Ch. 7].    The problem formulation is: given the system of differential equations,

$$\dot{x}^i = f^i(x_1, x_2, \ldots, x_n, u_1, u_2, \ldots, u_r, t) \tag{1}$$

$$x^i(t_0) = x_0^i \quad i = 1, 2, \ldots, n$$

minimize

$$S = \sum_1^n c^i x^i(t_f) \tag{2}$$

The case of fixed-time ($t_f$ given) with free right end conditions [$x^i(t_f)$ free] is considered. We define an auxiliary variable $p_i(t)$ so that

$$\dot{p}_i = -\sum_{j=1}^n p_i \frac{\partial f^j}{\partial x^i} \quad p_i(t_f) = -c^i \quad i = 1, 2, \ldots, n \tag{3}$$

The main object of the proof is to formulate the total variation $\Delta S$ of the performance function with respect to the total variation $\Delta x$ of the state variable. Note that $\Delta x$ depends upon both $\Delta u$ and $\Delta t$. For simplicity, we first consider $\Delta u$. We assume that an optimum control vector $\bar{u}(t)$ has been found which minimizes $S$. A variation $\Delta u(t)$ from $\bar{u}(t)$ will cause the variation $\Delta x(t)$ from the optimum state vector $\bar{x}(t)$. From (1), we have

$$\sum_1^n p_i \, \Delta \dot{x}^i = \sum_1^n p_i [f^i(\bar{x} + \Delta x, \bar{u} + \Delta u, t) - f^i(x, u, t)] \tag{4}$$

Multiplying both sides of (4) by $dt$ and integrating between $t_0$ and $t_f$

$$\int_{t_0}^{t_f} \sum_1^n p_i \, \Delta \dot{x}^i \, dt = \int_{t_0}^{t_f} \sum_1^n p_i [f^i(\bar{x} + \Delta x, \bar{u} + \Delta u, t) - f^i(x, u, t)] \, dt \tag{5}$$

Another way to obtain the left member of (5) is to consider the derivative of $\sum_1^n p_i \, \Delta x_i$

$$\frac{d}{dt} \sum_1^n p_i \, \Delta x^i = \sum_1^n \dot{p}_i \, \Delta x^i + \sum_1^n p_i \, \Delta \dot{x}^i \tag{6}$$

Multiplying both sides of (6) by $dt$ and integrating

$$\int_{t_0}^{t_f} \sum_1^n p_i \, \Delta \dot{x}^i \, dt = \left[ \sum p_i \, \Delta x^i \right]_{t_0}^{t_f} - \int_t^{t_f} \dot{p}_i \, \Delta x^i \, dt \tag{7}$$

By equating the right members of (5) and (7), we obtain

$$\left[ \sum_1^n p_i \, \Delta x^i \right]_{t_0}^{t_f} = \int_{t_0}^{t_f} \sum_1^n p_i \, \Delta x^i \, dt$$

$$+ \int_{t_0}^{t_f} \sum_1^n p_i [f^i(\bar{x} + \Delta x, \bar{u} + \Delta u, t) - f^i(\bar{x}, \bar{u}, t)] \, dt \qquad (8)$$

Now we shall relate the left member of (8) to the performance function. Since $x^i(t_0) = x_0^i$ is given, $\Delta x^i(t_0) = 0$. Also by (3), we have $p_i(t_f) = -c^i$. Then

$$[\sum p_i \, \Delta x^i]_{t_0}^{t_f} = \sum p_i(t_f) \, \Delta x^i(t_f) - \sum p_i(t_0) \, \Delta x^i(t_0) = -\sum c^i \, \Delta x^i(t_f) = -\Delta S \qquad (9)$$

Combining (9), (8) and (3)

$$\Delta S = -\int_{t_0}^{t_f} \sum_{i=1}^n \left( -\sum_{j=1}^n p_j \frac{\partial f^j}{\partial x^i} \right) \Delta x^i \, dt$$

$$- \int_{t_0}^{t_f} \sum_1^n p_i [f^i(\bar{x} + \Delta x, \bar{u} + \Delta u, t) - f^i(\bar{x}, \bar{u}, t)] \, dt \qquad (10)$$

The second term of the right-hand member of (10) is evaluated by separating the variations in $f^i$ due to $\Delta u$ from those due to $\Delta x$. The variation in $f^i$ due to $\Delta x$ is expanded in a Taylor series with a remainder term giving, if $f \in C^2$

$$f^i(\bar{x} + \Delta x, \bar{u} + \Delta u, t) - f^i(\bar{x}, \bar{u}, t) = f^i(\bar{x}, \bar{u}, \Delta u, t) - f^i(\bar{x}, \bar{u}, t)$$

$$+ \sum_{j=1}^n \frac{\partial f^i(\bar{x}, \bar{u}, \Delta u, t)}{\partial x^j} \Delta x^j + \frac{1}{2} \sum_{\substack{j=1 \\ s=1}}^n \frac{\partial^2 f^i(\bar{x} + \xi \, \Delta x, \bar{u} + \Delta u, t)}{\partial x^j \, \partial x^s} \Delta x^j \, \Delta x^s \qquad (11)$$

where

$$0 \le \xi \le 1$$

From (10) and (11), we obtain

$$\Delta S = -\int_{t_0}^{t_f} \sum p_i [f^i(\bar{x}, \bar{u} + \Delta u, t) - f^i(\bar{x}, \bar{u}, t)] \, dt$$

$$- \int_{t_0}^{t_f} \sum_{\substack{i=1 \\ j=1}}^n p_i \left\{ \frac{\partial}{\partial x^j} [f^i(\bar{x}, \bar{u} + \Delta u, t) - f^i(\bar{x}, \bar{u}, t)] \right\} \Delta x^j \, dt$$

$$- \frac{1}{2} \int_{t_0}^{t_f} \sum_{\substack{i=1 \\ j=1 \\ s=1}}^n p_i \frac{\partial^2 f^i(\bar{x} + \xi \, \Delta x, \bar{u} + \Delta u, t)}{\partial x^j \, \partial x^s} \Delta x^j \, \Delta x^s \, dt$$

$$+ \int_{t_0}^{t_f} \sum_{\substack{i=1 \\ j=1}}^n p_j \frac{\partial f^j}{\partial x^i} (\bar{x}, \bar{u}, t) \, \Delta x^i \, dt \qquad (12)$$

In order to show the process of derivation, we consider the simple case of a linear system in which $u(t)$ are separable in $f^i$. That is

$$f^i = \sum_{j=1}^n a_{ij}(t)x^j + \varphi(u, t) \tag{13}$$

Then the last three members of (12) vanish and

$$\Delta S = -\int_{t_0}^{t_f} [H(\bar{x}, \bar{u} + \Delta u, t) - H(\bar{x}, \bar{u}, t)] \, dt \tag{14}$$

where

$$H = \sum_1^n p_i f^i = (p_i, f^i) = p^T f \tag{15}$$

is the Hamiltonian function, and is equal to the dot product of the two vectors $p$ and $f$. In order for the performance function $S$ to be a minimum, it is required that any variation $\Delta S$ be positive ($\Delta S > 0$). Then from (14)

$$H(\bar{x}, \bar{u} + \Delta u, t) - H(\bar{x}, \bar{u}, t) < 0 \tag{16}$$

or

$$H(\bar{x}, \bar{u}, t) > H(\bar{x}, \bar{u} + \Delta u, t) \tag{17}$$

for all $t$, $t_0 \leq t \leq t_f$.

The preceding development is based on the variation of the control vector. Let us now consider the influence of the time variation. In the case considered, since $t_f$ is fixed, the overall time period cannot be varied. However, the control variation $\Delta u$ can occur in the interval $t_1$, $t_2$ ($t_2 > t_1$) completely adjustable in the interval $t_0$, $t_f$. We will assume a variation $\Delta u_k$ in only one component of the control vector $\bar{u}$. For this variation, $\Delta u_k$ is zero everywhere except in the interval $t_1$, $t_2$. If the interval $t_1$, $t_2$ is chosen so that the inequality

$$H(\bar{x}, \bar{u} + \Delta u, t) - H(\bar{x}, \bar{u}, t) > 0 \tag{18}$$

is satisfied, then we will have $\Delta S < 0$. A similar argument may be presented for all the control variables. Therefore, a necessary condition for a minimum of $S$ is

$$H(\bar{x}, \bar{u} + \Delta u, t) - H(\bar{x}, \bar{u}, t) \leq 0 \tag{19}$$

or

$$H(\bar{x}, \bar{u}, t) \geq H(\bar{x}, \bar{u} + \Delta u, t) \tag{20}$$

for all $t$, $t_0 \leq t \leq t_f$. Inequality (19) states that for a minimum of $S$, the Hamiltonian function, for a given $\bar{x}$, must be maximized with respect to $u$. This is the main result of the Maximum Principle.

Some comments are to be made now.

1. Equations (8) and (9) are the most important relations in this development. On one hand, the variation of $S$ is related to the expression $\sum p_i \, \Delta x^i$. On the other hand, Eq. (8) relates $\sum p_i \, \Delta x^i$ to the variation of $f$ which is calculated on the basis of $\Delta x$ and $\Delta u$.

2. To deal with nonlinear systems, we must evaluate the additional terms in (12), i.e., $f^i$, $\partial f^i/\partial x^j$, $\partial^2 f^i/\partial x^i \cdot \partial x^j$ in various circumstances. This is what is done in the Pontryagin's presentation.

3. The argument used above on $\Delta S$ for a minimum of $S$ is somehow analytical. We can also use geometrical arguments by examining the behavior of the vectors $p$ and $f$. This is also what is done in the Pontryagin's presentation.

4. An interesting interpretation of the relations (15) and (20) is that the control vector $\bar{u}(t)$ is employed so that the dot product of the state velocity vector $\dot{\bar{x}}(t)$ with the auxiliary vector $\bar{p}$ is a maximum for a minimum of $S$ (also, it is a minimum for a maximum of $S$.)

5. It must be noted that a basic difference between Kopp's presentation and Pontryagin's presentation is that in Kopp's $t_f$ is fixed, and in Pontryagin's $t_f$ is free and $x(t_f)$ is fixed. This condition is the basic difficulty in proving the Maximum Principle.

**4B.2 Pontryagin's Presentation** [G-30, Ch. 2].    **4B.2.1 Formulation of the Problem.** Consider the performance functional

$$J = \int_{t_0}^{t_1} f^0[x(t), u(t), t] \, dt \tag{21}$$

and the system of differential equations

$$\frac{dx^i}{dt} = f^i(x^1, x^2, \ldots, x^n, u^1, u^2, \ldots, u^r) = f^i(x, u) \qquad i = 1, 2, \ldots, n \tag{22}$$

or, in vector form

$$\frac{dx}{dt} = f(x, u) \tag{23}$$

where $f^0$, $f^i$ and $\partial f^i/\partial x^j$ are assumed given and continuous on $X \times \bar{U}$ where $\bar{U}$ is the closure of $U$ in $E_r$, $x \in X$ and $u \in \bar{U}$. The optimum problem is formulated in the following form.

In the $(n + 1)$-dimensional phase space $X$ of the variables $x^0, x^1, \ldots, x^n$ the point $x_0 = (0, x_0)$ and the line $\Pi$ are given. The line $\Pi$ is assumed to be parallel to the $x^0$ axis and to pass through then point $(0, x_1)$. Among all the admissible controls $u = u(t)$, having the property that the corresponding solution $x(t)$ of the system

$$\frac{dx^i}{dt} = f^i(x^1, \ldots, x^n, u) \qquad i = 0, 1, 2, \ldots, n \tag{24}$$

with initial condition $x(t_0) = x_0$ intersects $\Pi$, find one whose point of intersection with $\Pi$ has the smallest coordinate $x^0$ (Fig. 4B-1) [Note that $x$ is an $n$-vector, but $x$ is an $(n + 1)$-vector].

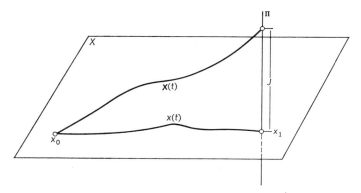

**Fig. 4B-1**  Geometric representation of the trajectories and the performance function.

*4B.2.2  Theorem of the Maximum Principle.* Consider the system of equations

$$\frac{dp_i}{dt} = -\sum_{j=0}^{n} \frac{\partial f^j(x, u)}{\partial x^i} p_j \qquad i = 0, 1, \ldots, n \tag{25}$$

for the auxiliary unknowns $p_0, p_1, \ldots, p_n$ and the function:

$$\mathscr{H}(p, x, u) = [p, f(x, u)] = \sum_{j=0}^{n} p_j f^j(x, u) \tag{26}$$

With the aid of (26), (24) and (25) may be written in the form of the Hamiltonian system

$$\frac{dx^i}{dt} = \frac{\partial \mathscr{H}}{\partial p_i} \qquad i = 0, 1, 2, \ldots, n \tag{27}$$

$$\frac{dp_i}{dt} = -\frac{\partial \mathscr{H}}{\partial x^i} \qquad i = 0, 1, 2, \ldots, n \tag{28}$$

(Note that for $x$, the superscripts indicate the components of the vector, the subscripts indicate the value of $x$ at a given time, e.g. $x_0 = x(t_0)$. For $p$, the subscripts indicate the components of the vector, while the given time is indicated as an argument.) For fixed values of $p$ and $x$, the function $\mathscr{H}$ becomes a function of the parameter $u \in U$. The least upper bound of the values of this function will be denoted by $\mathscr{M}(p, x)$

$$\mathscr{M}(p, x) = \sup_{u \in U} \mathscr{H}(p, x, u)$$

Theorem of the Maximum Principle. Let $u(t)$, $t_0 \leq t \leq t_i$, be an admissible. control such that the corresponding trajectory $x(t)$ which begins at the point $x_0$ at the time $t$ is defined on the entire interval $t_0 \leq t \leq t_1$, and passes, at the time $t_1$, through a point on the line $\Pi$. In order that $u(t)$ and $x(t)$ be optimal, it is necessary that there exist a nonzero absolutely continuous vector function

$$p(t) = [p_0(t), p_1(t), \ldots, p_n(t)]$$

corresponding to the functions $u(t)$ and $x(t)$ so that

1. The function $\mathscr{H}[p(t), x(t), u]$ of the variable $u \in U$ attains its maximum at the point $u = u(t)$ almost everywhere in the interval $t_0 \leq t \leq t_1$

$$\mathscr{H}[p(t), x(t), u(t)] \; (=) \; \mathscr{M}[p(t), x(t)] \tag{29}$$

2. At the terminal time $t_1$ the relations

$$p_0(t_1) \leq 0, \qquad \mathscr{M}[p(t_1), x(t_1)] = 0 \tag{30}$$

are satisfied. Furthermore, it turns out that if $p(t)$, $x(t)$, $u(t)$ satisfy the system (27), (28), and condition (21), the time functions $p_0(t)$ and $\mathscr{M}[p(t), x(t)]$ are constant. Thus, (30) may be verified at any time $t$, $t_0 \leq t \leq t_1$, and not just at $t_1$.

*4B.2.3 Proof.* As we have seen in Section 1, attention is first put on the product $p \cdot \delta x$. Let $\delta x(t) = [\delta x^0(t), \delta x^1(t), \ldots, \delta x^n(t)]$ be defined by the following variational equations

$$\frac{d(\delta x^i)}{dt} = \sum_{j=0}^{n} \frac{\partial f^i[x(t), u(t)]}{\partial x^j} \, \delta x^j \qquad i = 0, 1, \ldots, n \tag{31}$$

with initial condition $\delta x(t_0) = \xi_0$. From (31), we see that to each vector $\xi_0 = \delta x(t_0)$ there corresponds a family of vectors $\xi_t = \delta x(t)$ for $t > t$. Let $X_{t_0}$ and $X_t$ spaces having their origins at $x(t_0)$ and $x(t)$, and $\delta x(t_0) \in X_{t_0}$, $\delta x(t) \in X_t$. The correspondence between $\xi_0$ and $\xi_t$ can be throught of as a linear mapping $A_{tt_0}$ (Fig. 4B-2) giving

$$\xi_t = A_{tt_0}(\xi_0) = \delta x(t) \tag{32}$$

The vectors $A_{tt_0}(\xi_0)$ satisfy system (31), so that we have

$$\frac{d}{dt} [A_{tt_0}(\xi_0)]^i = \sum_{j=0}^{n} \frac{\partial f^i[x(t), u(t)]}{\partial x^j} [A_{tt_0}(\xi_0)]^j \tag{33}$$

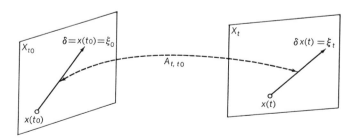

***Fig. 4B-2*** Mapping of the spaces $X_{t0}$ and $X_t$.

Let us compare the two linear and homogeneous systems (31) and (25). Their matrices having the form

$$\left\{ \frac{\partial f^i(x(t), u(t))]}{\partial x^j} \right\}, \quad \left\{ - \frac{\partial f^j[x(t), u(t)]}{\partial x^i} \right\}$$

are the negative transposes of one another. Systems (31) and (25) are adjoint and we can write

$$\frac{d}{dt} \sum_{j=0}^{n} p_j(t)\, \delta x^j(t) = \sum_{j=0}^{n} \frac{dp_j(t)}{dt}\, \delta x^j(t) + \sum_{j=0}^{n} p_j(t) \frac{d\, \delta x^j(t)}{dt}$$

$$= - \sum_{j,k=0}^{n} \frac{\partial f^k[x(t), u(t)]}{\partial x^j} p_k(t)\, \delta x^j(t) + \sum_{j,k=0}^{n} p_j(t) \frac{\partial f^j[x(t), u(t)]}{\partial x^k}\, \delta x^k(t) = 0 \quad (34)$$

So that the product

$$[p(t), \delta x(t)] = \sum_{j=0}^{n} p_j(t)\, \delta x^j(t) = \text{constant}$$

LEMMA 1.   If $p(t)$ is a solution of system (25), and $\xi_0$ is an arbitrary vector given at the point $x(t_0)$,

$$[p(t), A_{tt_0}(\xi_0)] = \text{constant} \quad (35)$$

on the entire interval $t_0 \leq t \leq t_1$. [See (32)]. Equation (35) has a geometric interpretation [Appendix 1B, Section 1B.1, Eq. (16)]. There is a hyperplane $L_t$ in $X_t$, passing through $n(t)$ and defined by the equation

$$\sum_{j=0}^{n} p_j(t)\, x^j = 0 \quad (36)$$

The vector $p(t)$ is normal to $L_t$.

Equations (31) through (36) clarify the relation between $p$ and $\delta x$ if they exist. The attention is put now on the method of obtaining $\delta x$. The variation of the state is in principle due to that of the control. Let us now define $\tau_1, \tau_2, \ldots$ = instants of time; $\delta t_1, \delta t_2, \ldots$ = non-negative time intervals; $v_1, v_2, \ldots$ = points in $U$, and $a = (\tau_i, v_i, \tau, dt, dt_i)$ a certain parameter characterizing the totality of the quantities $\tau_i, v_i, \tau, \delta t, \delta t_i$; we have the following lemma.

LEMMA 2.   If $a = \lambda'a' + \lambda''a'' + \cdots$ (where $\lambda' \geq 0$, $\lambda'' \geq 0, \ldots$), the corresponding vectors $\Delta x$ are given by the same linear relation

$$\Delta x_a = \lambda'\, \Delta x_{a'} + \lambda''\, \Delta x_{a''} + \cdots \quad (37)$$

If we take all the possible symbols $a$ (for fixed $\tau$), the vectors $\Delta x = \Delta x_a$ will fill out a set $K_\tau$ in $X_\tau$. The set $K_\tau$ is a convex cone in $X_\tau$ (see Appendix 1D. Section 1D.1). $K_\tau$ is also called the cone of attainability.

At this point, it is important to know how to compute the position $x_*(t)$ corresponding to the perturbed control $u_*(t)$ and with respect to $A_{\tau t_0}(\xi_0)$ (due to the

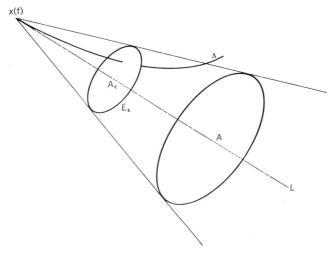

**Fig. 4B-3**   Possible trajectory $\Lambda$ of the state.

initial variation $\xi_0 = \delta x(t_0)$) and $\Delta x$ (the current variation). The following two relations are useful

$$x_*(\tau + \epsilon\, \delta t) = x(t) + \epsilon A_{\tau t_0}(\xi_0) + \epsilon\, \Delta x + O(\epsilon) \tag{38}$$

where $\Delta x$ is a vector which does not depend on $\epsilon$ [$\epsilon$ is a positive quantity and $O(\epsilon)$ a quantity of higher order] and which is defined by

$$\Delta x = f[x(\tau), u(\tau)]\, \delta t + \sum_{i=1}^{s} A_{\tau\tau_i}\{f[x(\tau_i), v_i] - f[x(\tau_i), u(\tau_i)]\}\, \delta t_i \tag{39}$$

LEMMA 3.   Let $\tau(t_0 < \tau < t_1)$ be a regular point* for the control $u(t)$, and let $x(t)$ be the trajectory corresponding to $u(t)$ which starts at point $x_0$. Let $\Lambda$ be a curve which starts at $x(\tau)$, and which has a tangent ray $L$ at this point. If $L$ belongs to the interior of the cone $K_\tau$, there exists a control $u_*(t)$ such that the corresponding trajectory $x_*(t)$, starting at the same point $x_0$, passes through a point $\Lambda$ [distinct from $x(\tau)$] (see Fig. 4B-3).

The purpose of this lemma is to show the geometric behavior of the possible trajectory with respect to $L$ and to the cone $K$.

LEMMA 4.   If the control $u(t)$ and the corresponding $x(t)$, $t_0 \le t \le t_1$, are optimal, then for any regular point $\tau$ $(t_0 < \tau < t_1)$, the ray $L_\tau$, starting at the point $x(\tau)$ and going in the direction of the negative $x^0$-axis, does not belong to the interior of the cone $K_\tau$ (i.e., $L_\tau$ passes either outside this cone, or along its boundary).

---

* See definition of a regular point in [G–30], Ch. 1, pp. 76–77.

This lemma gives the geometric argument for proving the optimality of the performance function $J$. Because, if $u(t)$ and $x(t)$ were not optimal, then there would exist another set of $u_*(t)$ and $x_*(t)$ which would lead to a smaller value of $x^0(t_1)$. This lemma constitutes the pivot-step of the proof, and corresponds to the step between Eqs. (15) and (16) of Section 4B.1. The problem is now to relate the geometric condition to the Hamiltonian function. For this, we argue that, because of the statement of the Lemma 4, since $L_\tau$ does not belong to the interior of $K_\tau$, there exists a hyperplane $\Gamma$ supporting $K_\tau$ at its vertex such that $K_\tau$ lies entirely in one of the two closed half spaces defined by $\Gamma$. The equation of $\Gamma$ (in $X_\tau$) is $\sum_{j=0}^{n} a_j x^j = 0$. Since we may multiply each $a_j$ by the same nonzero number without thereby changing $\Gamma$, we may consider that $K_\tau$ lies in the negative half space $\sum_{j=0}^{n} a_j x^j \leq 0$. In other words, for any vector $\Delta x$ defined by formula (39), the inequality

$$(a, \Delta x) \leq 0 \qquad (\Delta x \in K_\tau) \tag{40}$$

is satisfied. Setting $\delta t_1 = \delta t_2 = \cdots \delta t_s = 0$ in (39), we obtain $\Delta x = f[x(\tau), u(\tau)] \, \delta t$ and from (40),

$$\{a, f[x(\tau), u(\tau)] \, \delta t\} \leq 0$$

Since this inequality holds true for every $\delta t$

$$\{a, f[x(\tau), u(\tau)]\} \leq 0$$

or, in other words

$$\mathscr{H}[a, x(\tau), u(\tau)] = 0 \tag{41}$$

Let

$$p(t, a) = [p_0(t, a), p_1(t, a), \ldots, p_n(t, a)]$$

be the solution of the system (25) with initial condition

$$p(\tau, a) = a \tag{42}$$

The solution $p(t, a)$ is defined on the entire interval $t_0 \leq t \leq t_1$ since system (25) is linear. The following three lemmas lead to Eqs. (29) and (30) which are valid until $\tau$.

LEMMA 5.   If $a$ satisfies condition (40), then

$$\mathscr{H}[p(t, a), x(t), u(t)] = \mathscr{M}[p(t, a), x(t)] \tag{43}$$

at every regular point for $u(t)$ in the half-open interval $t_0 < t \leq \tau$. The proof of this relation is based on the fact that the inequality

$$\mathscr{H}[p(\tau_1, a), x(\tau_1), v_1] - \mathscr{H}[p(\tau_1, a), x(\tau_1), u(\tau_1)] \leq 0 \tag{44}$$

holds true for a given $\tau_1$ and for any $v_1 \in U$. Furthermore, (43) is also valid when $t = \tau$ then

$$\mathscr{H}[p(\tau, a), x(\tau), u(\tau)] = \mathscr{M}[p(\tau, a), x(\tau)] \tag{45}$$

Therefore (41) and (42) imply the following assertion.

LEMMA 6.   If the vector $a$ satisfies (40), then

$$\mathcal{M}[p(\tau, a), x(\tau)] = 0 \tag{46}$$

LEMMA 7.   If the absolutely continuous function $p(t)$ satisfies Eq. (25) and the relation

$$\mathcal{H}[p(t), x(t), u(t)] = \mathcal{M}[p(t), x(t)] \tag{47}$$

almost everywhere on some interval $I$ then the function $\mathcal{M}[p(t), x(t)]$ is constant on all of $I$. ($I$ is the half-open interval $t_0 < t \leq \tau$.)

The following two lemmas extend the validity of Eqs. (29) and (30) to the final time $t_1$.

LEMMA 8.   If $\tau$ and $\tau'$ are regular points for $u(t)$ and $t_0 < \tau' < \tau < t_1$, then $A_{\tau\tau'}(K_\tau) \subset K_\tau$, where $A_{\tau\tau'}$ is the mapping of $X_{\tau'}$, onto $X$.

LEMMA 9.   If the control $u(t)$ and the corresponding trajectory $x(t)$, $t_0 \leq t \leq t_1$, are optimal, the ray $L_{t1}$ issuing from $x(t_1)$ in the direction of the negative $x^0$ axis, does not belong to the interior of $K_{t_1}$.

# Relation Between the Variational Methods and the Continuous Dynamic Programming Method

We have seen in this chapter that the computational requirements derived from the variational methods and the dynamic programming method are quite different. However, here we shall show that the basic analytic expressions of these methods can be derived from each other. A number of works have been done on this question [G-25, Ch. 4; G-26, Ch. 5; G-29, Ch. 7; G-30, Ch. 1; G-34, Ch. 5; G-51, Ch. 8; G-58, Chs. 3 and 4]. Here we shall use the presentation by A. R. M. Noton [G-58] showing the relation between the Pontryagin's equations and the continuous dynamic programming.

*4C.1 The Continuous Dynamic Programming.* Let the problem be to minimize

$$J(u) = \int_0^T v(x, u, t)\, dt \tag{48}$$

subject to

$$\dot{x} = g(x, u, t) \qquad x(0) = c \tag{49}$$

The minimized form of the integral (48) is defined to be the cost function

$$f[x(0), 0] = \underset{u(t)}{\text{Min}}\, J(u) = \underset{u(t)}{\text{Min}} \int_0^T v(x, u, t)\, dt \tag{50}$$

We write the integral of (50) as the limiting form of a series

$$f[x(0), 0] = \lim_{h \to 0} \underset{u(0),\, u(h),\, \ldots,\, u(T-h)}{\text{Min}} \sum_{n=1}^{N-1} v(x, u, nh)h \tag{51}$$

so that we can use the Principle of Optimality and derive

$$f[x(t), t] = \lim_{h \to 0} \underset{u(t)}{\text{Min}}\, \{v(x, u, t)h + f[x(t + h), (t + h)]\} \tag{52}$$

where

$$x(t + h) = x(t) + hg(x, u, t) \tag{53}$$

by (49).

The term $f[x(t+h), (t+h)]$ in (52) is a function of $x_1, x_2, \ldots, x_m$ and $t$, and can be expanded as an $(m+1)$-dimensional Taylor series about $x(t)$ (by taking only two terms)

$$f[x(t+h), (t+h)] = f(x, t) + h\frac{\partial f}{\partial t} + \sum_{n=1}^{m} [x_n(t+h) - x_n(t)]\frac{\partial f}{\partial x_n} + \cdots \quad (54)$$

Substituting the $m$ equations of (53) into (54) and the resulting equations into (52) gives

$$f(x, t) = \lim_{h \to 0} \underset{u(t)}{\text{Min}} \left[ v(x, u, t)h + f(x, t) + h\frac{\partial f}{\partial t} + h\sum_{n=1}^{m} g_n(x, u, t)\frac{\partial f}{\partial x_n} \right] \quad (55)$$

By definition $f(x, t)$ is not a function of $u(t)$, since the latter is eliminated to form $f(x, t)$. The term cancels out therefore in (55), which becomes

$$\underset{u(t)}{\text{Min}} \left[ v(x, u, t) + \sum_{n=1}^{m} g_n(x, u, t)\frac{\partial f}{\partial x_n} \right] + \frac{\partial f}{\partial t} = 0 \quad (56)$$

where the term $\partial f/\partial t$ is separated from the minimization because $f$ is independent of $u$. Equation (56) represents the continuous form of dynamic programming and gives us now two problems:

1. We assume that $u$ takes on its optimal value $\bar{u}$, the minimization operator disappears in (56), and we have to solve the partial differential equation

$$v(x, \bar{u}, t) + \sum_{n=1}^{m} g_n(x, \bar{u}, t)\frac{\partial f}{\partial x_n} + \frac{\partial f}{\partial t} = 0 \quad (57)$$

in $x_1, x_2, \ldots, x_m$ and $t$ in order for obtaining the function $f(x, \bar{u}, t)$ ($v$ and $g_n$ given)

2. Once $f(x, \bar{u}, t)$ is obtained, we rewrite the left-hand member of (56) as a function of $u$ and proceed to minimization with respect to $u$

$$\underset{u(t)}{\text{Min}} \, \phi(u) = \underset{u(t)}{\text{Min}} \left[ v(x, u, t) + \sum_{n=1}^{m} g_n(x, u, t)\frac{\partial f}{\partial x_n} \right] + \frac{\partial f}{\partial t} \quad (58)$$

As we see, the key problem is the solution of the partial differential equation (57) which is not an easy matter. Here, we shall discuss two basically different methods: (a) method of characteristics, (b) method of $f$-expansion. Method (a) will lead to Pontryagin's equations. Method (b) will show an important aspect of the optimization problem: the relation between the continuous dynamic programming and *feedback* controls.

*4C.2  The Continuous Dynamic Programming and the Maximum Principle.*  The aim of the method of characteristics is chiefly to break down the one partial differential equation into a set of ordinary differential equations which can be integrated. According to this method, the partial differential equation

$$F(x_0, x_1, \ldots, x_m, z, p_0, p_1, \ldots, p_m) = 0 \quad (59)$$

where

$$p_n = \partial z / \partial x_n, \quad n = 0, 1, \ldots, m \tag{60}$$

can be replaced by the following set of ordinary differential equations.

$$\frac{dx_n}{dx_0} = \frac{\partial F}{\partial p_n} \bigg/ \frac{\partial F}{\partial p_0} ; \quad n = 1, 2, \ldots, m \tag{61a}$$

$$\frac{dp_n}{dx_0} = -\left(\frac{\partial F}{\partial x_n} + p_n \frac{\partial F}{\partial z}\right) \bigg/ \frac{\partial F}{\partial p_0} ; \quad n = 1, 2, \ldots, m \tag{61b}$$

$$\frac{dz}{dx_0} = \left(\sum_{n=1}^{m} p_n \frac{\partial F}{\partial p_n}\right) \bigg/ \frac{\partial F}{\partial p_0} \tag{61c}$$

Let us compare (59) to (57) and take $x_0 = t$, $z = f$ and $x = (x_1, x_2, \ldots, x_m)$, then

$$p_0 = \frac{\partial f}{\partial t} ; \quad p_n = \frac{\partial f}{\partial x_n} \quad n = 1, 2, \ldots, m \tag{62}$$

Let us put: (the Hamiltonian function)

$$H = v(x, u, t) + \sum_{n=1}^{m} g_n(x, u, t) p_n \tag{63}$$

Equation (59) is to be Eq. (57), thus

$$F \equiv p_0 + H(x, \bar{u}, p, t) \tag{64}$$

From (64), we can evaluate the derivatives required in (61)

$$\frac{\partial F}{\partial z} = 0, \quad \frac{\partial F}{\partial p_0} = 1, \quad \frac{\partial F}{\partial x_0} = \frac{\partial H}{\partial t} + \sum_{i=1}^{r} \frac{\partial H}{\partial \bar{u}_i} \cdot \frac{\partial \bar{u}_i}{\partial t} \tag{65}$$

$$\frac{\partial F}{\partial x_n} = \frac{\partial H}{\partial x_n} + \sum_{i=1}^{r} \frac{\partial H}{\partial \bar{u}_i} \cdot \frac{\partial \bar{u}_i}{\partial x_n} \tag{66a}$$

$$\frac{\partial F}{\partial p_n} = \frac{\partial H}{\partial p_n} + \sum_{i=1}^{r} \frac{\partial H}{\partial \bar{u}_i} \cdot \frac{\partial \bar{u}_i}{\partial x_n} \tag{66b}$$

An interesting point is the introduction of the restriction on $u(t)$ into the mathematical derivations. Let the $r$ components of $u$ be restricted, for example, to lying within or on a boundary in $U$. *Within* the boundary, because of the minimization

$$\frac{\partial H}{\partial \bar{u}_i} = 0 \tag{67}$$

and *on* the boundary $\bar{u}_i$ is constant; hence

$$\frac{\partial \bar{u}_i}{\partial x_n} = \frac{\partial \bar{u}_i}{\partial p_n} = 0 \tag{68}$$

Therefore, for all components $\bar{u}_i$, either (67) or (68) holds true and Eqs. (66) become

$$\frac{\partial F}{\partial x_n} = \frac{\partial H}{\partial x_n} \tag{69}$$

$$\frac{\partial F}{\partial p_n} = \frac{\partial H}{\partial p_n} \tag{70}$$

Substitution of these derivatives into (61 a and b) yields

$$\frac{dx_n}{dt} = \frac{\partial H}{\partial p_n} \tag{71}$$

$$\frac{dp_n}{dt} = -\frac{\partial H}{\partial x_n} \tag{72}$$

Equations (67), (71), and (72) are the Pontryagin's equations.

# The Continuous Dynamic Programming and Feedback Controls

In Chapter 4, we have shown mainly optimization techniques from which a set of numerical values of the optimal control vector are calculated. The values of the state vector appear as a consequence of the control action. In a number of control problems, it is desired to calculate the optimal control vector according to the value of the state vector. The mutual dependence of the two vectors appears then in a feedback and feed-forward configuration. Mathematically, the problem is to obtain the optimal control vector as an analytical function of the state vector $\bar{u} = \bar{u}(x, t)$. Although in cases where the system dynamics are nonlinear this problem is not always solvable, in a number of cases closed-form expressions have been obtained. These are notably the cases where the system dynamics are linear with constant or time-varying parameters, and where the performance function has a linear or a quadratic form. The problem of optimal feedback controls has been treated extensively in the literature [G-28, G-34, G-35, G-49, G-54, G-58]. Here we shall discuss only a small part of it [G-58, Ch. 3].

Let us consider a system described by $m$ first-order differential equations in the state variables $x_1, x_2, \ldots, x_m$

$$\dot{x} = Bx + Cu = g(x, u) \tag{73}$$

where $B$ is an $(m \times m)$ and $C$ an $(m \times r)$ matrix, and the $r$-component vector $u$ is to be chosen to minimize

$$J = \int v(x, u, t)\, dt = \int_0^T (x^T A x + u^T H u)\, dt, \qquad \underset{u(t)}{\text{Min}} J = f \tag{74}$$

where $A$ and $H$ are symmetric matrices. $B$, $C$, $A$, and $H$ can all be functions of time.

From Appendix 4C, we have seen that the continuous dynamic programming leads to the partial differential equation

$$\underset{u(t)}{\text{Min}} \left[ v(x, u) + \sum_{n=1}^{m} g_n(x, u) \frac{\partial f}{\partial x_n} \right] + \frac{\partial f}{\partial t} = 0 \tag{75}$$

which is to be solved for $f$.

The $f$-expansion method consists of assuming a form for $f$ and then of confirming that such a form will in fact satisfy (75). As a trial solution, let

$$f(x, t) = \operatorname*{Min}_{u(t)} \int_0^T (x^T A x + u^T H u)\, dt = x^T K x \qquad (76)$$

where $K$ is an $(m \times m)$ matrix which can always be taken to be symmetrical [See Appendix 1C, Section 1C.2, Eq. (26).] The matrix $K$ is a function only of $t$ and not of $x$. Thus

$$\partial f/\partial t = x^T \dot{K} x \qquad (77)$$

and

$$\begin{bmatrix} \partial f/\partial x_1 \\ \partial f/\partial x_2 \\ \cdot \\ \cdot \\ \cdot \\ \partial f/\partial x_n \end{bmatrix} = 2Kx \qquad (78)$$

The functional equation (75) can be written as

$$\operatorname*{Min}_{u(t)} [x^T A x + u^T H u + (Bx + Cu)^T K x + x^T K(Bx + Cu)] + x^T \dot{K} x = 0 \qquad (79)$$

In order to minimize (79) with respect to $u$, differentiate partially with respect to the $r$ components of $u$ and collect the resulting $r$ equations in one matrix equation. It is

$$Hu + C^T K x = 0 \qquad (80)$$

$H$ being symmetrical. Therefore

$$u = -H^{-1}C^T K x = -Dx \qquad (81)$$

To evaluate $D$, it is necessary to calculate $K$ ($H$ and $C$ known). For this, substitute (81) into (79)

$$x^T A x + x^T K C H^{-1} C^T K x + x^T B^T K x + x^T K B x - 2x^T K C H^{-1} C^T K x + x^T \dot{K} x = 0 \qquad (82)$$

or

$$x^T(A - KCH^{-1}C^T K + B^T K + KB + \dot{K})x = 0 \qquad (83)$$

Equation (83) can be satisfied for all $x$ if

$$\dot{K} = KCH^{-1}C^T K - A - B^T K - KB \qquad (84)$$

Equation (84) is a generalized matrix form of Ricatti's differential equation, this is an ordinary nonlinear differential equation, the solution of which will yield the matrix $K$, and hence $D$. From (81), we see that the control $u$ is obtained as a function (furthermore, linear) of $x$, and a feedback loop can thus be implemented.

*Example*

$$\dot{x}_1 = -x_1 + u$$
$$\dot{x}_2 = x_1$$

$$J = \int_0^T (x_2^2 + .1u^2)\, dt$$

$$A = \begin{bmatrix} 0 & 0 \\ 0 & 1 \end{bmatrix} \qquad B = \begin{bmatrix} -1 & 0 \\ 1 & 0 \end{bmatrix} \qquad C = \begin{bmatrix} 1 \\ 0 \end{bmatrix} \qquad H = [.1]$$

$$K = \begin{bmatrix} k_{11} & k_{12} \\ k_{21} & k_{22} \end{bmatrix} \quad \text{where} \quad k_{12} = k_{21}$$

By integrating numerically the corresponding equation (84), the values of $k_{11}$, $k_{12} = k_{21}$, $k_{22}$ are obtained. The control signal is then calculated by (81), which in this case becomes

$$u = -1.700x_1 - 3.155x_2$$

# REFERENCES

[4-1]   McCann, M. J., "Variational Approaches to Optimal Trajectory Problems," in [G-31].

[4-2]   Kirchmayer, I. K. and R. J. Ringlee, "Optimal Control of Thermalhydro System Operation," Second IFAC Congress, Basle, September 1963.

[4-3]   Lefkowicz, I. and D. P. Eckman, "Application and Analysis of a Computer Control System," *J. Basic Eng.*, December 1959, 569–577.

[4-4]   Gould, L. A. and W. Kipiniak, "Dynamic Optimization and Control of a Stirred-Tank Chemical Reactor," *Tr. AIEE, Appl. Ind.*, January 1961, 734–746.

[4-5]   Thomae, H. F., "A Technique for Optimization of Ascent Trajectories and Propellant Loadings of Multistage Space Vehicles," *Raumfahrtforschung*, Heft 3/64, 115–123.

[4-6]   Melbourne, W. G. and C. G. Sauer, Jr., "Payload Optimization for Power-Limited Vehicles," *Tech. Rep.* N° 32–118. Jet Propulsion Laboratory, 1961.

[4-7]   Melbourne, W. G., "Three Dimensional Optimal Thrust Trajectories for Power-Limited Propulsion Systems," *ARS J.*, December 1961, 1723–1728.

[4-8]   Melbourne, W. G. and C. G. Sauer, Jr., "Optimum Interplanetary Rendezvous with Power-Limited Vehicles," *AIAA J.*, **1**, n 1, January 1963, 54–60.

[4-9]   Ross, S., "Composite Trajectories Yielding Maximum Coasting Apogee Velocity," *ARS J.*, November 1959, 843–848.

**[4-10]**  Cavoti, C. R., Necessary and Sufficient Conditions for an Optimum in a Class of Flight Trajectories, *Z. Flugwissenschaften*, Heft 12, 1963, 485–495.

**[4-11]**  Cavoti, C. R., "Numerical Solutions on the Astronautical Problem of Thrust Vector Control for Optimum Solar Sail Transfer," *Rep.* R 63-SD 30, Space Sciences Laboratory, General Electric, March 1963.

**[4-12]**  Cavoti, C. R., "Method of Characteristic Constants and Retro-Thrust Optimality in a Gravitational Field-Necessary and Sufficient Conditions," *AIAA Aerospace Sciences Meeting*, New York 1964.

**[4-13]**  Graham, R. G., "The Effects of State Variable Discontinuities in the Solution of Variational Problems," *Rep.* TDR 269 (4550-20)-4, AD 602480, *Aerospace Corp.*, El Segundo, California, July 1964.

**[4-14]**  Bryson, Jr, A. E., W. F. Denham, and S. E. Dreyfus, "Optimal Programming Problems with Inequality Constraints," *AIAA J.*, **1,** n 11, November 1963, 2544–2550.

**[4-15]**  Denham, W. F., "Some Inequality Constraint Problems in the Calculus of Variations," *Tech. Rep.* n⁰ 436, AD 601 842, Cruft Laboratory, Harvard University, April 1964.

**[4-16]**  Chang, S. S. L., "A Modified Maximum Principle for Optimum Control of a System with Rounded Phase Space Coordinates," Second IFAC Congress, Basle, 1963.

**[4-17]**  Kaufmann A., La programmation dynamique, Collection "Information Scientifique," Bull, Paris, 1961.

**[4-18]**  Naslin P., Introduction élémentaire à la programmation dynamique, *Automatisme*, March 1965, 114–118; April 1965, 160–164.

**[4-19]**  Ruge, H., "On the Optimal Control of Hydroelectric Power Systems," second IFAC Congress, Basle, 1963.

**[4-20]**  Smith, F. T., "An Introduction to the Application of Dynamic Programming to Linear Control Systems," *Rand Rep.*, RM-3526-PR, February 1963.

**[4-21]**  Smith, F. T., "The Application of Dynamic Programming to a Linear Time-varying System," *Rand Rep.* RM-4099-PR, July 1964.

**[4-22]**  Burkhart, J. A. and F. T. Smith, "Application of Dynamic Programming to Optimizing the Orbital Control Process of a 24-Hour Communication Satellite," *AIAA J.*, November 1963, 2551–2557.

**[4-23]**  Kelley, H. J., "Gradient Theory of Optimal Flight Paths," *ARS J.*, October 1960, 947–953.

**[4-24]**  Bryson, Jr, A. E., W. F. Denham, F. J. Carroll, and K. Mikami, "Determination of Lift or Drag Programs to Minimize Re-Entry Heating," *J. Aerospace Sci.*, April 1962, 420–430.

**[4-25]**  Kelley, H. J., "Methods of Gradients," Ch. 6 in *Optimization Techniques*, G. Leitmann Ed., Academic Press, 1962.

**[4-26]**  Hadley, G., *Nonlinear and Dynamic Programming*, Addison-Wesley, 1964, Ch. 9.

[4-27]   De Backer, W., "The Generalized Gradient, Its Computational Aspects and Its Relations to the Maximum Principle," Internal Report, Scientific Data Processing Center-Euratom-CCR, Ispra: Italy, April 1963.

[4-28]   Meditch, J. S. and L. W. Neustadt, "An Application of Optimal Control to Midcourse Guidance," Second IFAC Congress, Basle, 1963.

[4-29]   Pearson, J. D., Theoretical Methods of Optimization: Dynamic Programming, in [G-31].

[4-30]   McIntyre, J. E., "Neighboring Optimal Terminal Control with Discontinuous Forcing Functions," *AIAA J.*, **4**, n 1 (January 1966), 141–148.

[4-31]   Kahne, S. J., "Optimal Cooperative State Rendezvous and Pontryagin's Maximum Principle," *Internat. J. Control* (G.B.), **2**, n 5 (November 1965), 425–431.

[4-32]   Shapiro, S., "Lagrange and Mayer Problems in Optimal Control," *Automatica* (G.B.), **3**, n 3–4 (January 1966), 219–230.

[4-33]   Rubio, J. E., "On the Uniqueness of the Solution of Some Equations of Optimal Control Theory," *Internat. J. Control* (G.B.), **3**, n 1 (1966), 69–78.

[4-34]   Mayne, D., "A Second-order Gradient Method for Determining Optimal Trajectories of Nonlinear Type Discrete-time Systems," *Intern. J. Control*, **3**, n 1 (1966), 85–95.

[4-35]   Mangasarian, O. L., "Sufficient Conditions for the Optimal Control of Nonlinear Systems," *SIAM J. Control*, **4**, n 1 (February 1966), 139–152.

[4-36]   Douglas, J. M., "The Use of Optimization Theory to Design Simple Multi-variable Control Systems," *Chem. Eng.*, *Sci.* (G.B.), **21**, n 6–7 (June–July 1966), 519–531.

[4-37]   Lastman, G. J., "Optimization of Nonlinear Systems with Inequality Constraints," Ph.D. Thesis, Texas Univ., Austin, N 67-8/23.

[4-38]   Schoenberza, M., "System Optimization with Fixed Control Structure" (Submarine diving), R278 AD-628112 (January 1966), 82.

# 5

---

# DYNAMICAL
# SUBOPTIMIZATION
# TECHNIQUES

## 5.1 INTRODUCTION

### 5.1.1 Optimization and Suboptimization

WE HAVE SEEN in the preceding chapter that all three categories of dynamical optimization techniques encounter computational difficulties in their application, especially in the case of nonlinear systems. A great number of works have been devoted to the solution of these difficulties. In one way or another, the suggested methods imply the concept of "approximation," justifying, therefore, the name of suboptimization techniques.

An overall examination shows that these techniques are still in the research stage. Few of them have been applied to concrete engineering problems leading to a technological implementation. Lacking valuable comparative computing experimentations, it is even difficult to judge the relative merits of the methods intended to solve the same problem.

Furthermore, like all fields in expansion, inevitable confusions arise:

1. Similar words as approximation, successive approximation, iterative process, linearization, quasilinearization, are used for different meanings.

2. Some techniques attack the computing difficulties by using analytical expressions; others attack the analytical part of the optimization procedure for solving indirectly the computing difficulties.

3. Some methods use one single strategy, others use a combination of several strategies.

4. Some strategies are really new, others are contained in part in the previous optimization techniques.

5. Several identical techniques appear different because of the use of different mathematical reasoning; other dissimilar techniques, on the contrary, solve the same problem.

In this chapter, we shall clarify all these questions in the following way:

1. We shall summarize the various stages of the three optimization procedures in order to (a) state more precisely the computational difficulties; (b) show explicitly the various mathematical entities (functions, functionals, equations) to which the approximation can be applied; and (c) show the level at which the suboptimization technique operates. This is done in this subsection with the help of Table 5.1.

2. We shall classify the various kinds of elementary strategies used alone, or in combined form, in suboptimization techniques.

3. We shall expose successively typical suboptimization techniques for (a) overcoming the $2PBVP$; (b) overcoming the dimensionality problem; and (c) improving the gradient method. This will be done in Sections 5.2, 5.3, and 5.4.

4. In Section 5.5, we shall give some concluding remarks and indications of research direction.

The following remarks must be made concerning Table 5.1: (1) The Pontryagin formulation is used in the column of Variational Calculus, the $2PBVP$ is the same for either classical or Pontryagin formulation; (2) side constraints are not considered in the indicated procedure, their introduction does not change essentially the computational difficulties; (3) the right member of $f_N(c)$, although the particular notation $\underset{u}{\text{Min}}$, is a functional and can be consequently manipulated (derivated, approximated, etc); (4) The computational difficulties in the Variational Calculus and the gradient method are apparently similar. It is, in fact, the case because they both concern a successive approximating process. However, there are differences in the number of initial guesses needed, and in the resulting functions of the iterative process.

### 5.1.2    Elementary Strategies

In order to solve the computational difficulties of the three optimization techniques mentioned previously, a great number of suboptimization techniques have been suggested. Some of them attack the analytical extremalizing stage of the basic procedures; others attack the numerical computing stage. Simple ones use some quite fundamental process that we

*Table 5.1* Computational Difficulties of Dynamical Optimization Techniques and Application Points of Dyamical Subopti-mization Techniques

| | Variational Calculus (Pontryagin's formulation) | Dynamic Programming | Gradient Method |
|---|---|---|---|
| Problem formulation stage | State function $x(t)$; $x \in X$<br>Control function $u(t)$; $u \in U$<br>System constraint equations<br><br>$\varphi = \dot{x} - g(x, u, t) = 0$  $x(t_0) = x_0$<br><br>Other constraint equations<br><br>$\varphi(x, u, t) = 0\big|_{t=t_0}^{t=t_f}$<br><br>Functional to be optimized<br><br>$S = \sum_{1}^{n} c_i x_i(t_f)$ | State function $x(t)$; $x \in X$<br>Control function $u(t)$; $u \in U$<br>System dynamics<br><br>$\dot{x} = g(x, u, t)$  $x(t_0) = c$<br><br>or<br><br>$x^{(n+1)} = x^{(n)} + \tau g(x^{(n)}, u^{(n)}, t^{(n)})$<br><br>Penalty functional<br><br>$v(x, u, t)$<br><br>Cost functional to be optimized<br><br>$J = \int_{0}^{T} v(x, u, t)\, dt$ | State function $x(t)$; $x \in X$<br>Control function $u(t)$; $u \in U$<br>System dynamics (equations)<br>$\dot{x} = g(x, u, t)$  $x(t_0) = x_0$<br>Constraints (equations)<br><br>$\varphi(x, u, t) = 0\big|_{t=t_0}^{t=t_f}$<br><br>Functional to be optimized<br><br>$J = \int_{t_0}^{t_f} F(x, u, t)\, dt$ |

**Table 5.1** *continued*

| | Variational Calculus (Pontryagin's formulation) | Dynamic Programming | Gradient Method |
|---|---|---|---|
| Analytical extremalizing stage | New equations $g_{n+1}$:<br><br>$\dot{x}_{n+1} = \dfrac{dS}{dt} = g_{n+1}; \; x_{n+1}(t_0) = 0$<br><br>Adjoint system:<br><br>$\dot{p}_i = -\displaystyle\sum_{k=1}^{n+1} p_k \dfrac{\partial g_k}{\partial x_i}; \; p_i(t_f) = -c_i$<br><br>$i = 1, 2, \ldots, (n+1)$<br><br>Hamiltonian functional:<br><br>$H = \displaystyle\sum_{1}^{n+1} p_i g_i(x, u, t)$<br><br>Extremalizing equations:<br><br>$\dot{p}_i = -\dfrac{\partial H}{\partial x_i} = -\displaystyle\sum_{k=1}^{n+1} p_k \dfrac{\partial g_k}{\partial x_i}$<br><br>$\dot{x}_i = \dfrac{\partial H}{\partial p_i} = g_i(x, u, t)$<br><br>$\dfrac{\partial H}{\partial u} = \displaystyle\sum p_i \dfrac{\partial g_i(x, u, t)}{\partial u} = 0$ | Recurrent equation<br><br>$f_1(c) = \underset{u}{\text{Min}} \, [v(c, u)\tau]$<br><br>$f_N(c) = \underset{u}{\text{Min}} \, [v(c, u)\,\tau$<br>$\qquad\qquad + f_{N-1}[c + g(c, u)\tau]$<br><br>$N = 2, 3, \ldots$<br><br>The quantity<br><br>$f_n(c) = \text{Min } J$<br><br>is function of $c$, $N$, $x(t)$, $u(t)$.<br>It is also subject to analytical approximations. | Adjoint system (equations)<br><br>$\dot{\lambda}_i = -\displaystyle\sum_{j=1}^{n} \dfrac{\partial g_j}{\partial x_j} \lambda_j$<br><br>Influence function $\lambda_i(t)$<br>Control influence function<br><br>$\lambda_u(t) = \displaystyle\sum_{i=1}^{n} \lambda_i(t) \dfrac{\partial g_i}{\partial u}$<br><br>Similar procedure for taking into account constraints $\psi$. |

| | | | |
|---|---|---|---|
| Numerical computing stages | (a) From $\partial H/\partial u = 0$, obtain<br>$$u = u(p, x).$$<br>(b) Substitute $u = u(p, x)$ in $p$ and $x$ so that<br>$\dot{p} = \dot{p}(p, x)\quad p(t_f)$ known<br>$\dot{x} = \dot{x}(p, x)\quad x(t_0)$ known<br>(c) Guess $p(t_0)$.<br>(d) Solve by numerical integration for $p(t)$ and $x(t)$.<br>(e) Compare<br>$$p(t_f)_{\text{calc.}} \text{ to } p(t_f)_{\text{known}}.$$<br>(f) Restart from step (c) if the two values of $p(t_f)$ are not equal.<br>(g) When the correct solutions of $p(t)$ and $x(t)$ are obtained, compute the control program $u(t)$ from<br>$$u = u(p, x).$$ | (a) Use the recurrent equations to calculate $f$ for all values of $c$ and $N$.<br>(b) Memorize sequences of<br>$$\left.\begin{array}{l} f_k(c) \\ u_k(c) \end{array}\right\}\ k = 1, 2, \ldots, N$$<br>(c) Exploit the chart for given values of $N$ and $c$.<br>(d) Final objective is $u_k(c)$. | (a) Calculate $\partial g_j/\partial x_i$ for obtaining the analytical form of the adjoint system.<br>(b) Guess $u(t) = u_0(t)$.<br>(c) Numerically integrate<br>$\dot{x} = g(x, u, t)$ knowing $x_0$ and $u_0(t)$ for obtaining $x(t)$.<br>(d) Guess $\lambda_i(t_0) = \lambda_{i0}$.<br>(e) Solve numerically<br>$$\dot{\lambda} = \dot{\lambda}(x, u, \lambda, t)$$<br>knowing $u_0(t)$, $x_0$, $x(t)$, $\lambda_0$.<br>(f) Calculate $\partial g/\partial u$ and $\lambda_u$.<br>(g) Calculate the gradient factor<br>$$k = \partial J \Big/ \int_0^T [\lambda_u(t)]^2\, dt.$$<br>(h) Iterate by modifying $u(t)$ according to<br>$$u_{\text{new}} = u_{\text{old}} + k\lambda_u(t).$$<br>(i) Stop when $\delta J \cong 0$. |
| Computational difficulties | (a) No help to make correctly the initial guess $p(t_0)$.<br>(b) High sensitivity of the oscillations with respect to the initial guess.<br>(c) Uncertain monotonicity and convergence.<br>(d) Intrinsically lengthy iterative procedure. | Great (even prohibitive) memory capacity needed in high-order dimensional problems. | (a) Possible singularities of the matrix $\partial g_j/\partial x_i$.<br>(b) Uncertain monotonicity and convergence.<br>(c) Difficulty for converging simultaneously $\lambda_{i0}$ and $u(t)$. |

might call "elementary strategies"; other ones, more complicated, use a combination of these strategies. Seven important elementary strategies, each having one distinct feature, can be recognized as follows:

(a) Adjoint strategy, or Lagrange-multiplier strategy.
(b) Imbedding strategy.
(c) Constrained-Gradient strategy.
(d) Approximation strategy.
(e) Approximating process strategy.
(f) Quasilinearization strategy.
(g) Grid-manipulation strategy.

The first three of these strategies are contained in germ form already in the three previously described fundamental optimization procedures; they are employed here for solving computational difficulties. Strategies (d) through (g) are all more or less related to the concept of "approximation." The distinct features come from the fact that it is applied to functions in some cases, and to functionals or equations in others; it operates in the time domain in some instances, and in the space domain in others. The distinction is also viewed in the result of the approximation: (1) a discrete sequence is obtained from a continuous time-function or a difference differential equation is obtained from a continuous differential equation; (2) a quantized space functional is obtained from a continuous space functional; (3) a linearized functional or equation, more tractable, is obtained from a nonlinear functional or equation. Covering all these distinct features, we can, however, cite a general feature: various forms of polynomial expansion and Taylor series expansion are commonly used for the approximation purpose.

For a better understanding of how they are used in suboptimization techniques, we describe in greater detail the contents of the seven elementary strategies.

(a) ADJOINT STRATEGY (or Lagrange-multiplier strategy). *Situation.* Presence of a performance functional $J = \int F \, dx$, system constraint $\varphi = 0$ and side constraints $\psi = 0$.

*Strategy Step 1.* Form the constrained performance functional (or the augmented functional) $F^* = F + \lambda_1 \varphi + \lambda_2 \psi$, $\lambda_1(t)$, and $\lambda_2(t)$ being arbitrary functions of time.

*Step 2.* Derive the Euler equations from $\delta F^* = 0$, so that there are the same number of equations [algebraic for $u(t)$ and differential for $x(t)$ and $\lambda(t)$] and of unknowns $u(t)$, $x(t)$, and $\lambda(t)$.

*Comments.*

1. In static optimization problems, this strategy is used for solving constrained optimization multidimensional problems.

2. The variables $\lambda(t)$ are slack-variables, or co-state variables; they are supposed to evolve jointly with the state variables.

3. In the suboptimization techniques, this strategy will be used to adjoint unknown initial conditions.

(b) IMBEDDING STRATEGY. *Situation.* 2 *PBVP* relative to the system $\dot{x} = \dot{x}(p, x)$, $\dot{p} = \dot{p}(p, x)$, $x(t_0)$, $p(t_f)$ known; $x(t_f)$ also known in some cases, $p(t_0)$ unknown.

*Strategy.* Optimize by dynamic programming. Euler equations and $p(t)$ disappear, as well as the 2*PBVP*. The only knowledge of $x(t_0) = c$ suffices to derive the optimal control $u(t)$ by the recurrent algorithm.

*Comments.*

1. The imbedding strategy changes the 2*PBVP* into a one-sided boundary problem easily solved. But the dimensionality problem is resulted. There is no true benefit.

2. See "Invariant imbedding," a better strategy for solving the 2*PBVP*.

(c) CONSTRAINED-GRADIENT STRATEGY. *Situation:* Determine $u(t)$ to optimize $J = \int F(x, u, t) \, dt$ constrained by $\dot{x} = g(x, u, t)$. Normally, without constraint, one would operate $\partial J/\partial u = 0$.

*Strategy Step 1:* Use influence functions $\lambda_i(t)$ and control influence function $\lambda_u(t)$ to evaluate the influence $\delta x(t_0)$ and $\delta u$ on $\delta J$.

*Step 2:* Use the gradient factor $K$ indicating the steepest direction of search.

*Comments.*

1. The $\lambda_i$-functions are obtained by solving the same adjoint system as in strategy (a), but their purpose is more teleological.

2. This strategy can be used in all suboptimization techniques involving an iterative or an approximating process.

(d) APPROXIMATION STRATEGY.[1] *Situations.*

(1) Existence of one- or multidimensional functions known only by experimental points, and which involve computational difficulties in an optimization procedure [e.g. $f_k(c)$, $n_k(c)$ in the dynamic programming procedure].

(2) Existence of one- or multidimensional analytically known functions or functionals involving analytical untractability in an optimization procedure [e.g. $\varphi$, $g$, $\psi$, $S$, $H$ in the Variational procedure; $g$, $v$, $J$, $f_N(c)$ in the dynamic programming procedure].

*Strategies.*

(1) Use a polynomial expansion, e.g. of the form $f(x) = \sum a_i p_i(x)$ where $a_i$ are unknown coefficients, $p_i(x)$ are known functions. Determine by identification the coefficients $a_i$ by using the known experimental points.

---

[1] For details on Taylor series, differentiation, and derivation of functionals, see [G-3, Chs. 3, 4, and 5] and [G-4, Ch. 1].

(2) Expand the function or functional into a Taylor series around some known points, e.g.

$$f(x + h) = f(x) + hf'(x) + \frac{h^2}{2!}f''(x) + \cdots + \frac{h^n}{n!}f^{(n)}(x)$$

where the coefficients are the derivatives of various degrees $d^k f/dx^k$, $k = 1$, $2, \ldots, n$, taken on the known point.

*Comments.*

1. The desired criteria of an approximation are conformity, unicity, and uniformity in the whole argument space. These qualities are determined by the form of the expansion, and the number of terms used.

2. The linearization of a function or a functional is an approximation where only terms of first degree are taken into account, e.g.

$$f(x + \Delta x) = f(x) + \Delta x f'(x); \qquad \Delta f = f(x + \Delta x) - f(x) = \Delta x f'(x)$$

while an approximation can have various degrees of accuracy according to the number of terms taken; therefore, it is not necessarily linear.

3. Clearly, the approximation is used practically in all suboptimization techniques.

(e) APPROXIMATING PROCESS STRATEGY (OR SUCCESSIVE APPROXI-MATION PROCESS). *Situation.* Existence of one or a set of algebraic equations, differential equations (e.g. adjoint equations, Euler equations), partial differential equations [e.g. $\partial H/\partial u = 0$, $f_N(c) = \underset{u}{\text{Min}} (-)$], the closed-form solution of which involves analytical difficulties.

*Strategy Step 1:* Make an initial guess.

*Step 2:* Establish an iterative algorithm.

*Step 3:* Define a stopping condition.

*Step 4:* Operate the iterations starting from the initial guess until the satisfaction of the stopping condition.

*Comments.*

1. The Gradient method did use an approximating process. The approximating process suggested for the $2PBVP$ suffers from the convergence slowness.

2. The desired criteria of an approximating process are: good initial guess, existence and high rate of convergence, unicity of solution.

3. The solution at each iteration constitutes an approximation of the desired solution.

(f) QUASILINEARIZATION STRATEGY [5-7, 5-8, 5-9]. *Situation.* Existence of a set of nonlinear differential equations. Difficulty of solution because of the nonlinearity and the $2PBVP$.

Example:   $\dot{x} = b(x, p)$    $x(0) = c$    $\dot{p} = d(x, p)$    $p(T) = 0$

*Strategy Step 1:* Expand $b$ and $d$ in a Taylor series in the functions $x(t)$

and $p(t)$ around $x_n$ and $p_n$ and only retain terms up to first order.

$$\dot{x}_{n+1} = b(x_n, p_n) + \frac{\partial b}{\partial x}\bigg|_{x=x_n} (x_{n+1} - x_n) + \frac{\partial b}{\partial p}\bigg|_{p=p_n} (p_{n+1} - p_n)$$

$$\dot{p}_{n+1} = a(x_n, p_n) + \frac{\partial d}{\partial x}\bigg|_{x+x_0} (x_{n+1} - x_n) + \frac{\partial d}{\partial p}\bigg|_{p=p_n} (p_{n+1} - p_n)$$

*Step 2:* These differential equations are linear in $x_{n+1}$ and $p_{n+1}$. Use the property of linear equations: the solution is the sum of the particular solution $r(t)$ plus the sum of weighted homogeneous solutions $c_1 s_2(t)$, $c_2 s_2(t)$

$$x(t) = x_r(t) + c_1 x_{s1}(t) + c_2 x_{s2}(t)$$

$$p(t) = p_r(t) + c_1 p_{s1}(t) + c_2 p_{s2}(t)$$

*Step 3:* Combine results of step 1 and step 2 for obtaining an iterative algorithm.

*Comments.*

1. This strategy uses (a) the approximation strategy (Taylor series); (b) linearization (first-order terms plus the property of linear differential equations; (c) approximating process strategy (iterative process).

2. This strategy will be used in the generalized Newton-Raphson method.

(g) GRID-MANIPULATION STRATEGY. *Situation.* Existence of multi-dimensional operating space in $x_i(t)$, $u_j(t)$, $p_i(t)$, $t$. $i = 1, 2, \ldots, n$; $j = 1, 2, \ldots, r$. The largeness of this space causes prospection or computation difficulties in various optimization techniques.

*Strategies.*

(1) *Coarse-fine strategy:* use few points of the space and narrow the optimal area successively.

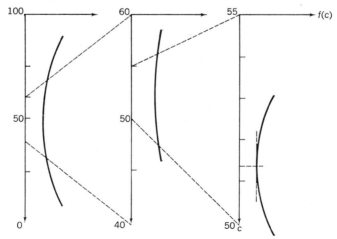

(2) *Tube strategy:* only consider in the phase-space a tube around a coarse solution.

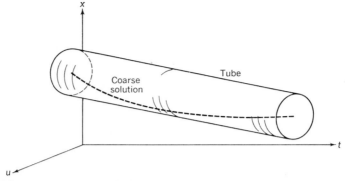

(3) *Bloc strategy:* Transit in the phase-space by small blocs.

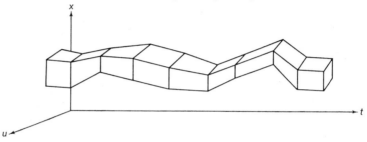

(4) *Discretization strategy:* Discretize the state-space so that the performance function $J(x)$ can be calculated more easily.

(5) *Hypersphere strategy:* Consider a hypersphere around the point $(p, x)_{t=t_0}$ so that it is more easily attainable.

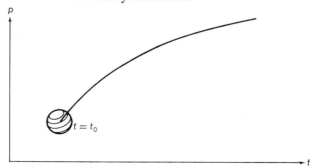

*Comments.*

1. These strategies are types of nonconventional approximations.
2. These strategies, except (4) and (5), are used for overcoming the dimensionality problem.

## 5.2   OVERCOMING THE 2*PBVP*

### 5.2.1   Approximating Process Around the Unknown Optimal Control Function

Instead of the lengthy hunting of the initial conditions $p_i(t_0)$, this sub-optimization technique hunts the unknown optimal control function $u^*(t)$. By doing so this technique has the possibility of using the Maximum Principle condition $\partial H/\partial u = 0$ as a natural stopping condition [5-1, 5-2, 5-3, 5-4][1].

As noted in Table 5.1, the three unknown optimal vectors $x$, $p$, and $u$ are solutions of the three systems

$$\dot{x} = \frac{\partial H(x, p, u)}{\partial p} = g(x, u, t) \qquad x(0) = x_0 \qquad (5\text{-}1)$$

$$\dot{p} = -\frac{\partial H(x, p, u)}{\partial x} = -\sum p \frac{\partial g(x, u, t)}{\partial x} \qquad p(t_f) = -c_i \qquad (5\text{-}2)$$

$$\frac{\partial H}{\partial u} = \sum p \frac{\partial g(x, u, t)}{\partial u} = 0 \qquad (5\text{-}3)$$

The first two systems consist of differential equations, while the third consists of algebraic equations in $x$, $p$, and $u$. These equations may be linear or nonlinear according to the original form of $g(x, u, t)$. In order to avoid iterations around the differential equations, one method is to try successive approximations of $u(t)$. The approximating process is:

1. Initial guess: $\tilde{u}$, an arbitrary function of $u(t)$.
2. Algorithm:
   (a) Solve with $\tilde{u}(t)$ and by using numerical integration method

$$\dot{\tilde{x}} = g(\tilde{x}, \tilde{u}, t) \quad \text{to yield} \quad \tilde{x}(t) \qquad (5\text{-}4)$$

[1] Regarding the use of approximating process strategy for solving the 2*PBVP*, the reader can find: (1) an example in [5-5] where some system properties have eased the initial guess of $p_i(t_0)$; (2) in [5-6] how upper and lower bounds of $p_i(t)$ can be determined by using primal-and-dual-problem concept similar to that exposed in Chapter 2 under linear programming methods.

(b) Solve with $\tilde{u}(t)$ and $\tilde{x}(t)$, and always by using numerical integration method

$$p = \sum p \frac{\partial g(\tilde{x}, \tilde{u}, t)}{\partial x} \quad \text{to yield} \quad \hat{p}(t) \qquad (5\text{-}5)$$

(c) Calculate the error or value-function with $\tilde{u}$, $\tilde{x}$, and $\tilde{p}$

$$M = \frac{\partial H(\tilde{p}, \tilde{x}, \tilde{u})}{\partial u} = \tilde{p} \sum \frac{\partial g(\tilde{x}, \tilde{u}, t)}{\partial u} \qquad (5\text{-}6)$$

(d) If $M = 0$ as required by Eq. (5-3), then $\tilde{u}$ is the optimal control $u^*(t)$. If $M \neq 0$, $\tilde{u}$ was not optimal, then choose new values of $u$ according to the magnitude of $M$.

It is interesting now to compare this technique with the procedure of the Gradient method (Table 5.1). If we replace $p$ by $\lambda_i$, we see immediately that $p$ is the influence function of the Gradient method. Also, by Eq. (5-3), we can easily identify $M$ with the control influence function $\lambda_u(t)$. Furthermore, we know that one of the $g$-functions, namely $g_{n+1}$ represents the performance function $S$ or $J$, so that the condition $\partial J/\partial u = 0$ is contained within $\partial H/\partial u = 0$. There are, however, two differences.

(1) Here, we do not have to guess the initial co-state $\lambda_i(t_0)$; this is an advantage.

(2) Here, new values of $u$ are chosen according to the *magnitude* of $M$ only, the steepest *direction* towards $\delta J = 0$ is ignored.

Therefore, obvious possible improvement of this suboptimization technique is to calculate the gradients $\partial J/\partial u$ and take them into account for the $u$-adjustments.

## 5.2.2  Combined Strategy Based on Invariant Imbedding

We have seen, when discussing elementary strategies, that the imbedding strategy transforms a difficult $2PBVP$ into an easy $1PBVP$. However, the dimensionality problem is also brought in. The present suboptimization technique uses the following:

1. An invariant imbedding strategy which transforms the $2PBVP$ related to differential equations, to be solved in the time domain, into a $1PBVP$ related to partial differential equations, to be solved in the space domain.

2. The quasilinearization strategy for implementing an iterative algorithm for solving the $1PBVP$ [5-7; 5-8; 5-9; 5-11; G-37, 135–145].

We consider again Eqs. (5-1) through (5-3). Assuming that Eq. (5-3) has been solved yielding $u$ in terms of $p$ and $x$, then Eqs. (5-1) and (5-2) can be written as

$$\dot{x} = b(x, p) \qquad x(0) = c \tag{5-7}$$

$$\dot{p} = d(x, p) \qquad p(T) = 0 \tag{5-8}$$

We try to imbed this 2*PBVP* in a more general one valid for any initial state $c$ and any initial time $\tau$. It is obvious that the correct choice of the unknown initial co-state $p$ will be a function of initial state and initial time. We note this dependence by

$$p(\tau) = r(c, \tau) \tag{5-9}$$

Let us now consider the process at time $\tau + \Delta$, approximating $\dot{p}$ by

$$\dot{p} = \frac{p(\tau + \Delta) - p(t)}{\Delta} \tag{5-10}$$

We see that the ordinate of the $p$-curve for the abscissa $\tau + \Delta$ is:

$$\begin{aligned} p(\tau + \Delta) &= p(\tau) + \dot{p}\Delta \\ &= r(c, \tau) + d[c, r(c, \tau), \tau]\Delta \end{aligned} \tag{5-11}$$

according to Eqs. (5-9) and (5-8). On the other hand, from Eq. (5-9), we see that

$$p(\tau + \Delta) = r\{c + b[c, r(c, \tau), \tau]\Delta, \tau + \Delta\} \tag{5-12}$$

The right-hand side of Eq. (5-12) is the value of the $p$-curve at time $\tau + \Delta$ corresponding to a value of the $x$-curve at time $\tau + \Delta$ of initial state $c + b[c, r(c, \tau), \tau]\Delta$. Equations (5-11) and (5-12) show that the function $r(c, \tau)$ satisfies the functional equation:

$$r\{c + b[c, r(c, \tau), \tau]\Delta, \tau + \Delta\} = r(c, \tau) + d[c, r(c, \tau), \tau]\Delta \tag{5-13}$$

In the limits as $\Delta$ tends to zero, this equation becomes the first-order quasi-linear partial differential equation

$$b(c, r, \tau)r_c + r_\tau = d(c, r, \tau) \qquad \tau < T \tag{5-14}$$

In addition $r$ satisfies, in view of Eq. (5-8)

$$r(\dot{c}, T) = 0 \tag{5-15}$$

Note that in Eq. (5-14), $b$ and $d$ are known functions, $r_c$ and $r_\tau$ are the partial derivatives of the unknown function $r(c, \tau)$ with respect to the arguments $c$ and $\tau$. We see that the preceding mathematical treatment, called invariant imbedding, has derived a partial differential equation, i.e., Eq. (5-14). The variable of the latter is $r(c, \tau)$, the imbedded missing $p(t_0)$, which is necessary for resolving Eqs. (5-7) and (5-8). In a similar way, we can also imbed the missing $x(T)$ by denoting

$$s(c, \tau) = x(T) \tag{5-16}$$

we find that the function $s$ satisfies the equation

$$s(c, \tau) = s\{c, b[c, r(c, \tau), \tau]\Delta, \tau + \Delta\}$$

from which it follows that

$$b(c, r, \tau)s_c + s_\tau = 0 \qquad (5\text{-}17)$$

and from Eq. (5-7)

$$s(c, T) = c \qquad (5\text{-}18)$$

Of course, it is sufficient to have one of the missing quantity $p(0)$ or $x(T)$ to resolve the $2PBVP$.

The question now arises of how to solve Eq. (5-14) for obtaining $r(c, \tau)$. The reader is reminded here that $\tau$ is the time variable, while $c$ denotes the initial state. The function $r(c, \tau)$ is, therefore, regarded as a function of time with a variable parameter $c$ just like $f_k(c)$ and $u_k(c)$ of the dynamic programming procedure. If Eq. (5-14) is linear in $r$, we may even have a closed-form analytical solution by using conventional methods. If Eq. (5-14) is not linear, we can (1) either use the numerical integration method, or (2) use a combination of the numerical integration method, the quasilinearization strategy and the double imbedding strategy.

The numerical integration method is similar to that of using the backwards iterative algorithm

$$x_{n-1} = x_n - g(x_n) \qquad (5\text{-}19)$$

starting from the known $x(T) = x_T$, for solving

$$\dot{x} = g(x) \qquad (5\text{-}20)$$

The difference, however, is that here the values of $r(c, \tau)$ will be calculated for a grid of $c$-values on the basis of a knowledge of $r(c, \tau + \Delta)$. As a consequence, we not only will have the missing $p(0)$ for the known initial state $c$, but a grid of $p(c, 0)$-values for a grid of $c$-values.

For values of $\tau$ sufficiently smaller than $T$ appreciable errors may build up by accumulation in the calculation of $r$ based on Eqs. (5-14) and (5-15). We can use a triple strategy as mentioned previously leading to the following steps (see Fig. 5-1):

*Step 1:*  A double imbedding is considered, the $c$-imbedding for eventual modification of the initial state, the $\tau$-imbedding for increasing computing accuracy.

*Step 2:*  Replace $\tau$ by $A$, $A$ can vary from $0$, $1\tau$, $2\tau$, ... , $(N-1)\tau$, $N\tau = T$. Start with $A = N\tau$.

*Step 3:*  Use the quasilinearization strategy, i.e. determine the iterative procedure by using one particular solution and two weighted homogeneous solutions. Note that the accuracy of this procedure depends on the value of

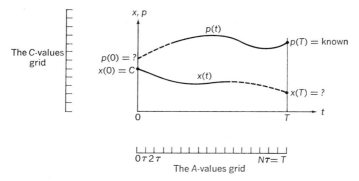

**Fig. 5-1**  Double invariant imbedding.

the initial guess. The result $r(c, \tau)$ of this step is accurate since $r(c, T) = 0$ is known.

*Step 4:*  Consider $A = (N - 1)\tau$, the value of $r(c, \tau)$ obtained from the previous step will be used as initial estimate for the new quasilinearization procedure. And so on until $A = 0$.

Such a suboptimization technique shows that (1) there are a number of strategies which can mutually help overcome various kinds of difficulties, and (2) the combined use of these strategies remains complicated, and its efficiency has to be proven by mathematical experimentations.[1]

### 5.2.3  Discretization of the Control Function—Greenspan's Approach

This optimization technique uses the discretization of the control function so that the resulting extremalizing equations are a set of algebraic nonlinear equations instead of the Euler nonlinear differential equations.[2]

Consider the problem of determining $u(t)$ to minimize

$$J = \int_a^b F(u, u', t) \, dt \tag{5-21}$$

$$u(a) = \alpha, \qquad u(b) = \beta$$

where $a$, $b$, $\alpha$, and $\beta$ are given constants. Let us now divide the interval $a \leq t \leq b$ into $n$ parts by means of the points $t_0, t_1, \ldots, t_{n-1}, t_n$, where

---

[1] An analytical example can be found in [5-9]: computing details of the quasilinearization strategy will be given in Subsection 5.2.4.
[2] See [5-12, 5-13]. Reference [5-13] pertains to rigorous theoretical study. Reference [5-12] contains a great number of illustrative examples.

$a = t_0$, $b = t_n$, $t_i < t_{i+1}$, $i = 0, 1, \ldots, n - 1$. Let $\Delta_i = t_{i-1}$; $i = 1, 2, \ldots$, $n$ and $u_i = u(t_i)$, $i = 1, 2, \ldots, n$. Then approximate the *functional J* by the *function*

$$J_n = \sum_{i=1}^{n} \left[ F\left(t_{i-1}, u_{i-1}, \frac{u_i - u_{i-1}}{\Delta_i}\right) \right] \Delta_i \tag{5-22}$$

Since $u_0 = \alpha$, $u_n = \beta$, $J_n$ is a function of only $u_1, u_2, \ldots, u_{n-1}$. To find an extremal of $J_n$, consider the equations

$$\frac{\partial J_n}{\partial u_i} = 0 \qquad i = 1, 2, \ldots, n - 1 \tag{5-23}$$

The solution of system (5-23) constitutes the approximation of the optimal control $u^*(t)$ at the points $t_1, t_2, \ldots, t_{n-1}$.

As illustrative example, consider the problem of finding a continuous differentiable $u(t)$ satisfying the boundary conditions $u(0) = 1$, $u(1) = \cosh 1$ and minimizing

$$J = \int_0^1 u[1 + (u')^2]^{\frac{1}{2}} \, dt \tag{5-24}$$

Let us divide $0 < t < 1$ into five *equal* parts by setting $\Delta_i \equiv 0.2$; $i = 1, 2, 3, 4, 5$ (Fig. 5-2). Replace $J$ by

$$J_5 = (.2) \sum_{i=0}^{4} \left\{ u_i \left[ 1 + \left(\frac{u_{i+1} - u_i}{.2}\right)^2 \right]^{\frac{1}{2}} \right\} \tag{5-25}$$

Into $J_5$, substitute $u_0 = 1$, $u_5 = \cosh 1 = 1.5431$. Consider then the four equations

$$\frac{\partial J_5}{\partial J_1} = 0, \qquad \frac{\partial J_5}{\partial u_2} = 0, \qquad \frac{\partial J_5}{\partial u_3} = 0, \qquad \frac{\partial J_5}{\partial u_4} = 0, \tag{5-26}$$

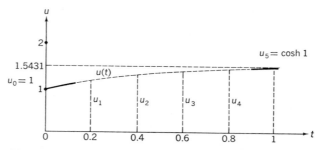

**Fig. 5-2**   Discretization of the unknown control function $u(t)$.

which are equivalent to:

$$\frac{u_1 - 1}{[.04 + (u_1 - 1)^2]^{\frac{1}{2}}} + \frac{.4 + (u_2 - u_1)^2}{[.04 + (u_2 - u_1)^2]^{\frac{1}{2}}} - \frac{u_1(u_2 - u_1)}{[.04 + (u_2 - u_1)^2]^{\frac{1}{2}}} = 0$$

$$\frac{u_1(u_2 - u_1)}{[.04 + (u_2 - u_1)^2]^{\frac{1}{2}}} + \frac{.04 + (u_3 - u_2)}{[.04 + (u_3 - u_2)^2]^{\frac{1}{2}}} - \frac{u_2(u_3 - u_2)}{[.04 + (u_3 - u_2)^2]^{\frac{1}{2}}} = 0$$

$$\frac{u_2(u_3 - u_2)}{[.04 + (u_3 - u_2)^2]^{\frac{1}{2}}} + \frac{.04 + (u_4 - u_3)^2}{[.04 + (u_4 - u_3)^2]^{\frac{1}{2}}} - \frac{u_3(u_4 - u_3)}{[.04 + (u_4 - u_3)^2]^{\frac{1}{2}}} = 0 \qquad (5\text{-}27)$$

$$\frac{u_3(u_4 - u_3)}{[.04 + (u_4 - u_3)^2]^{\frac{1}{2}}} + \frac{.04 + (1.54 - u_4)^2}{[.04 + (1.54 - u_4)^2]^{\frac{1}{2}}} - \frac{u_4(1.54 - u_4)}{[.04 + (1.54 - u_4)^2]^{\frac{1}{2}}}$$

The solution of system (5-27) will yield 4 values of the unknowns $u_1$, $u_2$, $u_3$, $u_4$ approximating the optimal control function $u^*(t)$ at points $t_1$, $t_2$, $t_3$, $t_4$. Of course, more points of $t$ must be used when higher accuracy is desired, the technique, however, remains the same. Two questions arise now: (1) how to solve the nonlinear algebraic system (5-27); (2) is the technique extendable to multidimensional problems.

Several computing techniques may be used for solving the system (5-27).

1. The generalized Newton-Raphson method which will be described in the next subsection.
2. Introduce parameters and additional equations so that in the resulting system radicals appear in a relatively simpler fashion; then use the Newton-Raphson method.
3. In the second part of technique (2), only apply the Newton-Raphson method to that part of the system which does not contain radicals.

The extension of this suboptimization technique to higher dimensional problems can be made naturally without conceptual difficulties. Notational cares must be taken however: (1) where several independent variables are present: $t$, $\tau$, $\sigma$, ... etc., form a lattice of points by discretizing these variables; (2) where several dependent variables are present: $u(t, \tau, \sigma, \ldots)$, $v(t, \tau, \sigma, \ldots)$, $w(t, \tau, \sigma, \ldots)$ etc. define functions $u$, $v$, $w$, ... in every point of the lattice; (3) form the approximate performance functional $J$ as a function of all the discretized dependent variables; (4) the extremalizing system is obtained by zeroing all the partial derivatives of $J$ with respect to all the dependent variables. It may seem that the resulting algebraic system will contain an prohibitive number of equations. Iterative algorithms suitable to digital computing can be found, however, avoiding the writing and the programming of many equations.

Several interesting points can be derived from this suboptimization technique: (1) the technique uses several elementary strategies: one grid-manipulation strategy (discretized $t$ and $u$); one approximation ($J$); one approximating process mechanized by the Newton-Raphson method. (2) the technique is similar to the piece-wise linearization method described in Subsection 2.4.1. (3) like the two preceding techniques, the final computing method needed is the generalized Newton-Raphson method.

### 5.2.4  Approximating Process using the Generalized Newton-Raphson Method

It is very significant to note that the preceding suboptimization techniques, although starting from different elementary strategies, and applied to different mathematical entities of the optimization procedure, all end at the need of using the Newton-Raphson numerical computing method. The reason is that this method constitutes a very powerful means of solving nonlinear equations, be these algebraic, differential, or partial differential. Because of its importance, this method will be discussed here in greater detail [5-14 to 5-17].

To introduce the basic concepts, we shall use a simple example: find the roots of a known algebraic nonlinear function $g(x)$. We write $g(x)$ under the form (See [G-55], pp. 178–183)[1]

$$g(x) = x - f(x) \qquad (5\text{-}28)$$

Since the roots $x_r$ are those values of $x$ satisfying the equation $g(x) = 0$, $x_r$ represent the point $R$ where the function $x$ is equal to $f(x)$ (Fig. 5-3a). Using this property, we can, for instance, implement the following approximating

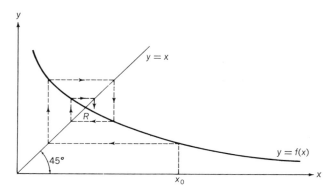

**Fig. 5-3a**   Nonefficient root search.

[1] If $g(x)$ does not contain $x$ explicitly, one can always write $x = g(x) + x = f(x)$. In cases where $g(x)$ has several roots, the shape of $f(x)$ in Fig. 5-3 will be such that it meets several times the 45° line.

process:

1. Starting from an initial guess $x_0$.
2. Compute successively

$$x_1 = f(x_0)$$
$$x_2 = f(x_1) \cdots x_{n+1} = f(x_n) \qquad (5\text{-}29)$$

3. Stop when $|x_{n+1} - x_n| \le \epsilon$.

It can be seen from Fig. 5-3a that this search process leads to a spiral path. The conditions for the search to be convergent are:

1. The function $f(x)$, then also the function $g(x)$, must be continuous in the search area.
2. The curve $f(x)$ must cross the line defined by $y = x$. This is equivalent to stating that the curve $g(x)$ must cross the line defined by $g(x) = 0$ (existence of roots).
3. The vertical distance between the curve $f(x)$ and the line $y = x$ must decrease at each step of iteration. This implies that $df/dx > -1$ in the search area, and that $dg/dx < 2$.
4. The condition 3 is only valid if $df/dx < 0$. If $df/dx > 0$, one must use $-f(x) = -x + g(x)$ instead of $f(x) = x - g(x)$

The speed of convergence depends on the absolute value of $dF/dx$ (Fig 5-3a). The flatter the curve $f(x)$ ($-1 < df/dx < 0$, but $df/dx$ close to 0), the slower the iterative process. This situation is similar to that of resolving the 2PBVP by using a simple approximating process around the $p(t_0)$ without supplementary strategies.

The Newton-Raphson method, using slope properties of $f(x)$, can accelerate the convergence. In effect, after two steps involving the computations of say $f(x_{n-1})$ and $f(x_n)$, the coordinates of the points $A$, $B$, and $C$ are known (Fig. 5-3b). If we assume $AC$ on $f(x)$ as a straight line, then the abscissa $x_{n+1}$ of the point $R$ can be determined geometrically. First, we have

$$\frac{RD}{AD} = \frac{CB}{AB} = -\frac{f(x_n) - f(x_{n-1})}{x_{n-1} - x_n} = \frac{f(x_n) - f(x_{n-1})}{x_n - x_{n-1}} = \alpha$$

Second, we have $DB = RD$.
Then

$$\frac{DB}{AD} = \frac{RD}{AD} = \alpha$$

Since $DB = x_{n-1} - x_n$, $AD = x_{n-1} - x_{n+1}$,
we have

$$\frac{x_{n+1} - x_n}{x_{n-1} - x_{n+1}} = \alpha$$

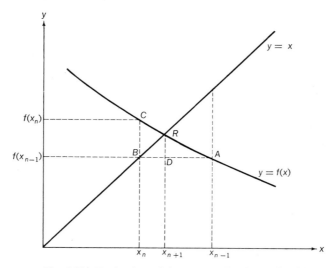

**Fig. 5-3b**   Derivation of the Newton-Raphson algorithm.

and

$$x_{n+1} = \frac{\alpha}{\alpha + 1}\, x_{n-1} + \frac{1}{\alpha + 1}\, x_n$$

Putting $q = \alpha/(\alpha + 1)$, we have $(1 - q) = 1/(\alpha + 1)$.
Then

$$x_{n+1} = qx_{n-1} + (1 - q)x_n$$

Instead of the algorithm (5-29), we can use the 3-point algorithm:

1. Guess $x_{n-1}$.
2. Compute   $x_n = f(x_{n-1})$ and $f(x_n)$.
3. Compute   $\alpha = [f(x_n) - f(x_{n-1})]/(x_n - x_{n-1})$
   $q = \alpha/(\alpha + 1)$
   $x_{n+1} = qx_{n-1} + (1 - q)x_n$                    (5-30)
4. Stop when $|x_{n+1} - x_n| \leq \epsilon$.

For showing how the Newton-Raphson method can be generalized, we shall use directly the 2PBVP as it was stated, without other manipulations as done in the preceding suboptimization techniques. However, instead of two sets of $\dot{x} = \partial H/\partial p$, $\dot{p} = -\partial H/\partial x$, we shall define one total set

$$\dot{x}_i = F_i(x^{(1)}, x^{(2)}, \ldots, x^{(q)}), \qquad i = 1, 2, \ldots, q$$

$q$ being twice the number of state-vector components of the original 2PBVP. For the generalization, we find first the following points:

(a) We have a vector of $q$ components $x = (x^{(1)}, x^{(2)}, \ldots, x^{(q)})$ instead of a scalor $x$ of the preceding example. In the state space of $q$ dimensions, this vector is represented by a point.

(b) This vector is a function of time, the solution $x(t)$ from $t = 0$ to $t = T$ is represented by a trajectory such as $AMB$ (Fig. 5-4a).

(c) If an approximating process is to be used, the initial guess must be a trajectory $x_0(t)$, e.g., $AOB$. The subsequent approximations $x_1(t)$, $x_2(t)$, $\ldots$, $x_n(t)$ (trajectories $A1B$, $A2B$, $\ldots$, $ANB$) transform $AOB$ to $AMB$.

(d) In the 2PBVP, terminal conditions at $t = 0$ and $t = T$ are not completely known. Therefore, the terminal points of $x_0(t)$ are two arbitrary points $A'$ and $B'$ (Fig. 5-4c). The approximating process must then transform the guess trajectory $A'OB'$ into the correct solution $AMB$.

(e) In the scalar case, the stopping condition is $|x_{n+1} - x_n| \leq \epsilon$. In the vector case, this condition must be extended twice: first in the space domain, from $x_1$ to $x_q$; second in the time domain from $t = 0$, to $t = T$ (Fig. 5-4d). The following condition is generally used

$$\rho = \sum_{j=1}^{q} \max_{t} |x_{n+1}^{(j)} - x_n^{(j)}| \leq \epsilon \qquad (5\text{-}31)$$

which means: at each iteration step $n$, compare to $\epsilon$ the value of $\rho$ which is the

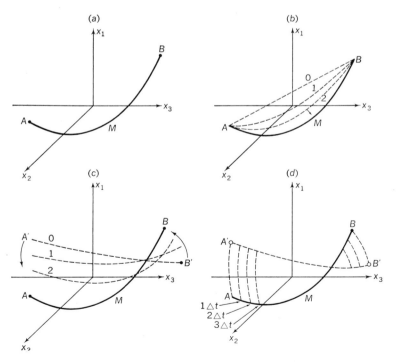

**Fig. 5-4**  Approximating process applied to time functions.

$q$-sum of all the differences $|x_{n+1}^{(j)}(t) - x_n^{(j)}(t)|$, the difference for a given vector $x^{(j)}$ is the greatest of this vector in the range of time $0 \le t \le T$.

Once these points clarified, we can show now how the Newton-Raphson method is generalized.

| Solve the algebraic equation | Solve the differential equation |
|---|---|
| $g(x) = x - f(x) = 0$ | $G(x) = \dot{x} - F(x) = 0$ |

(a) Guess an initial value $x_0$ of the variable $x$.

(b) Compute $x_1$ from the equation

$$g(x_1) = x_1 - f(x_0) = 0$$

This implies a very simple numerical computation by writing:

$$x_1 = f(x_0)$$

(a) Guess an initial solution $x_0(t)$ of the vector function $x(t)$

(b) Compute $x_1(t)$ by solving

$$G[x_1(t)] = \dot{x}_1(t) - F[x_0(t)] = 0$$

This implies the solution of a set of differential equations. This is much more difficult than the algebraic case. One can use, for instance, the numerical integration method, by writing

$$x_1(t_j) - x_1(t_{j-1}) = \Delta t F[x_0(t_{j-1})]$$
$$j = 1, 2, \ldots, r \, \Delta t$$
$$r \, \Delta t = T$$

(c) Compare $x_1$ to $x_0$ by using the criterium

$$|x_1 - x_0| \le \epsilon$$

(c) Compare $x_1(t)$ to $x_0(t)$ by using the criterium of the form (5-31).

(d) If the condition in (c) is satisfied, the problem is solved.

(e) If not, solve

$$g(x_2) = x_2 - f(x_1) = 0$$

and so on, until the condition in (c) is satisfied. This is the normal iterative process.

(f) In the accelerating iterative process, one uses the slope coefficient

$$\alpha = \frac{f(x_n) - f(x_{n-1})}{x_n - x_{n-1}}$$

(d) If the condition in (c) is satisfied, the problem is solved.

(e) If not, solve

$$G[x_2(t)] = \dot{x}_2(t) - F[x_1(t)] = 0$$

and so on, until the condition in (c) is satisfied. This is the normal iterative process.

(f) In the accelerating iterative process, one uses the generalized slope coefficient: the Jacobian

$$J_j = \sum_{j=1} \frac{\partial F_i}{\partial x_j} \quad i = 1, 2, \ldots, u$$

if the system $G$ is independent, $q = u$.

(g) The iterative algorithm then becomes

$$q = \alpha/(1 + \alpha)$$
$$x_{n+1} = q x_{n-1} + (1 - q) x_n$$

(g) The iterative algorithm then becomes

$$\dot{x}_{n+1}(t)$$
$$= J[x_n(t)][x_{n+1}(t) - x_n(t)] + F[x_n(t)]$$

This is always a simple numerical computation.

Note that this is a 3-point algorithm with respect to the variable $x$.

(h) The stopping condition is similar to (c).

This is again a system of differential equations which can be solved, for instance, by a numerical integration method.

Note that this is a 2-vector algorithm with respect to the vector $x(t)$.

(h) The stopping condition is similar than in (c).

From the preceding, we can see that in the application of the generalized Newton-Raphson method for solving differential equations, the main difficulty lies in the solution of the iterative differential equations. This difficulty is even increased in the case of the 2*PBVP*, in which the incompleteness of the terminal conditions causes extra troubles. Here we show how this method is applied to the 2*PBVP* by using the minimum-time example discussed in Subsection 4.2.2.2.

**Method**

(a) *Problem*

$$\dot{x} = F(x, t) \quad t \in [0, t_f]$$
$$x = \{x^{(1)}, x^{(2)}, \ldots, x^{(q)}\}$$
$$F = \{f^{(1)}, f^{(2)}, \ldots, f^{(q)}\}$$
$$f^{(j)} = f^{(j)}(x^{(1)}, x^{(2)}, \ldots, x^{(q)})$$
$$j = 1, 2, \ldots, q$$

End conditions:

1. half given at $t = 0$
2. half given at $t = t_f$

**Example**

(a) *Minimum-time Problem*

$$\dot{x}_1 = f_1 = x_2$$
$$\dot{x}_2 = f_2 = \frac{x_3^2}{x_1} - \frac{k}{x_1^2} + a\frac{x_5}{(x_5^2 + x_6^2)^{\frac{1}{2}}}$$
$$\dot{x}_3 = f_3 = -\frac{x_2 x_3}{x_1} + a\frac{1}{(x_5^2 + x_6^2)^{\frac{1}{2}}}$$
$$\dot{x}_4 = f_4 = x_5\left(\frac{x_3^2}{x_1^2} - \frac{k}{x_1^3}\right) - x_6\frac{x_2 x_3}{x_1^2}$$
$$\dot{x}_5 = f_5 = x_6\frac{x_3}{x_1} - x_4$$
$$\dot{x}_6 = f_6 = x_6\frac{x_2}{x_1} - 2x_5\frac{x_3}{x_1}$$
$$a(t) = T/(m_0 + \dot{m}t)$$

$$x_1(0) = 1 \qquad x_1(t_f) = 1.525$$
$$x_2(0) = 0 \qquad x_2(t_f) = .0$$
$$x_3(0) = 1 \qquad x_3(t_f) = .8098$$
$$x_4(0) = ? \qquad x_4(t_f) = ?$$
$$x_5(0) = ? \qquad x_5(t_f) = ?$$
$$x_6(0) = ? \qquad x_6(t_f) = ?$$

| **Method** | **Example** |
|---|---|

Minimize $t_f$

Minimize $t_f$

$$T = 0.1405$$
$$m_0 = 1$$
$$\dot{m} = -.07487$$
$$k = 1$$

(b) *Iteration algorithm*
$$\dot{x}_{n+1} = J(x_n, t)[x_{n+1} - x_n] + (F \cdot c_n, t)$$

$$J(x_n, t): \text{Jacobian}$$

at each $n$th iteration, this system of differential equation is to be solved. Since $J$ and $F$ are known, this is a linear system of the form

$$\dot{x} = C(t)x(t) + D(t)$$

$$F(x_n, t) = \{f^{(1)}, \ldots, f^{(6)}\}$$

$$J_j = \sum_{i=1}^{6} \frac{\partial f^{(i)}}{\partial x_j}$$

$$J_1 = \frac{1}{x_1^2} [x_2(x_3 - x_6) + x_3(2x_5 - x_6)]$$
$$+ \frac{1}{x_1^4} [k(2x_1 + 3x_5 + x_1 x_3(2x_2 x_6$$
$$- x_3(2x_5 - x_1)]$$

$$J_2 = \frac{1}{x_1^2} [x_1(x_6 - x_3) - x_3 x_6]$$

$$J_3 = \frac{1}{x_1^2} \{x_1[2(x_3 - x_5) - x_2 + x_6]$$
$$+ 2x_3 x_5 - x_2 x_6\}$$

$$J_4 = -1$$

$$J_5 = \frac{a(t)(x_6^2 - x_5)}{(x_5^2 + x_6^2)^{3/2}}$$
$$+ \frac{1}{x_1^3} [x_1 x_3(x_3 - 2x_1) - k]$$

$$J_6 = -\frac{a(t)x_6(x_5 - 1)}{(x_5^2 + x_6^2)^{3/2}}$$
$$+ \frac{1}{x_1^2} [x_1(x_2 + x_3) - x_2 x_3]$$

(c) *Computational scheme*
(c1) Choose an initial guess $x_0(t)$

(c1) Initial guess $x_0(t)$

$$x_0^{(1)}(t) = r_0 + [(r_f - r_0)/t_{f0}]t$$
$$x_0^{(2)}(t) \equiv 0$$
$$x_0^{(3)}(t) = \left[\frac{k}{x_0^{(1)}(t)}\right]^{\frac{1}{2}}$$
$$x_0^{(4)}(t) = 1$$
$$x_0^{(5)}(t) = \begin{cases} .52 & t \in [0, \frac{1}{2}t_{f0}] \\ -.5 & t \in [\frac{1}{2}t_{f0}, t_{f0}] \end{cases}$$
$$x_0^{(6)}(t) = \begin{cases} .3 & t \in [0, \frac{1}{2}t_{f0}] \\ 0.000 & t \in [\frac{1}{2}t_{f0}, t_{f0}] \end{cases}$$

| **Method** | **Example** |
|---|---|

(c2) Generate by numerical integration a set

$$\{x^{(q/2+i)}(t)\}, \quad i = 1, 2, \ldots, \frac{q}{2}$$

of solutions of the homogeneous system

$$\dot{x} = C(t)x(t)$$

with initial conditions

$x^{(q/2+1)}(t_0)$

$\quad = \{0, 0, \ldots, 0, x_{q/2+1} = 1, 0, \ldots, 0\}$

$x^{(q/2+2)}(t_0)$

$\quad = \{0, 0, \ldots, 0, 0, 1, \ldots, 0\}$

$\quad\quad\quad\quad\quad \cdot$

$\quad\quad\quad\quad\quad \cdot$

$\quad\quad\quad\quad\quad \cdot$

$x^{(q)}(t_0) = \{0, 0, \ldots, 0, 0, 0, \ldots, 1\}$

(c2) Generate three solutions of the homogeneous system

$$\dot{x} = J(t)x(t)$$

Compute first

$$J_j = J_j[x_0^{(1)}(t), x_0^{(2)}(t), \ldots, x_0^{(6)}(t)]$$

$$f_0 = f_0[x_0^{(1)}(t), x_0^{(2)}(t), \ldots, x_0^{(6)}(t)]$$

Then

$x_{H1}^{(j)}[(n + 1)\,\Delta t]$

$\quad = x_{H1}^{(j)}(n\,\Delta t) + \Delta t J_j x_{H1}(n\,\Delta t)$

$x_{H2}^{(j)}[(n + 1)\,\Delta t]$

$\quad = x_{H2}^{(j)}(n\,\Delta t) + \Delta t J_j x_{H2}(n\,\Delta t)$

$x_{H3}^{(j)}[(n + 1)\,\Delta t]$

$\quad = x_{H3}^{(j)}(n\,\Delta t) + \Delta t J_j x_{H3}(n\,\Delta t)$

$$j = 1, 2, \ldots, 6 \quad n = 0, 1, \ldots, N$$

$$t_{f0} = N\,\Delta t$$

with initial conditions

$$x_{H1}(t_0) = \{0, 0, 0, 1, 0, 0\}$$

$$x_{H2}(t_0) = \{0, 0, 0, 0, 1, 0\}$$

$$x_{H3}(t_0) = \{0, 0, 0, 0, 0, 1\}$$

(c3) Generate a particular solution $x^{(P)}(t)$ of the nonhomogeneous system $\dot{x} = C(t)x(t) + D(t)$ with initial conditions.

$x^{(p)}(t_0)$

$\quad = \{x_{10}, x_{20}, \ldots, x_{(q/2)0}; k_1,$

$\quad\quad\quad\quad\quad\quad\quad k_2, \ldots, k_{q/2}\}$

where $k_i$ are arbitrary.

(c3) Generate one particular solution by numerical integration

$x_p^{(j)}[(n + 1)\,\Delta t]$

$\quad = x_p^{(j)}(n\,\Delta t) + \Delta t[x_p^{(j)}(n\,\Delta t)J_j + f_j]$

$$j = 1, 2, \ldots, 6 \quad n = 0, 1, \ldots, N$$

with initial conditions

$$x_p(t_0) = \{0, 0, 0, k_1, k_2, k_3\}$$

| **Method** | **Example** |
|---|---|

(c4) The solution with the prescribed boundary conditions is then given by

$$x(t) = c_{(q/2+1)}x^{(q/2+1)}(t)$$
$$+ c_{(q/2+2)}x^{(q/2+2)}(t)$$
$$+ c_q x^{(q)}(t) + x_p(t)$$

where the $q/2$ constants $c_{(q/2+i)}$, $i = 1, 2,$ $\ldots, \dfrac{q}{2}$ are determined from the boundary conditions at $t = t_f$ by the solution of $q/2$ simultaneous linear algebraic equations.

(c4) Solve for the $c$'s constants by the set:

$$c_1 x_{H1}^{(i)} + c_2 x_{H2}^{(i)} + c_3 x_{H3}^{(i)}$$
$$= x_0^{(i)}(t_f) - x_p^{(i)} \quad i = 1, 2, 3$$

once $c_1$, $c_2$, $c_3$ are computed, the set $x_1(t)$ result of the first iteration is obtained by

$$x_1(t) = c_1 x_{H1}(t) + c_2 x_{H2}(t)$$
$$+ c_3(x_{H3}(t) + x_p(t)$$

(c5) Computation of

$$\rho = \sum_t \max |x_{n+1}(t) - x_n(t)|$$

(c5) Compute the criterion $\rho$

$$\rho = \sum_1^6 \max_t |x_1^{(j)}(t) - x_0^{(j)}(t)|$$

for all the 6 components of $x_1(t)$ and $x_0(t)$

(c6) Special criterion must be used for the time minimal problem.

(c6) Use a complex criterion for the iterations:

$$\bar{\rho} = \rho + \frac{1}{b}|t_{fk+1} - t_{fk}| \leq \eta$$

$$t_{fk+1} = t_{fk} + \frac{t_{fk} - t_{fk-1}}{r(t_{fk}) - r(t_{fk-1})}|r_f - r(t_{fk})|$$

is a weighting factor

*Comments.*

1. Although derived from different arguments, the quasilinearization and the generalized Newton-Raphson method result in the same operating procedure.

2. Great similarity also exists between the generalized Newton-Raphson method and the Gradient method. In both cases, the slope property, i.e., the Jacobian, is used. Unicity of the convergence toward the correct initial conditions is guaranteed by the form of the stopping condition (5-31).[1] The rate of convergence is not necessarily increased, because the sensitivity to the initial condition is not taken into account during the iterations. This point will be discussed in more details in Section 5-4.

[1] In [5-18], the elementary strategy $g - v$ is used so that the hunting of the initial conditions is made easier.

## 5.3  OVERCOMING THE DIMENSIONALITY PROBLEM

Since the origin of the dimensionality problem is the high number of points to be stored, the idea which comes immediately is to decrease this by using coarser grid. Such a strategy may succeed if functions $f_k(c)$ and $u_k(c)$ are unimodal. Otherwise, the true optimum may escape from the too large links of the coarse grid.

### 5.3.1  Lagrange-multiplier Method

The state parameter $c$ of functions $f_k(c)$ and $u_k(c)$ is a many-components vector in the multidimensional problems. In some cases, the Lagrange-multiplier method can be used to reduce the number of these components by one. As a consequence, memory requirement will be decreased, while computing time will be increased.[1] To see how this method operates, we reformulate the original problem with an additional constraint: Let $x$, $u$ be two scalor variables; find $u$ to minimize

$$J(u) = \int_0^T F(x, u)\, dt \qquad (5\text{-}32)$$

subject to

$$\dot{x} = g(x, u), \qquad x(0) = c \qquad (5\text{-}33)$$

and

$$\int_0^T H(x, u)\, dt = q \qquad (5\text{-}34)$$

Constraints of the type (5-34) can represent any of the kinds discussed in Chapter 4: terminal constraints, inequality boundary constraints, also one of the system constraints in cases where $g(x, u)$ is multidimensional. To treat this problem by means of dynamic programming techniques, we note that the value of the minimum now depends upon $c, q$, and $T$. Hence, we introduce the function of the three variables

$$f(c, q, T) = \operatorname*{Min}_u J(u) \qquad (5\text{-}35)$$

which means that the functions $f_k(c)$ and $u_k(c)$ become $f_k(c, q)$ and $u_k(c, q)$, two arguments $c$ and $q$ instead of only one, $c$.

Now, instead of the formulation (5-32) and (5-34), let us consider the

[1] See [G-25, Ch. 6] and [G-18, pp. 435–439]. The method does not consist of a mere substitution by $\lambda$ of one of the $x$-components. The new variable $\lambda$ is connected to a constraint such as discussed in elementary strategy (a).

problem of finding $u$ to minimize the new functional

$$J_1(u) = \int_0^T F(x, u)\, dt + \lambda \int_0^T H(x, u)\, dt \tag{5-36}$$

subject only to

$$\frac{dx}{dt} = g(x, u) \qquad x(0) = c \tag{5-37}$$

where $\lambda$ is an unknown parameter. The important point to observe is that for any fixed value of $\lambda$, the variational problem can be treated in terms of sequences of functions of one variable, which is $c$ in this case. The computational approach is, therefore, the following. We allow $\lambda$ to assume a set of values $\lambda_1, \lambda_2, \ldots, \lambda_k$. For each fixed value of $\lambda$, say $\lambda_i$, we use the recurrent function of the dynamic programming technique to obtain the corresponding optimal values $u$ and $x$. With the known values of $x$, $u$, and $\lambda_i$ we compute the integral

$$\int_0^T H(x, u)\, dt = \int_0^T H[x(t, \lambda_i), u(t, \lambda_i)]\, dt = q_i \tag{5-38}$$

We thus obtain the solution of the original problem for a set of $q$-values among which must be situated the one given in Eq. (5-34). The two-dimensional memory problem has been converted to a succession of one-dimensional computing problems. We are trading time to memory.

Several interesting points must be noted concerning this suboptimization technique: (1) the discussion here completes Subsection 4.4.3 showing how the dynamic programming technique treats the various kinds of constraints; (2) the way of adjoining the constraints by Lagrange multipliers is absolutely similar to that used in the classical variational approach; (3) the number of state parameters can be decreased by $2, 3, \ldots$ if $2, 3, \ldots$ Lagrange multipliers are used.

### 5.3.2   Functional Approximation

One important category of suboptimization techniques for overcoming the dimensionality problem consists of using approximation of various functionals, which are:

1. The penalty functional $v(x, u)$.
2. The cost or performance functional $J(u)$.
3. The recurrence computing functional $f_N(c)$.
4. And, in some cases, the functional $\partial f_N(c)/\partial u$.

The diversity of points at which the approximation strategy can be applied indicates how complicated the results may be. The complication is increased, furthermore, by two other reasons. First, many kinds of approximations are used, for instance [5-20, 5-23, 5-30, G-25 Chs. 12 and 18]:

1. Polynomial approximation of the form

$$f(x) = k_0 + k_1 x + k_2 x^2 + \cdots$$

2. Taylor series expansion by taking first-degree terms only or first plus second-degree terms.
3. Linearized expansion of multidimensional functionals, etc.

Second, exploitation of the results are different from one case to another.

1. In rare cases, a closed-form analytical solution of $u(t)$ is obtained.
2. In other cases, a set of differential equations are obtained, the solution of which again needs some approximating process.

A typical example of techniques in this direction will be discussed here. Let us consider again the recurrent functional equation

$$f_N(c) = \underset{u}{\text{Min}} \{v(c, u)\tau + f_{N-1}[c + g(c, u)\tau]\} \tag{5-39}$$

The continuous version of this functional equation is the following:

$$f(c) = \underset{u}{\text{Min}} \{v(c, u) + f[c + g(c, u)]\} \tag{5-40}$$

Assuming that $g(c, u)$ is small with respect to $c$, we can expand $f[c + g(c, u)]$ into a Taylor series and linearize by taking only first-degree terms

$$f[c + g(c, u)] \cong f(c) + g(c, u)f'(c) + \cdots \tag{5-41}$$

Substituting this into Eq. (5-40), we have

$$f(c) = \underset{u}{\text{Min}} [v(c, u) + f(c) + g(c, u)f'(c)] \tag{5-42}$$

Since $f(c)$ exists in both sides we can write

$$0 = \underset{u}{\text{Min}} [v(c, u) + g(c, u)f'(c)] \tag{5-43}$$

This is a partial differential equation with respect to $c$ and $u$. Let us define the following function

$$H(c, u) = v(c, u) + g(c, u)f'(c) \tag{5-44}$$

According to Eq. (5-43), we have

$$0 = \underset{u}{\text{Min}} H(c, u) \tag{5-45}$$

leading to the necessary condition (a more familiar form of partial differential equation

$$\frac{\partial H}{\partial u} = \frac{\partial v(c, u)}{\partial u} + \frac{\partial g(c, u)}{\partial u} \frac{\partial f}{\partial c} = 0 \tag{5-46}$$

Here, we have assumed that $f'(c) = \partial f/\partial c$ is not function of $u$; the operating line remains the same if it was.

The preceding method has deliberately changed the computational exploitation of the optimality principle into an analytical exploitation. The analytical exploitation will succeed if we know how to solve Eq. (5-43) or Eq. (5-46) in terms of $u$. Unfortunately, if the functionals $v$ and $g$ were given and known, the functional $f$ was not. Consequently some kind of approximating numerical computing process must be used. For this, we shall consider a special kind of approximation to $f(c)$

$$f(c) \simeq f^*(c, \beta) \tag{5-47}$$

where the vector $\beta$ represents a set of adjustable parameters. Once $f$ is known, all the terms in Eq. (5-46) can then be computed in terms of $\beta$ and the problem becomes that of adjusting $\beta$ so that $\partial H/\partial u$ is equal to zero. Or, for greater precision, we must state that $\beta$ is adjusted to that $H(c, u, \beta)$ is minimized with respect to $u$ according to Eq. (5-45). The various steps of the approximating process are:

1. Use $f^*(c, \beta)$ as an approximation of $f(c)$ and give an initial guess of $\beta$.
2. Compute the partial derivative $\partial f^*(c, \beta)/\partial c$ and extract the control vector $u^*(t)$ so that

$$H^*(c, \beta) = v(c, u^*) + g(c, u^*) \cdot \partial f^*/\partial c \tag{5-48}$$

is minimized (if $H^*$ is positive for all $c$ and $\beta$, the minimum is equal to zero).
3. If $H^* \neq 0$, use the increment law

$$\beta_{new} = \beta_{old} - \nu A \psi \tag{5-49}$$

where $\nu$ is some constant number, $A$ the Jacobian

$$A = \frac{\partial H^*[\beta, c(t_i)]}{\partial \beta_i} \tag{5-50}$$

and $\psi$ the stopping vector

$$\psi = \{H^*[\beta, x(t_1)], H^*[\beta, x(t_2)], \ldots, H^*[\beta, x(t_n)]\} \tag{5-51}$$

and use the stopping condition

$$\|\psi^2\| = \text{minimum} \tag{5-52}$$

As an application, let us take the problem

$$\dot{x} = g(x, u) \qquad \dot{x}_1 = \tan h\,(x_2) \qquad \dot{x}_2 = x_1 + x_2 + u$$

Choose $u$ to minimize

$$J = \tfrac{1}{2} \int_0^\infty (x_1^2 + u^2)\, dt$$

$$v = \tfrac{1}{2}(x_1^2 + u^2)$$

Consider the arbitrary particular structure for $f^*(c, \beta)$:

$$f^* = \beta_1 x_1^2 + \beta_2 x_2^2 + \beta_3 x_1 x_2 + \beta_4 x_1^4 + \beta_5 x_2^4$$

Choose five time points ($t_1 = 0, 0.75, 1.50; 2.25, 3.25$) leading to

$$H^* = \tfrac{1}{2}(x_1^2 - p_1^2) + p_1 \tan h(x_2) + p_2(x_1 + x_2)$$

where $p_1$ and $p_2$ are defined as $\partial f^*/\partial x_1$, $\partial f^*/\partial x_2$:

$$p_1 = 2\beta_1 x_1 + \beta_3 x_2 + 4\beta_4 x_1^3$$
$$p_2 = 2\beta_2 x_2 + \beta_3 x_1 + 4\beta_5 x_2^3$$

and where the suboptimal control policy is given by

$$u^*(x, \beta) = -p_2(x, \beta)$$

Using the algorithm (5-49), after iteration, $\|\psi\| \to 0$ with resulting parameters

$$\beta_1 = 1.212, \qquad \beta_2 = 1.629, \qquad \beta_3 = 2.409, \qquad \beta_4 = 0.0190, \qquad \beta_5 = -0.0143$$

*Comments.* (1) The increment law, by using $A$, takes into account the gradient or the sensitivity $\partial H/\partial \beta$. (2) The approximating process is very much similar to that discussed in Subsection 5.2.1. (3) Only mathematical experimentation can indicate for a special practical case the best alternative among all those possible approximation and approximating processes.

### 5.3.3  Approximation of Computed Functions

The suboptimization technique discussed in Subsection 5.3.1 attacks the dimensionality problem right at the formulation stage of the Dynamic Programming procedure. The techniques discussed in Subsection 5.3.2 attack the same problem at the analytical extremalizing stage of the procedure. Here, we shall discuss one direction of techniques attacking the same problem at the numerical computing stage of the procedure [5-19; 5-20; 5-31; G-25, pp. 244–245].

The dimensionality problem would disappear if we knew the analytical expression of the computed functions $f_k(c)$ and $u_k(c)$. For instance, if we had

$$u(c) = a_0 + b_0 c^2$$

the only requirement for the computer would be the storage of two constants $a_0$ and $b_0$, plus a very small program for calculating $u(c)$ when $c$ is given. The natural idea is, therefore, to use an artificial or approximate analytical expression. For instance, we can use the polynomial expansion for $u(c)$ which writes in the one-dimensional case

$$u(c) = a_0 + a_1 c + a_2 c^2 + \cdots + a_k c^k \qquad (5\text{-}53)$$

and in the two-dimensional case

$$u(c_1, c_2) = \sum_{i,j=0}^{k} a_{ij} c_1^i c_2^j \qquad (5\text{-}54)$$

The insertion of this approximation in the dynamical programming leads to:

(a) Adopt the number of terms, e.g. $k = 10$.

(b) Compute the coefficients $a_k$ or $a_{ij}$ [$\frac{1}{2}(k + 1)(k + 2)$ coefficients] from the known data of $c$ and $u(c)$. (A set of 10 linear algebraic equations is to be solved if $K = 10$.)

(c) Store the values of $a_k$ or $a_{ij}$.

(d) At the following step, compute $u(c)$ with the new $c$ and the stored coefficients according to Eq. (5-53) or Eq. (5-54).

The value of such a suboptimization technique depends on a number of factors: (1) the accuracy and the uniformity of the approximation; (2) the ease with which the coefficients are calculated; and (3) the case for calculating the approximated function each time the argument is given.

The polynomial expansion is not the best one with regard to these factors, because the accuracy of the approximation is not uniform in the argument space. Other types of expansion, namely the Legendre polynomials, trigonometric functions, and Cebycev polynomials offer more advantages, not only because of the uniformity of the accuracy, but also because of the ease of computing the various terms of the expansion. Once again, the practical value of one or another expansion can only be judged after mathematical experimentations on concrete problems. Here, we give the results of a comparative study of several expansions applied to the same problem [5-19].

The problem is the following. Given as system dynamics

$$\dot{x}_1 = x_2 + u$$
$$\dot{x}_2 = x_1$$

determine $u(t)$ bounded as $|u(t)| \leq 2$ so to drive $\{x_1(t), x_2(t)\}$ from the point $\{0, -0.915\}$ to the point $\{0, 0\}$ and to minimize

$$J = \int_0^{2.6} [x_1^2(t) + u^2(t)]\, dt$$

For comparison, of the exactness of the solutions, Pontryagin's method is applied leading to

$$\dot{x}_1 = x_2 + \frac{p_1}{2} \qquad \dot{x}_2 = x_1$$

$$\dot{p}_1 = 2x_1 - p_2 \qquad \dot{p}_2 = -p_1 - p_2$$

where $p_1/2$ is the optimal control obtained by maximizing the Hamiltonian function. The solution of this $2PBVP$ is not too difficult. Figure 5-5 shows the optimal trajectory. For comparison of computation requirements, the standard dynamic programming procedure is used in two versions.

*Version 1.* Grid of eleven values for $x_1$, $x_2$, and $u$; the time interval 2.6 sec divided into 13 $\Delta t$, of 0.2 each; words of 10 digits, and 4-point interpolation when the new $c$-value is not on the grid points.

*Version 2.* Same as Version 1, except that $\Delta t = 0.1$ sec instead of 0.2 sec. Three kinds of approximation are tested.

1. The cost function $f(x_1, x_2)$ and the policy function $u(x_1, x_2)$ are approximated with Legendre polynomials orthogonal in the considered interval $(-1, 1)$.

$$f_n(x) = \sum_{k=0}^{R} a_{k,n} P_k(x) \tag{5-55}$$

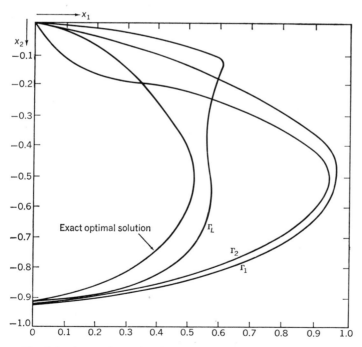

**Fig. 5-5** Comparison of various methods of approximation—Part I.

where $R$ is some predetermined value, the coefficients $a_{k,n}$ may be determined by the relation

$$a_{k,n} = \frac{2k+1}{2} \int_{-1}^{1} f_n(x) \cdot P_k(x)\, dx \tag{5-56}$$

which can be replaced exactly by

$$\int_{-1}^{+1} f_n(x) P_k(x)\, dx = \sum_{j=0}^{s} b_j f_n(x_j) P_k(x_j) \tag{5-57}$$

when $f_n(x) P_k(x)$ is a polynomial of degree $2s+1$ and the $x_i$'s are the zeros of the Legendre polynomial of degree $s+1$ and the $b_j$'s are the Christoffel numbers. In two dimensions

$$f(x_1, x_2) = \sum_{k=0}^{R} \sum_{l=0}^{R} b_{kl} P_k(x_1) P_l(x_2) \tag{5-58}$$

where

$$b_{kl} = \frac{(2k+1)(2l+1)}{4} \int_{-1}^{1} \int_{-1}^{1} f(x_1, x_2) P_k(x_1) P_l(x_2)\, dx_1\, dx_2 \tag{5-69}$$

The grid points, placed at the zeros of the eleventh degree Legendre polynomial, $b_{kl}$ may be calculated as

$$b_{kl} = \frac{(2k+1)(2l+1)}{4} \sum_{i=1}^{11} \sum_{j=1}^{11} a_i a_j f(x_{1i}, x_{2j}) P_k(x_{1i}) \cdot P_l(x_{2j}) \tag{5-60}$$

The coefficients $a_i$ and $a_j$ and the values of $P_k(x_{1i})$ and $P_l(x_{2j})$ are determined initially. The Legendre polynomials are evaluated by the recursion formula

$$P_0(x) = 1$$
$$P_1(x) = x$$

$$\begin{array}{c} . \\ . \\ . \end{array} \tag{5-61}$$

$$P_{n+1}(x) = \frac{2n+1}{n+1} x P_n(x) - \frac{n}{n+1} P_{n-1}(x)$$

2. Functions orthogonal on a disk. The region of interest in state space is transformed into a disk. The cost and policy function can be represented by

$$f(r, \varphi) = \sum_{m=0}^{R} \sum_{n=0}^{R} a_{mn} P_m(2r-1) P_n\!\left(\frac{\varphi}{\pi}\right) \tag{5-62}$$

where

$$a_{mn} = \frac{(2m+1)(2n+1)}{4} \int_{0}^{1} \int_{-\pi}^{+\pi} f(r, \varphi) P_m(2r-1) P_n\!\left(\frac{\varphi}{\pi}\right) \alpha\varphi\, dr \tag{5-63}$$

3. The method which requires for less computational time and storage is to use the product of polynomials in each state variable

$$f(x_1, x_2, \ldots, x_n) = g_1(x_1)g_2(x_2) \cdots g_n(x_n) \tag{5-64}$$

where $g_i(x_i)$ is a polynomial in $x_i$. In the case of 2 dimensions

$$f(x_1, x_2) = g_1(x_1)g_2(x_2) \tag{5-65}$$

In turn Legendre polynomials can be used to approximate $g_i(x_i)$

$$f(x_1, x_2) = \sum_{n=0}^{R} a_n P_n(x_1) \sum_{n=0}^{R} b_m P_m(x_2) \tag{5-66}$$

The approximate optimum trajectory for each of the preceding methods has been computed by using the same computer (IBM 7070, add time 60 $\mu$s, multiply time 3 ms). The computing time and storage requirements are summarized in Table 5.2 for easy comparison. Approximate trajectories are

**Table 5.2** Comparison of Computing Time and Storage Requirements of Several Approximation Methods

| Method | Computing Time and Storage Requirements | Name and Location of Computed Optimum Trajectory |
|---|---|---|
| (a) Standard dynamic programming: $x_1$, $x_2$, grids of 11 points; 13 $\Delta t$ of 0.2 sec. | 2300 memory positions. 7 minutes. | $\Gamma_1$ Fig. 5-5 |
| (b) Same as in (a) except 26 $\Delta t$ of 0.1 sec. | 3945 memory positions. 14 minutes | $\Gamma_2$ Fig. 5-5 |
| (c) Legendre polynomial approximation, 26 $\Delta t$ of .1 sec. | 2000 memory positions. 130 minutes | $\Gamma_L$ Fig. 5-5 |
| (d) Functions orthogonal on a disk of 69 points. | 2000 memory positions. $\simeq$70 minutes. | $\Gamma_D$ Fig. 5-6 |
| (e) Product of polynomials in each state variable, refined interpolation method. | No storage problem. $\simeq$60 minutes. | $\Gamma_P$ Fig. 5-6 |

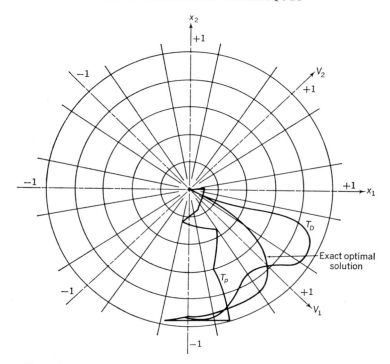

*Fig. 5-6*    Comparison of various methods of approximation—Part 2.

shown in Figs. 5-5 and 5-6, and compared to the theoretical optimum trajectory $\Gamma_0$. It can be concluded easily that in all cases, the accuracy of the approximations is far from sufficient. This category of suboptimization techniques can only supply very coarse solutions.

### 5.3.4  State-Increment Dynamic Programming—Larson's Approach

In the dynamic programming procedure treating a multidimensional problem, there are memory difficulties because it is necessary to store functions of state points covering the *entire* hyperplane defined by the state vector. If, instead of the hyperplane, we only consider a small hyperrectangle around the probable optimal point, the reduction in the storage requirement will be proportional to the reduction of surfaces (Fig. 5-7). Consider now at each iterative step, not the hyperrectangle but a block which consists of this hyperrectangle and of $\Delta T$, a portion of the time dimension, containing several elements of time $\Delta t$. The storage requirement will be increased slightly but this is still less stringent than if the hyperplane was considered.

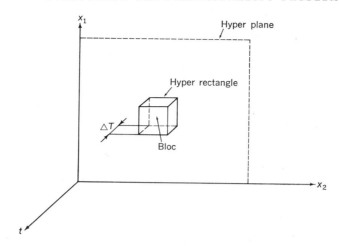

*Fig. 5-7*   Reduction of storage requirement by using bloc strategy.

For getting some concrete idea, let us suppose that the maximum of $x_1$ and $x_2$ is 100 units, each unit being equal to one $\Delta x$; that the hyperrectangle is of $(12\,\Delta x) \times (12\,\Delta x)$, and that $\Delta T = 5\,\Delta t$, the storage requirements are:

hyperplane          $100 \times 100 = 10000$

hyperrectangle      $12 \times 12 = 144$

one block           $5 \times 144 = 720$

At each iterative step, the reduction is roughly from 15 to 1. This is obviously an advantage. Now the question arises how this advantage can be kept throughout the procedure. This seems difficult, because the next optimal point may be located anywhere on the hyperplane. Here intervenes the time increment $\delta t$. As shown in Fig. 5-8*a*, if $\delta t$ was given, then the location of the next optimal point in the state space (not the phase space) will be remote or close ($B_1'$ or $B_1$) according to the value of the slopes ($\alpha'$ or $\alpha$). However, we can always fix the location of the next block $B_1$ or $B_1'$ very close to $B_0$ in the state space and use a variable time increment $\delta t$ or $(\delta t)'$ according to the values of the slopes ($\alpha$ or $\alpha'$). In this manner, the storage-reduction advantage will be kept throughout the optimization procedure. Furthermore, in many practical problems, the preferable direction can often be deduced by some physical understanding of the system, rendering the block-to-block transition even easier.

This is the Larsen's approach for overcoming the dimensionality problem

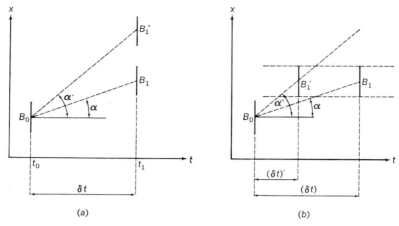

**Fig. 5-8** Dependence between $\delta t$ and the location of the next bloc.

[5-21, 5-22] where the elementary strategy g-3 (p. 250) is used among others. The complete procedure is now described.

**Step 1.** Formulate the discrete version of the problem: (multi-dimensional and nonlinear).

$$x(t + \delta t) = x(t) + g(x, u, t) \delta t \qquad (5\text{-}67)$$

$$J = \int_t^{t_f} v(x, u, \sigma) \, d\sigma \qquad (5\text{-}68)$$

$$f(x, t) = \min_{\substack{u(\sigma)\in U \\ t \leq \sigma \leq t_f}} \left\{ \int_t^{t_f} v[x(\sigma), u(\sigma), \sigma] \, d\sigma \right\} \qquad (5\text{-}69)$$

$$f(x, t) = \min_{u\in U} \{ v(x, u, t) + f[x + g(x, u, t) \delta t, t + \delta t] \} \qquad (5\text{-}70)$$

**Step 2.** Define the blocks in the phase space. Each state variable is quantized in uniform increments within its allowable range.

$$-\beta_i \leq x_i \leq \beta_i \qquad i = 1, 2, \ldots, n \qquad (5\text{-}71)$$

$$x_i = -\beta_i + j_i \Delta x_i \qquad j_i = 0, 1, \ldots, N \qquad (5\text{-}72)$$

$$N_i \Delta x_i = 2\beta_i \qquad i = 1, 2, \ldots, n \qquad (5\text{-}73)$$

A block is numbered by $j_i$ and contains $w_i$ units of $\Delta x_i$ (Fig. 5-9) in the $x_i$-axis. In the $t$-axis, a block is numbered by $m$ and covers $\Delta T$ seconds each containing

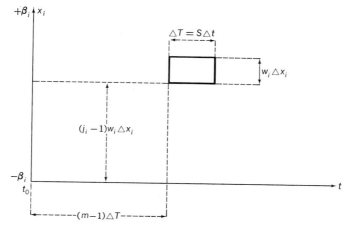

**Fig. 5-9** Definition of a bloc.

$s$ units of time:

$$(j_i - 1)w_i\,\Delta x_i \le x_i \le j_i w_i\,\Delta x_i \qquad j_i = 1, 2, \ldots, J_i \qquad (5\text{-}74)$$

$$(m - 1)\,\Delta T \le t - t_0 \le m\,\Delta T \qquad m = 1, 2, \ldots, M \qquad (5\text{-}75)$$

$$J_i w_i\,\Delta x_i = 2\beta_i = N_i\,\Delta x_i \qquad (5\text{-}76)$$

$$M\,\Delta T = t_f - t_0 \qquad (5\text{-}77)$$

$$t = t_0 + (m - 1)\,\Delta T + s\,\Delta t \qquad s = 0, 1, \ldots, S \qquad (5\text{-}78)$$

$$S\,\Delta t = \Delta T \qquad (5\text{-}79)$$

The number $w_i$ is taken to be a small integer, usually between 2 and 5. Typical values of $S$ range from 5 to 15.

**Step 3.** Starting from a known terminal, using initial arbitrary values of $\delta t_0$, $\Delta x_i(t_0)$, compute the optimal $u(t_0)$ from Eq. (5-70), then compute $g$ according to Eq. (5-67).

**Step 4.** Compute the time increment $\delta t_1$ according to the general formula

$$\delta t = \operatorname*{Min}_{i=1,2,\ldots,n}\left\{\left|\frac{\Delta x_i}{g_i(x, u, t)}\right|\right\} \qquad (5\text{-}80)$$

by using values obtained from step 3. An important consequence of this choice of $\delta t$ is that the next state always lies within a small neighborhood of the present state.

***Step 5.*** Apply the standard procedure [Eq. (5-70)] by using $\delta t_1$. Operate in a similar way until the boundary of the present block is reached. Proceed to transit to a neighboring block . . . until $t_f$ is reached.

As an illustration, consider the problem of computing minimum fuel trajectories for Mach-3 aircraft. The system equations are

$$\frac{dV}{dt} = \frac{g}{\omega}(T - D) - g \sin \gamma \tag{a}$$

$$\frac{dh}{dt} = V \sin \gamma \tag{b}$$

$$\frac{dr}{dt} = V \cos \gamma \tag{c}$$

where $r$, $V$, $h$ are the state variables: range, velocity, and amplitude; $T$ and $\gamma$ are the control variables: thrust and flight path angle; $w$ is the weight, $D$ is the drag. The performance function is the integral of fuel flow function $v$

$$J = \int_{t_0}^{t_f} v(V, h, T, \gamma, \omega, \sigma) \, d\sigma \tag{d}$$

By using Eq. (c), the $dt$ in Eqs. (a) and (b) can be traded with $dr$, resulting in two system equations with two state variables $V(r)$ and $h(r)$.

The total range of altitude and velocity is $0 \leq h \leq 90,000$ and $0 \leq V \leq 3.00$ M. The increment sizes are $\Delta h = 1,000$ feet and $\Delta V = .02$ M. The block size is $2\,\Delta h$ by $2\,\Delta V$ by $\Delta r = 10$ nmi. The preferred direction of motion for this problem is well defined; as the aircraft consumes fuel and loses weight, it tends to fly at a higher altitude and a higher velocity. The memory requirement for a single block is 54 locations. This value is so low that the entire row of blocks covering the same altitude is stored, needing 1100 fast access storage locations. The total computing time is such that optimal trajectories over a 1,000 nmi range can be generated in one hour of computing time on an IBM 7090 computer.

### 5.3.5   Successive Approximating Dynamic Programming—Durling's Approach

We have seen that when the suboptimization techniques discussed in Subsection 5.3.2 are used, we are always led to an approximating process around the control function. The problem is then to find (to guess) an initial feasible solution $u_0(t)$ over the entire time domain so that the iteration can be initiated. The techniques discussed in Subsection 5.3.3 do not need an initial guess of $u(t)$; however, not only does the obtained optimal function $u^*(t)$ have a very low accuracy, but the computing time is also relatively high. The idea comes naturally, therefore, to use this latter function, although not very accurate, as an initial guess. Then by some approximating process, the coarse solution is to be made approaching to an accurate solution. This is the approach suggested by Durling [5-19].

The approximating process used by Durling does not, however, involve one of the analytical development given in Subsection 5.3.2. The dynamic programming procedure is used at each iteration. However, to avoid the high storage requirement, the elementary strategy g-2, the tube strategy is used. Furthermore, to avoid the high computing time, the elementary strategy g-1, the coarse-fine strategy, is used within each tube. As a result, Durling's approach has a triple advantage: (a) low storage requirement, (b) fast computation, (c) accurate solution. The complete procedure of this technique can be stated as follows:

*Step 1.* Obtain a coarse solution $u_0(t)$ and $x_0(t)$ by using a simple method: a coarse-grid covering the whole domain, or a polynomial expansion of $f_k(c)$ and $u_k(c)$.

*Step 2.* Consider a tube around $x_0(t)$ resulting in a moving grid, having the same size but different location at successive times. Apply the dynamic programming procedure with a normal grid size.

*Step 3.* If the new state trajectory or control trajectory does touch the boundary of the tube, modify the location of the tube and apply the dynamic programming procedure again with the same grid size.

*Step 4.* If the new trajectories do not touch the boundary of the tube, then reduce the grid size and apply the dynamic programming procedure again.

*Step 5.* When the trajectories do not touch the boundary any more, and when the difference between successive performance values becomes negligible with the finest grid, stop the computation.

Two examples have been treated by this procedure: the two-dimensional example discussed in Subsection 5.3.3, and an eight-dimensional problem. In Table 5.3, we summarize the results obtained from the first example; the corresponding optimal trajectories are shown in Fig. 5-10.

*Comments.* (1) The cost decreases monotonically, after pass 3-2, no significant gain is obtained. (2) The state is not brought to (.0) exactly at time $t_1$; however, the error is interior to the grid size in each case. (3) The definite solution obtained after pass 3-2, does not coincide completely with the exact solution; the resulting control is only suboptimal.

**Table 5.3** The Two-dimensional Example

| Pass Number | Grid Size $\Delta x$ | $\Delta u$ | Total Cost | $x_1(t_1)$ | $x_2(t_1)$ |
|:---:|:---:|:---:|:---:|:---:|:---:|
| 1–1 | .2 | .4 | 4.3290 | .0191 | −.1080 |
| 2–1 | .04 | .08 | 4.3117 | .0310 | −.0291 |
| 2 | | | 4.3006 | .0288 | −.0297 |
| 3 | | | 4.2967 | .0295 | −.0283 |
| 4 | | | 4.2947 | .0298 | −.0318 |
| 5 | | | 4.2941 | .0306 | −.0312 |
| 3–1 | .008 | .016 | 4.2932 | .0089 | −.0143 |
| 2 | | | 4.2932 | .0113 | −.0147 |

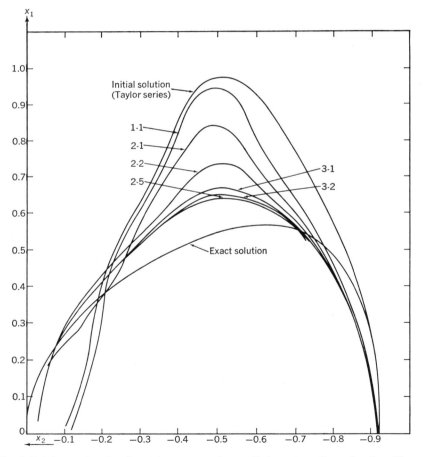

**Fig. 5-10** Approximating dynamic programming applied to a two-dimensional problem—state trajectories.

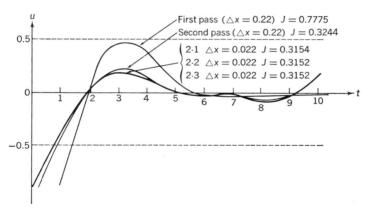

**Fig. 5-11** Approximating dynamic programming applied to an eight-dimensional problem—optimal control function.

The second example follows. System dynamics:

$$
\begin{bmatrix} x_1 \\ x_2 \\ x_3 \\ x_4 \\ x_5 \\ x_6 \\ x_7 \\ x_8 \end{bmatrix} = \begin{bmatrix} -1 & 1 & & & & & & \\ & -1 & 1 & & & & & \\ & & -1 & 1 & & & & \\ & & & -1 & 1 & & & \\ & & & & -1 & 1 & & \\ & & & & & -1 & 1 & \\ & & & & & & -1 & 1 \\ & & & & & & & -1 \end{bmatrix} \begin{bmatrix} x_1 \\ x_2 \\ x_3 \\ x_4 \\ x_5 \\ x_6 \\ x_7 \\ x_8 \end{bmatrix} + \begin{bmatrix} 0 \\ 0 \\ 0 \\ 0 \\ 0 \\ 0 \\ 0 \\ 1 \end{bmatrix} u; \quad \begin{bmatrix} x_1(0) \\ x_2(0) \\ x_3(0) \\ x_4(0) \\ x_5(0) \\ x_6(0) \\ x_7(0) \\ x_8(0) \end{bmatrix} = \begin{bmatrix} 0 \\ 0 \\ 0 \\ 0 \\ 0 \\ 0 \\ 0 \\ 1 \end{bmatrix}
$$

Performance function:

$$
J = \sum_{1}^{8} |x_i(10)| + \frac{1}{10} \int_0^{170} \{[x_1(t)]^2 + [u(t)]^2\} \, dt
$$

The $x$ and $u$-grids consist of 11 points along each coordinate axis. The time from 0 to 10 seconds are partitioned into 50 equal increments. The total program requires 9,669 positions of storage and a run time of 17.5 minutes (IBM 7070). The resulting control functions are plotted in Fig. 5-11.

## 5.4 IMPROVED GRADIENT METHODS

### 5.4.1 A Relaxation Method—Merriam III's Approach

One improvement of the method of Gradients has been suggested by Merriam III [G-34, Ch. 10]. This improvement consists of using a kind of relaxation method, and of implementing an algorithm which avoids the iterations around the initial conditions $\lambda_i(t_0)$ and which guarantees the convergence of the $u$-iterations. First, it is necessary to define more precisely what the relaxation method means. For this, we rewrite the extremalizing equations as follows:

$$\dot{x} = g[x(\sigma), u(\sigma), \sigma] \qquad x(0) = 0 \qquad (5\text{-}81)$$

$$\dot{p} = R[x(\sigma), p(\sigma), u(\sigma), \sigma] \qquad p(T) = 0 \qquad (5\text{-}82)$$

$$u(\sigma) = Q[x(\sigma), p(\sigma), \sigma] \qquad 0 < \sigma < T \qquad (5\text{-}83)$$

associated with the performance functional to be minimized

$$J = \int_0^T F[x(\sigma), u(\sigma)] \, d\sigma \qquad (5\text{-}84)$$

The main reason that the $2PBVP$ arises is because there is coupling between Eqs. (5-81) and (5-82) so that the numerical integration cannot be used. If we now assume the $u^{(i)}(\sigma)$ is a known time-independent function, then the corresponding state vector, denoted as $x^{(i)}(\sigma)$ is found by integrating

$$\dot{x}^{(i)}(\sigma) = g[x^{(i)}(\sigma), u^{(i)}(\sigma), \sigma] \qquad (5\text{-}85)$$

in the forward-time direction from the known boundary condition $x^{(i)}(0) = 0$. After this has been done, the costate vector $p^{(i)}(\sigma)$ can be found by integrating

$$\dot{p}^{(i)}(\sigma) = R[x^{(i)}(\sigma), p^{(i)}(\sigma), u^{(i)}(\sigma), \sigma] \qquad (5\text{-}86)$$

in the backward-time direction from the known boundary conditions $p^{(i)}(T) = 0$. The assumed function $u^{(i)}(\sigma)$ and the solutions $x^{(i)}(\sigma)$ and $p^{(i)}(\sigma)$ generally do not satisfy Eq. (5-83). If now another function $u^{(i+1)}(\sigma)$ can always be found so that system performance is improved, then the $2PBVP$ can be solved to any degree of accuracy desired by solving a sequence of one-point boundary-value problems. This method is called a relaxation procedure. In this respect, the suboptimization technique discussed in Subsection 5.2.1, and the Gradient method discussed in Section 4.5 are all relaxation methods of the preceding definition.

The difference between the Merriam III's approach and other methods lies in the way of finding the new approximating function $u^{(i+1)}(\sigma)$ from the results obtained from the old one $u^{(i)}(\sigma)$. To see this, let us consider the following integral

$$J^{(i+1)} = \int_0^T [F^{(i+1)} + p^{(i+1)}(g^{(i+1)} - \dot{x}^{(i+1)})] \, d\sigma \qquad (5\text{-}87)$$

and use it as a criterion for examining the convergence of the approximating process. In this integral, we have:

$$F^{(i+1)} = F[x^{(i+1)}(\sigma), u^{(i+1)}(\sigma), \sigma] \qquad g^{(i+1)} = g[x^{(i+1)}(\sigma), u^{(i+1)}(\sigma), \sigma] \qquad (5\text{-}88)$$

Let us now expand the integral (5-87) in a Taylor series about the previous iteration. Using

$$U^{(i)}(\sigma) = u^{(i+1)}(\sigma) - u^{(i)}(\sigma); \qquad X^{(i)} = x^{(i+1)}(\sigma) - x^{(i)}(\sigma) \quad (5\text{-}89)$$

$$F^{(i+1)} \cong F^{(i)} + \frac{\partial F^{(i)}}{\partial x^{(i)}} X^{(i)} + \frac{\partial F^{(i)}}{\partial u^{(i)}} U^{(i)} \quad (5\text{-}90)$$

$$g^{(i+1)} \cong g^{(i)} + \frac{\partial g^{(i)}}{\partial x^{(i)}} X^{(i)} + \frac{\partial g^{(i)}}{\partial u^{(i)}} U^{(i)} \quad (5\text{-}91)$$

$$p^{(i+1)} \cong p^{(i)} + \frac{\partial p^{(i)}}{\partial x^{(i)}} X^{(i)} \quad (5\text{-}92)$$

$$\dot{X}^{(i)} = \dot{x}^{(i+1)} - \dot{x}^{(i)} \quad (5\text{-}93)$$

$$\dot{x}^{(i)} = g^{(i)} \quad (5\text{-}94)$$

the integral of (5-87) is written, all second-degree terms in $U^{(i)}$ and $X^{(i)}$ being neglected:

$$J^{(i+1)} \cong J^{(i)} + \int_0^T \left[ -p^{(i)}\dot{X}^{(i)} + \left( \frac{\partial F^{(i)}}{\partial x^{(i)}} + p^{(i)}\frac{\partial g^{(i)}}{\partial x^{(i)}} \right) X^{(i)} \right.$$
$$\left. + \left( \frac{\partial F^{(i)}}{\partial u^{(i)}} + p^{(i)}\frac{\partial g^{(i)}}{\partial u^{(i)}} \right) U^{(i)} \right] d\sigma \quad (5\text{-}95)$$

The backward-time equation which corresponds to Eq. (5-86), is taken to be

$$-\dot{p}^{(i)} = \frac{\partial F^{(i)}}{\partial x^{(i)}} + p^{(i)}\frac{\partial g^{(i)}}{\partial x^{(i)}} \quad (5\text{-}96)$$

Substituting this into Eq. (5-96), we have

$$J^{(i+1)} \cong J^{(i)} + \int_0^T (-p^{(i)}\dot{X}^{(i)} - \dot{p}^{(i)}X^{(i)} + S^{(i)}U^{(i)}) \, d\sigma \quad (5\text{-}97)$$

where

$$S^{(i)} = \frac{\partial F^{(i)}}{\partial u^{(i)}} + p^{(i)}\frac{\partial g^{(i)}}{\partial u^{(i)}} \quad (5\text{-}98)$$

denotes a combination of performance sensitivity and system sensitivity with respect to the control variations. Since, in Eq. (5-97), the terms in $p$ and $X$ form a total differential, we have

$$J^{(i+1)} \cong J^{(i)} - [p^{(i)}X^{(i)}]_0^T + \int_0^T S^{(i)}U^{(i)} \, d\sigma \quad (5\text{-}99)$$

Because of the conditions $X^{(i)}(0) = 0$ and $p^{(i)}(T) = 0$, this expression is reduced to

$$J^{(i+1)} \cong J^{(i)} + \int_0^T S^{(i)}U^{(i)} \, d\sigma \quad (5\text{-}100)$$

This is the equation sought for the purpose of selecting $u^{(i+1)}$.

Up to here, the Merriam III's approach just supplies an approximating process in which the way of selecting the next-step operator is given. However, the sequence of $u^{(i+1)}$ so selected does not guarantee the convergence towards the minimum of $J$. In other words, nothing proves that the sequence of $J^{(i+1)}$ would satisfy the relation

$$J^{(0)} > J^{(1)} > \cdots J^{(i)} > J^{(i+1)} > J^{(i+2)} > \cdots \qquad (5\text{-}101)$$

However, if we choose the sequence of $u^{(i+1)}$ so that Eq. (5-101) is satisfied, then that sequence will also guarantee the convergence of $J^{(i+1)}$ towards its minimum. We can do this by rewriting Eq. (5-100) in the following form:

$$J^{(i+1)} - J^{(i)} \cong \int_0^T S^{(i)} U^{(i)} \, d\sigma \leq 0 \qquad (5\text{-}102)$$

Such a form means that $u^{(i+1)}$ must be chosen so that the "equal to or smaller than" zero condition must be satisfied, and that the equality holds only when $S^{(i)} = 0$, everywhere on the interval $0 \leq \sigma \leq T$. The equality $S^{(i)} = 0$, by Eq. (5-98), means that the control variations cause neither $F$-variations, nor $g$-variations.[1] A sufficient condition for satisfying (5-102) is selecting $U^{(i)}$ so that the integrand is negative or zero everywhere. This condition leads to[2]

$$U^{(i)} = -\delta^{(i)} \operatorname{sgn} S^{(i)} \qquad (5\text{-}103)$$

which can be written, by Eq. (5-90), in a more familiar way

$$u_{\text{new}} = u_{\text{old}} - \delta^{(i)} \operatorname{sgn} S^{(i)} \qquad (5\text{-}104)$$

where $\delta^{(i)}$ is defined as $\delta^{(i)} = \delta(x^{(i)}, u^{(i)}, p^{(i)}, \sigma)$ subject to the nonnegative condition

$$\delta(x^{(i)}, u^{(i)}, p^{(i)}, \sigma) \geq 0 \qquad (5\text{-}105)$$

The function $S^{(i)}$ plays the role of a switching function. The constant or function $\delta^{(i)}$ plays the role of a weighting factor which can be made proportional, for instance, to the magnitude of the gradients, or to any other significant factors of the optimized system. Figure 5-12 shows a flow-diagram for digital computation of the relaxation procedure based on first variation.

It is very interesting to show that if we continue the line of arguments started in the preceding relaxation procedure, we will find the Newton-Raphson procedure discussed in Subsection 5.3.4. For this, consider the function $H$ defined as

$$H = F(x^{(i)}, u, \sigma) + p^{(i)} g(x^{(i)}, u, \sigma) \qquad (5\text{-}106)$$

---

[1] To be more exact, we must determine that this concerns only first variations of $F$ and $g$, since we have only taken the first-degree terms of Taylor series. The same kind of arguments can be followed if second variations are considered. See [G-34, Ch. 10, 5-44, 5-45].

[2] It is interesting to compare the search law represented by Eq. (5-103) to the search law discussed in Subsection 3.6.2. The two laws, quite similar, are derived differently.

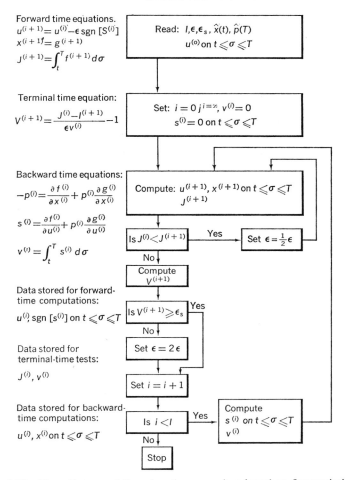

Forward time equations.
$$u^{(i+1)} = u^{(i)} - \epsilon \, \text{sgn} \, [S^{(i)}]$$
$$x^{(i+1)} = g^{(i+1)}$$
$$J^{(i+1)} = \int_t^T f^{(i+1)} d\sigma$$

Terminal time equation:
$$V^{(i+1)} = \frac{J^{(i)} - I^{(i+1)}}{\epsilon v^{(i)}} - 1$$

Backward time equations:
$$-\dot{p}^{(i)} = \frac{\partial f^{(i)}}{\partial x^{(i)}} + p^{(i)} \frac{\partial g^{(i)}}{\partial x^{(i)}}$$

$$s^{(i)} = \frac{\partial f^{(i)}}{\partial u^{(i)}} + p^{(i)} \frac{\partial g^{(i)}}{\partial u^{(i)}}$$

$$v^{(i)} = \int_t^T s^{(i)} d\sigma$$

Data stored for forward-time computations:
$$u^{(i)}, \, \text{sgn} \, [s^{(i)}] \text{ on } t \leqslant \sigma \leqslant T$$

Data stored for terminal-time tests:
$$J^{(i)}, \, v^{(i)}$$

Data stored for backward-time computations:
$$u^{(i)}, \, x^{(i)} \text{ on } t \leqslant \sigma \leqslant T$$

Read: $I, \epsilon, \epsilon_s, \hat{x}(t), \hat{p}(T)$
$u^{(0)}$ on $t \leqslant \sigma \leqslant T$

Set: $i = 0 \; j = \infty, \; v^{(i)} = 0$
$s^{(i)} = 0$ on $t \leqslant \sigma \leqslant T$

Compute: $u^{(i+1)}, x^{(i+1)}$ on $t \leqslant \sigma \leqslant T$
$J^{(i+1)}$

Is $J^{(i)} < J^{(i+1)}$   Yes → Set $\epsilon = \frac{1}{2}\epsilon$

No ↓

Compute $V^{(i+1)}$

Is $V^{(i+1)} \geqslant \epsilon_s$   Yes

No ↓

Set $\epsilon = 2\epsilon$

Set $i = i + 1$

Is $i < I$   Yes → Compute $s^{(i)}$ on $t \leqslant \sigma \leqslant T$ , $v^{(i)}$

No ↓

Stop

*Fig. 5-12*  Flow-diagram of the relaxation procedure based on first variations.

This function is similar to the Hamiltonian function or augmented function used in Chapter 4; however, $H$ is treated here as a function of $u$ and $\sigma$ only, because $x^{(i)}$ and $p^{(i)}$ are assumed fixed from the previous iteration. Also the notation

$$H^{(i)} = F(x^{(i)}, u^{(i)}, \sigma) + p^{(i)} g(x^{(i)}, u^{(i)}, \sigma) \qquad (5\text{-}107)$$

is used here in conjunction with $H$. The function $H$, in a case where it is unimodal, is sketched in Fig. 5-13 and shown as a strictly convex function. The role of the slope $S^{(i)}$ becomes apparent and suggests immediately selecting $u^{(i+1)}$ in order to correspond to the minimum point of $H$ where the slope is

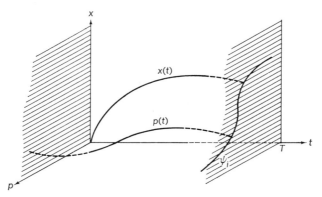

**Fig. 5-13**   Sketch of a general optimization problem.

zero. Some refinement can be added such as using the partial derivative $\partial S^{(i)}/\partial u^{(i)}$ (the rate of the slope) for determining the rate at which the condition $S^{(i)} = 0$ is being achieved. For this, consider the second-degree Taylor series about $u^{(i)}$

$$H \cong H^{(i)} + S^{(i)}(u - u^{(i)}) + \frac{1}{2}\frac{\partial S^{(i)}}{\partial u^{(i)}}(u - u^{(i)})^2 \tag{5-108}$$

When $u$ is adjusted to minimize the right-hand side of Eq. (5-108), the result is

$$u - u^{(i)} = \frac{-S^{(i)}}{\dfrac{\partial S^{(i)}}{\partial u^{(i)}}} \tag{5-109}$$

Finally, the step size is restricted by taking some fraction of this difference, thereby obtaining the iteration algorithm

$$U^{(i)} = -\alpha S^{(i)}/(\delta S^{(i)}/\delta u^{(i)}) \tag{5-110}$$

or, rather

$$\delta^{(i)} = \alpha \, |S^{(i)}|/(\delta S^{(i)}/\delta u^{(i)}) \tag{5-111}$$

This iteration algorithm is equivalent to Newton-Raphson method associated with trial-and-error minimization procedures such as discussed in Subsection 5.2.4.

### 5.4.2   Approximating Process Around the Unknown Initial Conditions

We have seen in Subsection 4.2.5, when discussing the *2PBVP*, that the slowness of the iterative procedure is because we hunt the unknown initial conditions $p_i(t_0)$ or $\lambda_i(t_0)$. In the gradient method, we also had to make an initial guess of $\lambda_i(t_0)$; this, however, gives less difficulty because there are two hunted objects: $\lambda_i(t_0)$ and $u(t)$. The $\lambda_i(t_0)$-achievement is somehow unrolled within the $u(t)$-achievement. The question which arises now is why the hunting of $\lambda_i(t_0)$ is difficult. In normal approximating process, using the gradient method, the success is due to:

1. The existence of gradients, or a sensitivity function which indicate not only the performance variation, e.g. $\partial J$, with respect to the parameter variation, e.g., $\partial u$, but also the direction of $\partial u$ so that $J$ is minimized ($\partial J = 0$).

2. The existence of a stopping condition, e.g. $J^{(n+1)} - J^{(n)} \leq \epsilon$ directly related to the performance functional.

In the situation of $\lambda_i(t_0)$, there is no sensitivity function of the form $\partial J/\partial \lambda_i(t_0)$; furthermore, the stopping condition is not directly related to $\lambda_i(t_0)$. This is the reason why the approximating process around $\lambda_i(t_0)$ is slow.

A special technique has been suggested by M. D. Levine and others [5-27, 5-29, 5-30] which consists of creating an artificial function $V$ relating $\lambda_i(t_0)$ to the other variables of the system so that the sensitivity function and then the steepest direction can be found after each iteration. For exposing this technique, we shall rewrite the extremalizing equations, based on the Pontryagin formulation, of a quite general optimization problem (Fig. 5-13).

$$\dot{x} = g(x, p, t) \qquad x(0) = 0 \quad x(T) \text{ unknown} \qquad (5\text{-}112)$$

$$\dot{p} = h(x, p, t) \qquad p(0) = a \qquad \text{unknown} \qquad (5\text{-}113)$$

$p, x : n$-vectors

$p(T)$ given by the constraints; ($T$ unknown)

$$\psi_j(x, p, t) = 0, \qquad j = 1, 2, \ldots, n+1 \qquad (5\text{-}114)$$

First, we choose one of the $\psi_j$ constraints as the stopping condition of each integration, i.e.

$$W(x, p, t) = \psi_m(x, p, t) = 0 \qquad (5\text{-}115)$$

leading to one component less in Eq. (5-114)

$$\psi_j(x, p, t) = 0, \qquad j = 1, 2, \ldots, n \qquad (5\text{-}116)$$

Using these terminal conditions, we can construct an artificial scalar terminal error criterion for the approximating process

$$V(T) = V[x(T, a), p(T, a), T(a)] = \sum_{i=1}^{n} [\psi_j(T)]^2 \qquad (5\text{-}117)$$

This is a positive semidefinite function[1] with a minimum value of zero. For

---

[1] The convergence problem is a stability problem. The use of Lyapunov terminology, such as $V$-function, semidefinite function, . . . is deliberate.

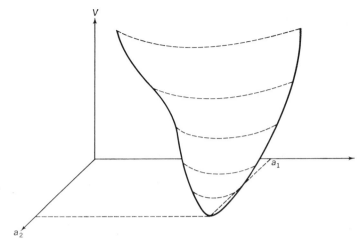

***Fig. 5-14***   The artificial error criterion $V$ in a two-dimensional problem.

example, if $n = 2$, $V$ will appear as a valley as shown in Fig. 5-14. For each value of $a$, we integrate Eqs. (5-112) and (5-113) until $W = 0$; this defines a $t = T$ for which $V(T)$ may be determined from Eq. (5-117) since $\psi_j$ is given. We may then plot this value as a point on the $V$-hypersurface. The next value of the vector $a$ must be on the direction ending at the point where $V(T)$ vanishes. For doing this in the *quickest* or the *steepest* way, we require the gradient of the hypersurface of *any* point on it. In other words, for a given $a$, a knowledge of $V_a = \partial V/\partial a$ is desired. Then we may choose

$$\frac{da}{d\sigma} = -KV_a \tag{5-118}$$

where $\sigma$ is a parameter characterizing the step size, and $K$ a constant characterizing the weight of the gradient taken in each direction. For digital computations, Eq. (5-118) is written

$$a_{\text{new}} = a_{\text{old}} - KV_a(T) \tag{5-119}$$

assuming unity step $d\sigma = 1$. The question arises now of how to calculate the partial derivative $V_a$. The general form is

$$V_a(T) = V_x(T)x_a(T) + V_p(T)p_a(T) + V_T(T)T_a \tag{5-120}$$

where $V_x$, $V_p$, and $V_T$ may be found directly by differentiation from Eq. (5-117). The difficulty lies in the evaluation of analytical expressions of $x_a(T)$, $p_a(T)$, and $T_a$.[1]

---

[1] Because this implies the differentiability of $g$ and $h$, and needs lengthy mathematical arguments for the existence proof. See appendix A of [5-27].

The evaluation of $x_a(T)$ and $p_a(T)$ can be made by using some transformations. For this, we note first that they are Jacobian matrices of the form

$$
x_a(T) = \begin{bmatrix} x_{1_{a1}}(T) & \cdots & x_{1_{an}}(T) \\ x_{2_{a1}}(T) & & x_{2_{an}}(T) \\ \cdot & & \cdot \\ \cdot & & \cdot \\ \cdot & & \cdot \\ x_{n_{a1}}(T) & \cdots & x_{n_{an}}(T) \end{bmatrix} \tag{5-121}
$$

$$
p_a(T) = \begin{bmatrix} p_{1_{a1}}(T) & p_{1_{a2}}(T) & \cdots & p_{1_{an}}(T) \\ p_{2_{a1}}(T) & & & p_{2_{an}}(T) \\ \cdot & & & \cdot \\ \cdot & & & \cdot \\ p_{n_{a1}}(T) & \cdot & \cdot & \cdot & p_{n_{an}}(T) \end{bmatrix} \tag{5-122}
$$

Now if we set $y = x_a$ and $z = p_a$, it can be shown that $y$ and $z$ are solutions of the differential system (called accessory equations)

$$
\frac{dy}{dt} = g_x y + g_p z \qquad y(0) = 0, \qquad y(T) = x_a(T) \tag{5-123}
$$

$$
\frac{dz}{dt} = h_p z + h_x y \qquad z(0) = I, \qquad z(T) = p_a(T) \tag{5-124}
$$

To determine $T_a$, consider Eq. (5-115); differentiating it with respect to $a$ yields:[1]

$$
W_x x_a + \frac{dW}{dt} T_a + W_p p_a = 0 \tag{5-125}
$$

But

$$
\frac{dW}{dt} = W_a \frac{dx}{dt} + W_p \frac{dp}{dt} + W_T \tag{5-126}
$$

Substituting Eq. (5-126) into Eq. (5-125), we get

$$
T_a = \frac{-(W_x x_a + W_p p_a)}{W_x \dfrac{dx}{dt} + W_p \dfrac{dp}{dt} + W_T} \tag{5-127}
$$

[1] If the terminal time $T$ is fixed, $T$ is independent from $a$, then $T_a = 0$.

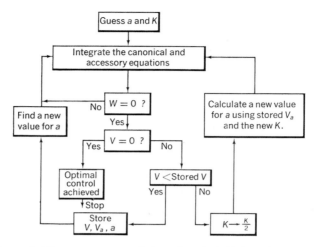

**Fig. 5-15**  Approximating process around the unknown initial conditions—flow diagram.

or using $y(T)$ and $z(T)$ from Eqs. (5-123) and (5-124)

$$T_a = - \frac{W_x(T)y(T) + W_p(T)z(T)}{W_x(T)\dfrac{dx}{dt}(T) + W_p(T)\dfrac{dp}{dt}(T) + W_T(T)} \tag{5-128}$$

Substituting Eqs. (5-123), (5-124), and (5-128) into Eq. (5-120), it is now possible to calculate $V_a$ with a knowledge of $x$, $p$, $T$, $y$, $z$.

The computational technique may be outlined with reference to the flow-diagram in Fig. 5-15.

**Step 1.**  Guess a value of $a$ and choose $k$.

**Step 2.**  Integrate Eqs. (5-112), (5-113), (5-123), and (5-124) until $W = 0$. This involves $2(n^2 + n)$ equations.

**Step 3.**  Calculate the value of $V(T)$ in Eq. (5-117) and ascertain if $V = 0$. If it is, then the problem has been solved.

**Step 4.**  If $V \neq 0$, check if the new value is smaller than the stored value of $V$. If it is not, halve the value of $K$ and find a new value of $a$ using the stored $V_a$. This is repeated until the new $V$ determined falls below the value of the stored $V$.

**Step 5.**  If step 4 has been satisfied, calculate $V_a(T)$ by Eq. (5-120) and update $a$ by using Eq. (5-119).

***Step 6.***  Return to step 2 and repeat steps 2, 4, 4, 5, and 6 until $V = 0$.

For time optimal problems, we must steepest descend by choosing $a$ and $T$, the optimal time. This requires

$$\frac{dV(T)}{dT} = V_x(T)\frac{dx(T)}{dt} + V_p(T)\frac{dp(T)}{dt} + V_T(T)$$

The time $T$ is updated at each stage (along with $a$) by choosing

$$T_{\text{new}} = T_{\text{old}} - K\frac{dV(T)}{dt} \tag{5-129}$$

In this case, we must choose an "artificial" stopping condition on the first run and then use

$$W = t - T \tag{5-130}$$

on all successive runs.[1]

***Illustrative Example.***  Find $u(t)$ which transfer the state $(x_1, x_2)$ from the point $(-5, -1)$ to the origin in minimum time.

$$\frac{dx_1}{dt} = x_2$$

$$\frac{dx_2}{dt} = -.1x_1 + u;$$

$$\frac{dp_1}{dt} = .1p_2, \qquad p_1(0) = a_1$$

$$\frac{dp_2}{dt} = -p_1, \qquad p_2(0) = a_2$$

$$u(t) = \text{sgn}\,[p_2(t)] \qquad \text{(bang-bang control)}$$

$$g_1 = x_2 \qquad g_2 = -.1x_1 + \text{sgn}\,(p_2)$$

$$h_1 = .1p_2 \qquad h_2 = -p_1$$

$$\frac{dy_{11}}{dt} = y_{21} \qquad y_{11}(0) = 0$$

$$\frac{dy_{12}}{dt} = y_{22} \qquad y_{12}(0) = 0$$

$$\frac{dy_{21}}{dt} = -.1y_{11} + \frac{2(-1)^j\,\text{sgn}\,(a_2)\delta_{t,y_j}z_{21}}{dp_2/dt} \qquad y_{21}(0) = 0$$

[1] It is interesting to compare the procedure used here with the one used in Subsection 5.3.4.

$$\frac{dy_{22}}{dt} = -.1y_{12} + \frac{2(-1)^{j}\,\text{sgn}\,(a_{2})\delta_{t.\gamma_{j}}z_{22}}{dp_{2}/dt} \qquad y_{22}(0) = 0$$

$$\frac{dz_{11}}{dt} = .1z_{21} \qquad z_{11}(0) = 1$$

$$\frac{dz_{12}}{dt} = -1z_{22} \qquad z_{12}(0) = 0$$

$$\frac{dz_{21}}{dt} = -z_{11} \qquad z_{21}(0) = 0$$

$$\frac{dz_{22}}{dt} = -z_{12} \qquad z_{22}(0) = 1$$

The artificial error function is defined as

$$V = x_1(T)^2 + x_2(T)^2$$

$$a_{1\text{new}} = a_{1\text{old}} - 2K[x_1(T)y_{11}(T) + x_2(T)y_{21}(T)]$$

$$a_{2\text{new}} = a_{2\text{old}} - 2K[x_1(T)y_{12}(T) + x_2(T)y_{22}(T)]$$

$$T_{\text{new}} = T_{\text{old}} - 2K[x_1(T)g_1(T) + x_2(T)g_2(T)]$$

To begin the computation

$$a_1 = .2 \qquad a_2 = 1 \qquad K = 10$$

Artificial stopping condition $W = t - (1 + t_1)$ where $t = t_1$ is defined by $y_{12}(t_1) - .01 = 0$. The solution is found after 33 steps, when $V < .005$. Figure 5-16a, b, c, d, and e show respectively the curves of the optimal $x$, $p$, $u$, $z$, and $y$.

*Comments*

1. The strategy which consists of using an artificial error function connecting directly the approximated object $a$, to a system characteristic, $\psi_j$, is new with respect to the techniques previously discussed.

2. This technique can be applied to the method of gradients. However, the $\lambda_i(t_0)$-approximating is only a minor problem there. The question is then whether it is worthwhile the great analytical and computational complications. The answer to this question needs mathematical experimentations.

### 5.4.3   Integrations and Use of Hybrid Computations

We have seen that, in the gradient method, we need the knowledge of the influence function $\lambda_i$. For this, we have to guess $\lambda_i(0)$ and then solve the adjoint differential system for obtaining $\lambda_i$. We have seen also that in all

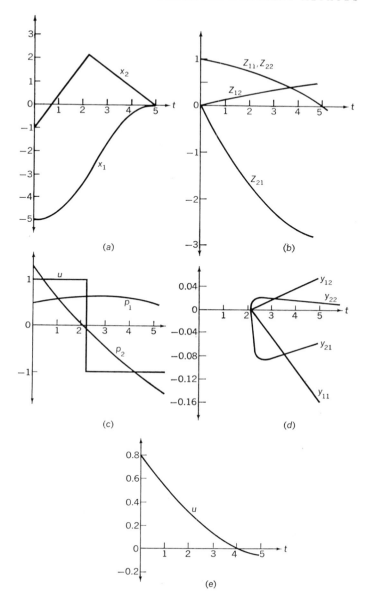

**Fig. 5-16** Solution of the example problem. (*a*) The state variables. (*b*) The *z* accessory extremals. (*c*) The adjoint variables and the control function. (*d*) The *y* accessory extremals. (*e*) The control function.

other approximating processes, we meet, at one point or at another, the
need of solving a set of differential equations. Throughout the discussions, we
have always assumed that this was done by numerical integration using digital
computations. Such a procedure generally leads to lengthy runs. This is not
only expensive, but precludes the application of the iterative procedure to
on-line control. Since integrations are typically best operated by analogue
computations, a number of works have been devoted to the application of
hybrid computers for implementing iterative processes derived from
optimization problems [5-25, 5-26, 5-43, 5-46]. Here, we shall expose how
this can be done by using a time optimal problem: the harmonic oscillator
problem, which was an illustrative example in the preceding subsection. At
the same time, we shall indicate another way of creating an artificial error
criterion when initial conditions $p(0) = a$ are approximated.[1]

The general time optimal problem can be formulated as follows. Determine
$u(t)$ to bring the state $x(t)$ to zero in the least positive time using an allowable
control $|u_k| \leq 1$. The system is assumed linear.

$$\dot{x} = A(t)x + B(t)u \tag{5-131}$$

where $x$ is an $n$-vector, $u$ an $r$-vector, $A(t)$ and $B(t)$ are $n \times n$ and $n \times r$
matrices respectively. The initial conditions $x_i(0)$ are known. Other characteristics are as usual. Hamiltonian:

$$H = p\dot{x} + 1 \tag{5-132}$$

Adjoint system:

$$\dot{p} = -\frac{\partial H}{\partial x} = -A(t)^T x \tag{5-133}$$

Bang-bang optimal control:

$$u_k^0 = -\text{sgn } p_i B_{ik} \tag{5-134}$$

Switching times for the bang-bang control are determined by zero crossings
of the function $p_i B_{ik}$.[2] The control law will be known if the functions $p_i(t)$
are known.

Let the unknown initial conditions $p_i(0)$ for the adjoint system be called $a_i$.
Neustadt's method provides an iterative procedure for finding the initial
state $a_i$. The method depends on the generation of a function[3]

$$Z_i(t, a) = -\int_0^t X_{ij}^{-1}(\tau) B_{jk}(\tau) u_k(\tau, a) \, d\tau \tag{5-135}$$

---

[1] Credited to L. W. Neustadt. See [5-33, 5-34].
[2] See similar control in the example given in Subsection 4.2.4.
[3] Note the difference between $Z_i$ here and $V$ of the preceding subsection. They play the
same role, but use different functions.

where $X^{-1}(\tau)$ is the inverse of the fundamental matrix solution to Eq. (5-131), $B_{jk}(\tau)$ is given by the problem, and $u_k(\tau, a)$ given by Eq. (5-134) and is only a function of $a$, not of $p(t)$. The iteration procedure requires a choice of a vector $a$ and then the integration to generate $Z(t, a)$. The integration is stopped at the time when

$$Z(t, a) \equiv a \cdot [Z(t, a) - x(0)] = 0 \qquad (5\text{-}136)$$

this time is denoted by $F(a)$, i.e.

$$a \cdot \{Z[F(a), a] - x(0)\} = 0 \qquad (5\text{-}137)$$

The vector $Z[F(a), a] - x(0)$ plays the role of a sensitivity function or a gradient. It is recorded and provides the basis for the corrections to $a$

$$a^{(i+1)} = a^i - K \frac{Z[F(a), a] - x(0)}{\|Z[F(a), a] - x(0)\|} \qquad (5\text{-}138)$$

where $K$ is a weighting constant. From one of these later equations, we can see that the function $Z$, which is dependent on $a$, plays the role of a dummy $x(0)$, and represents the value of $x(0)$ when $a$ is not the correct $p(0)$.

The object of the previous calculation is to find the $a$ that maximizes $F(a)$. The $a$-vector that does this also causes the boundary condition to be satisfied.

*Example.*   Harmonic oscillator problem

$$\dot{x}_1 = x_2$$

$$\dot{x}_2 = -\omega^2 x_1 + u \qquad |u| \le 1$$

$$u_0 = \text{sgn}\left[ -q_1 \frac{\sin \omega t}{\omega} + a_2 \cos \omega t \right]$$

$$Z_1(t) = -\int_0^t \frac{1}{\omega} (\sin \omega\tau) u(\tau, a)\, d\tau$$

$$Z_2(t) = \int_0^t (\cos \omega\tau) u(\tau, a)\, d\tau$$

$$a_1[Z_1(t, a) - x_1(0)] + a_2[Z_2(t, a) - x_2(0)] \le 0$$

Value of $a$ for the first trial

$$a_i = -\frac{x_i(0)}{\|x(0)\|}$$

In view of implementing the iterative procedure, a hybrid computer consisting of an analog computer and a digital computer is used (Fig. 5-17). The analog computer integrates the equations for $Z_1$ and $Z_2$; it stops when $g = a \cdot [Z(t, a) - x(0)] = 0$. The analog computer is put into the "Hold" mode,

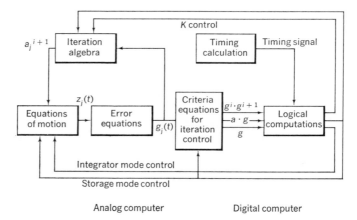

*Fig. 5-17*   Gradient method using hybrid computation.

and a new *a*-vector is obtained from the iteration rule. These calculations are made in the digital computer.

*Comments*

1. This hybrid computation method can be extended to all other approximating processes discussed in this chapter.

2. It appears that there are many possibilities of improving the iteration around the unknown costate initial conditions.

## 5.5  CONCLUSIONS

In this chapter, we have discussed a certain number of suboptimization techniques which, in principle, are independent, or at least complementary. However, although there are many distinct features in some minor points, they are similar in general lines and can be divided into two main categories. The first category contains those techniques in the line of the generalized Newton-Raphson method with some sophisticated gradient procedure. The sophistication concerns either greater care for solving the convergence problem, or suitable computing means for a definite kind of operations. The second category continues the line of dynamic programming procedure, and, by using a combination of grid-manipulating strategies and approximating-process strategy may arrive at a triple optimum of high accuracy, reasonable storage requirement, and reasonable computing time.

The general feeling remains, however, that there does not exist one method or one procedure suitable for treating all nonlinear dynamical optimization problems. This will never be the case for the simple reason that nonlinear problems cannot be grouped within one single class. As a consequence, the previously described techniques will only give the whole of their potential power if they are applied to each particular case, tested, compared, reorganized, etc. This is the mathematical experimentation mentioned several times in this book. This is also the practical and the unique approach for filling the gap between theory and application.

Another way of looking at these techniques is the one used by an always unsatisfied man. Truly, in all these suboptimization techniques, we have only considered a very small domain of analysis; only used two main categories of strategies: approximation and approximating process, of and around some functions and functionals; only applied one kind of property: derivatives, variations, and gradients. There are certainly other fields of analysis (functional analysis, logical analysis), other strategies( association, pattern recognition, various computer organization and programming), and other properties (total variation, order, trees) to be explored and used for optimization purposes. The field of short-range and long-range research looks immense. The techniques for solving a definite class of problem are never immutable, they always evolve.

# REFERENCES

[5-1]  Peterson, E. L. and F. X. Remond, "Computational Methods of Multi-dimensional Nonlinear Processes," Report RM 61 TMP-52, August 1961, General Electric Company, Santa Barbara, California.

[5-2]  Peterson, E. L., "Theory, Methods, and Application of Optimization Techniques," Seminar on Orbit Optimization and Advanced Guidance Instrumentation. Organized by AGAD/NATO, Düseldorf, October 1964, GECO, Santa Barbara.

[5-3]  Dreyfus, S. E., "A Computational Technique Based on Successive Approximations in Policy Space," Appendix III in [G-26].

[5-4]  Kelley, H. J., "A Successive Approximation Scheme Employing the Min Operations," Section 6-6, pp. 248–251 in [G-29].

[5-5]  Armstrong, E. S. and J. H. Suddath, "Application of Pontryagin's Maximum Principle to the Lunar Orbit Rendez-Vous Problem," *Progr. Astron. Aeron.*, **13,** Guidance & Control II, 1964, 865–885.

[5-6]  Pearson, J. D., "A Successive Approximation Method for Solving 2*PBVP* Arising in Control Theory," *Proc. 4th JACC*, 1963, Mineapolis, Minn, 36–41.

[5-7]    Bellman, R. and R. Kalaba, "Dynamic Programming, Invariant Imbedding and Quasilinearization: Comparisons and Interconnections," in [G-37], pp. 135–145.

[5-8]    Bellman, R. E. and R. Kalaba, "Dynamic Programming, Invariant Imbedding and Quasilinearization. Comparisons and Interconnections," Rand Memorandum RM-4038-PR, March 1964, AD 431866.

[5-9]    Bley, K. B., "Three New Computational Methods for Solving 2PBVP," Rand Report P 2878, AD 433279.

[5-10]   Bellman, R. and T. A. Brown, "On the Computational Solution of 2PBVP," Rand Memorandum RM-3857-PR April 1964, AD 436753.

[5-11]   Bellman, R. E., H. H. Kagiwada, and R. E. Kalaba, "Numerical Studies of a Nonlinear TPBVP Using Dynamic Programming, Invariant Imbedding and Quasilinearization," Rand Memorandum RM-4069 Pr, March 1964, AD 434561.

[5-12]   Greenspan, D., "On Approximating Extremals of Functionals Part I: The Method and Examples for Boundary Value Problems," MRC Report N° 466, March 1964, AD 437288.

[5-13]   Schlechter, S., "Iterative Methods for Nonlinear Problems," Trans. A.M.S. 104, 1962, 179–189.

[5-14]   Bellman, R., H. Kagiwada, and R. Kalaba, "Nonlinear Extrapolation and 2PBVP, Rand Memorandum RM 4079-PR, April 1964, AD 435765.

[5-15]   Kenneth, P. and R. McGill, "Two-Point Boundary Value Problem Techniques," in [G-50].

[5-16]   McGill, R. and P. Kenneth, "A Convergence Theorem on the Iterative Solution of Nonlinear 2PBVP Systems," XIVth International ASTRONAUTICAL Congress, Paris, September 1963.

[5-17]   McGill, R. and P. Kenneth, "Solution of Variational Problems by Means of a Generalized Newton-Raphson Operator, AIAA J., 2, N° 10, October 1964, 1761–1766.

[5-18]   Plant, J. B. and M. Athans, "An Iterative Technique for the Computation of Time Optimal Controls," Third IFAC Congress, London, June 1966.

[5-19]   Durling, A. E., "Computational Aspects of Dynamic Programming in Higher Dimensions," Technical Report TR-64-3, Electrical Engineering Department, Syracuse University, May 1964.

[5-20]   Peterson, E. L. and F. X. Remond, "Investigation of Dynamic Programming Methods for Flight Control Problems," General Electric Co. Santa Barbara, FDL-TDR-64-11, March 1964.

[5-21]   Larson, R. E., Dynamic Programming with Reduced Computational Requirements, IEEE Tr. on Automatic Control AC-10, N° 2, April 1965, 135–143.

[5-22]   Richardson, D. W., E. P. Hechtman, and R. E. Larson, "Control Data Investigation for Optimization of Fuel on Supersonic Transport Vehicle," Phase 1 Rep., Contract AF 33(657)-8822. Project 9056, Hughes Aircraft Co, Culver City, Calif, January 1963. Phase II Rep., R. R. Larson, "Dynamic Programming with Continuous Independent Variable." Tech Rept. 6302-6 Contracts Nb 225/24 NR 373, 360, SEL 64-019 Stanford University, Stanford, Calif. June 1964.

[5-23]  Durbeck, R. C., "An Approximation Technique for Suboptimal Control," *IEEE Tr* **AC-10**, n 2, April 1965, 144–149.

[5-24]  Bellman, R., "Some Directions of Research in Dynamic Programming," Memo RM 3661-PR, April 1964, AD 438 438.

[5-25]  Paiewonsky, B. et al, "Synthesis of Optimal Controllers Using Hybrid Analog-Digital Computers," pp. 285–303 in [G-37].

[5-26]  Gilbert, E. G., "The Application of Hybrid Computers to the Iterative Solution of Optimal Control Problems," pp. 261–283 in [G-37].

[5-27]  Levine, M. D., "A Steepest Descent Method for Synthesizing Optimal Control Program," in [G- 2].

[5-28]  Jazwinski, A. H., "Optimal Trajectories and Linear Control of Nonlinear Systems," *AIAA J.* **2**, n 8, August 1964, 1371–1379.

[5-29]  Wachman, M., "On the 2*PBVP*, with Application to Celestial Mechanics," General Electric Co, Space Sciences Laboratory, Technical Information Series, R 64 SD 57 August 1964.

[5-30]  Knapp, C. H. and P. A. Frost, "Determination of Optimum Control and Trajectories Using the Maximum Principle in Association with a Gradient Technique," *IEEE Tr*, vac-10 n 2, April 1965, 189–194.

[5-31]  Pearson, J. D., "Theoretical Methods of Optimization: Dynamic Programming," in [G-31].

[5-32]  Aoki, M., "Dynamic Programming and Numerical Experimentation as Applied to Adaptive Control," Doctoral Thesis, University of California at Los Angeles, November 1959.

[5-33]  Neustadt, L. W., "Synthesizing Time Optimal Control Systems," *J. Math. Anal. Appl.* **1** (1960) 484–493.

[5-34]  Neustadt, L. W., "On Synthesizing Optimal Controls," Second IFAC Congress, Basle 1963.

[5-35]  Longmuir, A. G. and V. Bohn, "The Synthesis of Suboptimal Feedback Control Laws," *IEEE Tr.*, **AC-12**, n 6 (December 1967), 755–8.

[5-36]  Fujisawa T. and Y. Yosuda, "An Iterative Procedure for Solving the Time Optimal Regulator Problem," *SIAM J. Control*, **5**, n 4, (November 1967), 501–512.

[5-37]  Naki, N. E. and L. A. Wheeler, "Optimum Terminal Control of Continuous Systems via Successive Approximation to the Reachable Set," *IEEE Tr.*, **AC-12**, n 5 (October 1967), 515–521.

[5-38]  Bingulac S. P. and S. J. Kahne, "On the Solution of 2*PBVP*," *IEEE Tr.*, **AC-10**, n 2 (April 1965), 208.

[5-39]  Knapp, C. H. and P. A. Frost, "Determination of Optimum Control and Trajectories using the Maximum Principle in Association with a Gradient Technique," *IEEE Tr*, **AC-10**, n 2 (April 1965), 185–193.

[5-40]  Tchamran A., "On Bellman's Functional Equation and a Class of Time-Optimal Control Systems," *J. Frankl. Inst.*, **280**, n 6 (December 1965), 493–505.

**[5-41]**   Noton, A. R. M., "Optimal Control and the 2*PBVP*," *Proc. Inst. Mech. Engrs.* (G.B.), **179,** pt 3H (1965), 181–188.

**[5-42]**   Levine, M. D., "A Steepest Descent Method for Synthesizing Optimal Control Programmes," *Proc. Inst. Mech. Engrs.* (G.B.), **179,** pt 3H (1965), 143–152.

**[5-43]**   Richard Lee Maybach, "The Solution of Optimal Control Problems on an Iterative Hybrid Computer," Ph.D. Thesis, n 67-6940, Arizona Univ., Tucson (1967), 179 pp.

**[5-44]**   Bullock, T. E., "Computation of Optimal Controls by a Method based on Second Variations," *NASA* CR 84847, SUDAAR 297 (December 1966), 97 pp.

**[5-45]**   Tracz, G. S. and B. Bernholz, "An Integral Equation Approach to Second Variation Techniques for Optimal Control Problems," *Technical Report, National Research Council of Canada* (1967), 34 pp.

**[5-46]**   Darcy, V. J., "An Application of an Analog Computer to Solve the 2PBVP for 4th order Optimal Control Problems," M.S. Thesis (June 1966), 65 pp, GGC/EE/66-7, AD639578.

**[5-47]**   Kang, G. Researches in Optimal and Sub-optimal Control Theory, Rept. 68-10, *NASA* CR-93737 (March 1968), 93 pp.

# General References

[G-1]   Zadeh, L. A. and C. A. Desoer, *Linear System Theory—The State Space Approach*, McGraw-Hill, 1963, 628 pp.

[G-2]   Kopal, Z., *Numerical Analysis*, Chapman and Hall, 1955, 556 pp.

[G-3]   Wilson, E. B., *Advanced Calculus*, Dover, 1958, 566 pp.

[G-4]   Volterra, V., *Theory of Functionals and of Integral and Integro-Differential Equations*, Dover, 1959, 226 pp.

[G-5]   Hancock, H., *Theory of Maxima and Minima, Dover*, 1960, 193 pp.

[G-6]   Elsgolc, L. E., *Calculus of Variations*, Pergamon, 1961, 176 pp.

[G-7]   Forsyth, A. R., *Calculus of Variations*, Dover, 1960, 656 pp.

[G-8]   Bolza, O., *Lectures on the Calculus of Variations*, Dover, 1961, 268 pp.

[G-9]   Bliss, G. A., *Lectures on the Calculus of Variations*, Chicago Univ. Press, 1946, 289 pp.

[G-10]  Valentine, F. A., *Contributions to the Calculus of Variations*, Chicago Univ. Press, 1937.

[G-11]  Cesari, L., J. La Salle, and S. Lefschetz (Eds.), *Contributions to the Theory of Nonlinear Oscillations*, Princeton Univ. Press, 1961, 286 pp.

[G-12]  Saady, Th. L. and J. Bram, *Nonlinear Mathematics*, McGraw-Hill, 1964, 381 pp.

[G-13]  Akhiezer, N. I. and I. M. Glazman, *Theory of Linear Operators in Hilbert Space*, Frederick Vugan Publ. Co.

[G-14]  Dantzig, G. B., *Linear Programming and Extension*, Princeton Univ. Press, 1963, 632 pp.

[G-15]  Kaufmann, A., *Méthodes et modèles de la recherche opérationnelle*, Vol. 1, 1959; Vol. 2, 1961; Dunod, Paris.

[G-16]  Hadley, G., *Linear Algebra*, Addison-Wesley, 1961.

[G-17]  Hadley, G., *Linear Programming*, Addison-Wesley, 1963, 520 pp.

[G-18]  Hadley, G., *Nonlinear and Dynamic Programming*, Addison-Wesley, 1964, 484 pp.

[G-19]  Churchman, C. W., R. L. Ackoff, and E. L. Arnoff, *Introduction to Operation Research*, John Wiley, 1957, 645 pp.

[G-20]  Charnes, A. and W. W. Cooper, *Management Models and Industrial Applications of Linear Programming*, 2 Vols., John Wiley, 1961, 859 pp.

**[G-21]**    Graves, R. L. and P. H. Wolfe (Eds.), *Recent Advances in Mathematical Programming*, McGraw-Hill, 1963, 347 pp.

**[G-22]**    Dennis, J. B., *Mathematical Programming and Electrical Networks*, The MIT Press, 1959, 186 pp.

**[G-23]**    Wilde, D. J., *Optimum Seeking Methods*, Prentice-Hall, 1964, 202 pp.

**[G-24]**    Bellman, R., *Dynamic Programming*, Princeton Univ. Press, 1957, 341 pp.

**[G-25]**    Bellman, R., *Adaptive Control Processes—A Guided Tour*, Princeton Univ. Press, 1961, 255 pp.

**[G-26]**    Bellman, R. E. and S. E. Dreyfus, *Applied Dynamic Programming*, Princeton Univ. Press, 1962, 363 pp.

**[G-27]**    Bellman, R. E. (Ed.), *Mathematical Optimization Techniques*, Univ. of Calif. Press, 1963.

**[G-28]**    Kipiniok, W., *Dynamic Optimization and Control*, The MIT Press, 1961, 231 pp.

**[G-29]**    Leitmann, G. (Ed.), *Optimization Techniques with Applications to Aerospace Systems*, Academic Press, 1962, 453 pp.

**[G-30]**    Pontryagin, L. S. et al, *The Mathematical Theory of Optimal Processes*, Interscience, Wiley, 1962, 360 pp.

**[G-31]**    Wescott, J., *An Exposition of Adaptive Control*, Pergamon, 1962, 100 pp.

**[G-32]**    *Proc. Symp. Optimal Control*, Imperial College, Department of Electrical Engineering, London, April 1964.

**[G-33]**    *Proc. 1st Intern. Symp. Optimizing Adaptive Control*, Instrument Society of America, Pittsburgh, 1963, 311 pp.

**[G-34]**    Merriam III, C. W., *Optimization Theory and the Design of Feedback Control Systems*, McGraw-Hill, 1964, 391 pp.

**[G-35]**    Fang, L. T. and C. S. Wang, *The Discrete Maximum Principle*, John Wiley, 1964, 158 pp.

**[G-36]**    Aris, R., *Discrete Dynamic Programming*, Blaisdell, 1964, 148 pp.

**[G-37]**    Balakrishnan, A. V. and L. W. Neustadt (Eds.), *Computing Methods in Optimization Problems*, Academic Press, 1964, 327 pp.

**[G-38]**    Good, H. H. and R. E. Machol, Systems Engineering, McGraw-Hill, 1957, 551 pp.

**[G-39]**    Eckman, D. P., *Automatic Process Control*, John Wiley, 1958, 368 pp.

**[G-40]**    Campbell, D. P., *Process Dynamics*, John Wiley, 1958, 316 pp.

**[G-41]**    Mishkin and Brown, *Adaptive Control Systems*, McGraw-Hill, 1961, 533 pp.

**[G-42]**    Grabbe, E. et al (Eds.), *Handbook of Automation, Computation and Control*, 3 Vols. John Wiley, 1958–1961.

**[G-43]**    Singer, S. F. (Ed.), *Progress in the Astronautical Sciences*, Vol. 1, North-Holland Publ. Co, 1962, 416 pp.

**[G-44]**    *Progress in Astronautics and Rocketry*, Series by American Rocket Society (Vol. 1–Vol. 9), 1960–1963 and American Institution of Aeronautics and Astronautics (Vol. 10–Vol. 13), 1963–1964.

**[G-45]** Battin, R. H., *Astronautical Guidance*, McGraw-Hill, 1964, 400 pp.

**[G-46]** Kirchmayer, L. K., *Economic Operation of Power Systems*, John Wiley, 1958, 260 pp.

**[G-47]** Kirchmayer, L. K., *Economic Control of Interconnected Systems*, John Wiley, 1959, 207 pp.

**[G-48]** Peterson, E. L., *Statistical Analysis and Optimization of Systems*, John Wiley, 1961, 190 pp.

**[G-49]** Chang, S. S. L., *Synthesis of Optimum Control Systems*, McGraw-Hill, 1961, 1961, 380 pp.

**[G-50]** Leondes, C. T. (Ed.), *Advances in Control Systems Theory and Applications*, Vol. 1, Academic Press, 1964, 365 pp.

**[G-51]** Peschon, J. (Ed.), *Disciplines and Techniques of Systems Control*, Blaisdell, 1965, 547 pp.

**[G-52]** Quade, E. S. (Ed.), *Analysis for Military Decisions*, Rand McNally North-Holland, 1964, 382 pp.

**[G-53]** Isaacs, R., *Differential Games*, John Wiley, 1965, 384 pp.

**[G-54]** Athans, M. and P. L. Falb, *Optimal Controls*, McGraw-Hill, 1966, 896 pp.

**[G-55]** Ledley, R. S., *Digital Computer and Control Engineering*, McGraw-Hill, 1960, 835 pp.

**[G-56]** Howard, R., *Dynamic Programming and Markovian Processes*, Technology Press and Wiley, 1960.

**[G-57]** Derosso, P. M., R. J. Roy, and C. M. Close, *State Variables for Engineers*, John Wiley, 1965, 608 pp.

**[G-58]** Norton, A. R. M., *Introduction to Variational Methods in Control Engineering*, Pergamon, 1965, 122 pp.

**[G-59]** Blakelock, J. H., *Automatic Control of Aircraft and Missiles*, John Wiley, 1965, 348 pp.

**[G-60]** Carpentier, J. and H. Garelly (Eds.), *Identification, Optimalisation et Stabilité des Systèmes Automatiques*, Actes du Congrés d'Automatique théorique, Paris 1965, Dunod, 1967, 345 pp.

**[G-61]** Roberts, S. M., *Dynamic Programming in Chemical Engineering and Process Control*, Academic Press, 1964, 457 pp.

**[G-62]** Dreyfus, S. E., *Dynamic Programming and the Calculus of Variations*, Academic Press, 1965, 248 pp.

**[G-63]** Fel'dbaum, A. A., *Optimal Control Systems*, Academic Press, 1965, 452 pp (Translated from Russian by A. Kraiman).

**[G-64]** Sworder D., *Optimal Adaptive Control Systems*, Academic Press, 1966, 187 pp.

**[G-65]** Leondes, C. T. (Ed.), *Advances in Control Systems—Theory and Applications*, Vol. 1, 1964, 365 pp., Vol. 2, 1965, 313 pp., Vol. 3, 1966, 346 pp., Vol. 4, 1966, 320 pp., Vol. 5, 1967, 426 pp., Vol. 6, 1968, 321 pp., Academic Press.

**[G-66]** Caioniello, E. R. (Ed.), *Functional Analysis and Optimization*, Academic Press, 1967, 225 pp.

**[G-67]** Savas, E. S., *Computer Control of Industrial Processes*, McGraw-Hill, 1965, 400 pp.

**[G-68]** Shinsky, F. G., *Process-Control Systems, Application, Design, Adjustment*, McGraw-Hill, 1967, 367 pp.

**[G-69]** Favret, *Introduction to Digital Computer Applications*, Reinhold, 1966.

**[G-70]** Miller, *Digital Computer Applications to Process Control*, Plenum, 1965, 400 pp.

**[G-71]** Kunzi, H. P. and W. Krelle, *Nonlinear Programming*, Blaisdell, 1966, 200 pp.

**[G-72]** Abadie, J. (Ed.), *Nonlinear Programming*, John Wiley, 1967, 317 pp.

**[G-73]** Eveleigh, V. W., *Adaptive Control and Optimization Techniques*, McGraw-Hill, 1967, 434 pp.

**[G-74]** Leitmann, G., *Optimal Control*, McGraw-Hill, 1966, 264 pp.

**[G-75]** Leondes, C. T., (Ed.), *Modern Control Systems Theory*, McGraw-Hill, 1965, 486 pp.

**[G-76]** Hsu, J. and A. Meyer, *Modern Control Principles and Applications*, McGraw-Hill, 1968, 769 pp.

**[G-77]** Boudarel, Delmas, and Guichet, *Commande Optimale des Processus* (in French), Dunod, Paris, Tome I, 1967, 308 pp; Tome II, 1968, 310 pp; Tome III, 1969, 257 pp.

**[G-78]** Wilde, D. J. and Ch. S. Beighter, Foundation of Optimization, Prentice-Hall, 1967, 480 pp.

# INDEX